Frontiers
of
Comparative Plant Ecology

Edited by I. H. RORISON, J. P. GRIME,
R. HUNT, G. A. F. HENDRY and D. H. LEWIS

A Symposium to mark the Silver Jubilee
of the establishment of the Unit of Comparative Plant Ecology (NERC)
held at the University of Sheffield on 22–24 September 1986

(Reprint of *The New Phytologist*, **106** (Supplement), 1987)

Published for the New Phytologist Trust

by ACADEMIC PRESS
LONDON   ORLANDO   SAN DIEGO   NEW YORK
AUSTIN   BOSTON   SYDNEY   TOKYO   TORONTO

ACADEMIC PRESS LIMITED
24/28 Oval Road
London NW1
(Registered Office)

US edition published by
ACADEMIC PRESS INC.
Orlando, Florida 32887

(Reprint of *The New Phytologist*, **106** (Supplement), 1987)

© 1987 The New Phytologist
ISBN 0 12 595960 5

Printed in Great Britain by the University Press, Cambridge

CONTENTS

New Phytol. (1987) **106** (Suppl.), 1–2

INTRODUCTION TO SYMPOSIUM

By I. H. RORISON

Unit of Comparative Plant Ecology (NERC), Department of Botany, The University, Sheffield S10 2TN, UK

In the early 1960s, the Nature Conservancy launched a number of units within Universities. One of these was the Grassland Research Unit at Sheffield. Supported directly by the Natural Environment Research Council since 1973, it adopted the name Unit of Comparative Plant Ecology (NERC) in 1974.

At a time of increasing specialization and fragmentation of botanical research, the Unit attempts to maintain a broad, integrated approach to the study of plant distribution and vegetation. Its methods involve field surveys, screening of the responses of plants to biotic and abiotic variables under standardized laboratory conditions, and detailed physiological and biochemical measurements on plants of contrasted ecology.

The results have included a range of laboratory measurements (e.g. aspects of germination, growth, mineral nutrition and cytological characteristics) which are used to predict aspects of the behaviour of plants in the field. This has led to the recognition of recurrent patterns of specialization in plants and to the development of strategy concepts which seek to bridge the gap that has for too long existed between physiological ecologists and population biologists.

The 25th Anniversary of the founding of the Unit fell on 1 October, 1986 and the occasion was marked by a Jubilee Symposium. The topics were a reflection of the Unit's current interests and included overviews of recent work and predictions for the future, both from past and present members of the Unit and from invited speakers. The 17 papers presented here were delivered to 160 delegates from 15 countries and there were 35 posters, the titles of which are listed at the end of this volume.

The main theme of the papers which follow is interspecific variation in the form and function of plants, particularly in relation to environmental factors. Plants are considered at various stages of their development from seed fall to flowering, and processes of capture and utilization of resources are discussed from the points of view of mineral nutrition through to carbon economy. Emphasis is placed on experimental and evolutionary perspectives, and on the value of interdisciplinary approaches, the aim being a fuller understanding of the general principles underlying plant distribution and the dynamics of vegetation.

At the meeting itself, the emergence of ecology as a truly predictive science was confirmed. Both the origins and the dynamics of change were debated and interest in functional relationships was paramount. There were timely reminders of the need to consider *limits* of tolerance and responses to *fluctuating* conditions. There was also a growing awareness of the need for measurements of *irregular* and *unexpected* change. The need for synthesis and for continuing refinement of general principles was acknowledged but the importance of studying exceptions to general rules was also stressed. We saw signs of plant ecologists becoming more controversial and competitive, as exponents of the much-needed holistic approach vied both with each other and with reductionists for centre stage.

Our gratitude for the success of the Symposium extends not only to our enthusiastic contributors but also to our four chairmen for their skilled direction of the sessions (Fig. 1). The symposium was funded both by our parent body, the Natural Environment Research Council, and by our host organization, the University of Sheffield. Our most generous single benefactor, however, was the *New Phytologist Trust*. It undertook to publish the entire proceedings and the incisive pen of its Executive Editor played a significant role in achieving the impact of the final presentation. We are also most grateful to our anonymous referees for their contribution to the standard achieved.

All present members of the staff of UCPE (Fig. 2) entered into the spirit of the Symposium with the same degree of enthusiasm and dedication which they have shown throughout the development of the Unit. This volume is a tribute to them, to former members of the Unit and to our University colleagues with whom we have enjoyed a mutually beneficial relationship throughout our 25 years.

If we are to single out one person to thank, it must be Roy Clapham. It was his foresight, and his influence on ecological thinking, which laid the foundation both of our research and of our Unit. His presence at the Jubilee Symposium was a special pleasure.

Finally, it is fitting to return to NERC to thank them above all for allowing us the freedom to explore our subject and to develop the most promising, and often unexpected, lines of enquiry. In this present time of need, when contractual work of an applied nature is vital to survival, we, and the scientific world in general, should be grateful for NERC's determination to continue backing fundamental research in order to safeguard the country's intellectual future. We have always worked on the premiss that, if you understand the fundamentals first, you can then apply them to problems both seen and unforeseen, and it is on this note that we move into the next 25 years.

Fig. 1. Professor A. R. Clapham, C.B.E., F.R.S., founder of the Unit, flanked by contributors to, and chairmen of, the Jubilee Symposium. Back row (from left to right): John Raven, Quentin Kay, Alastair Fitter, Philip Grime, Dick Verkaar, Michael Bennett, Roderick Hunt, Tony Bradshaw, John Hodgson, Ian Rorison. Centre row (from left to right): George Hendry, Michael Fenner, Phyllis Coley, John Grace, Arthur Willis, Barry Osmond, Derek Anderson. Front row (from left to right): Alan Baker, Tom Givnish, John Cooper, Roy Clapham, Ken Thompson, George Stewart.

Fig. 2. The Unit of Comparative Plant Ecology, September 1986. back row (from left to right): Lal Gupta, Bruce Campbell, Peter Moody, Andy Neal. Second row (from left to right): Colin Birch, Rod Hunt, Frank Sutton, John Hodgson, George Hendry. Third row (from left to right): Rita Spencer, Philip Grime, Ian Rorison, Nuala Ruttle, Joanna Mackey. Front row (from left to right): Jane Brocklebank, John Rose, Kate Ewart, Adam Price. Insets: Stuart Band, Judith Fletcher.

New Phytol. (1987) **106** (Suppl.), 3–21

COMPARISON – ITS SCOPE AND LIMITS

By A. D. BRADSHAW

Department of Botany, University of Liverpool, Liverpool L69 3BX, UK

SUMMARY

At any point in time, any discipline in science tends to be seized by a particular methodology or enthusiasm and other approaches get 'dumped'. So it is in ecology – although, because of obstinate individualism, nothing is ever completely forgotten. We are currently in a reductionist, population-dominated era. Twenty-five years ago, our interests lay primarily in comparisons of species. In between, the holistic approach has had a good run with studies of ecosystems everywhere. This introductory paper reflects on this history and particularly, because of the occasion, examines the value of the process of comparison.

Key words: Holism, reductionism, autecology, synecology, inference, ecological laws, evolution, species, intraspecific variation, ecological amplitude.

INTRODUCTION

In my Summary, I use the words 'enthusiasm' and 'dumped' advisedly to imply that what happens in science is more to do with emotion than with logic. Once we spend a lot of time looking at behaviour of species, then ecosystems become important. Now, nearly any ecologist who feels he must be respectable will work on populations. Although an historian may be able to see a proximal cause for what happens – the arguments of a persuasive scientist or a novel and interesting discovery – the ultimate causes may be little other than those which give the world its flourishing fashion industry.

We have seen, therefore, a wide variety of different approaches succeeding one another in ecology since the subject was given the respectability of a title by Haeckel. Is it because of obstinate individualism, or perhaps persevering wisdom, that no approach has ever been dumped completely? The enthusiastic young will think, I am sure, that the retention of old approaches is due to obstinate, even unimaginative, individualism. The older may see wisdom in this course. This meeting celebrates the 25th anniversary of the Unit of Comparative Plant Ecology (UCPE). Few research units survive this long, so some wisdom must have been seen in its activities by the august Conservancy and Research Council which have supported it – particularly since, despite some diversity, the Unit has indeed followed the single theme of comparative ecology which forms both its title and the topic of this symposium.

It is my intention, therefore, to look at approaches to ecology, to see where they have taken us and to try to understand why comparative ecology remains a vigorous and valuable enterprise which has not been swept aside by the dictates of scientific fashion.

ON APPROACHES TO ECOLOGY

There are now a number of excellent analyses of the history of ecology. The most perceptive are those of Harper (1982) and McIntosh (1985). For the moment, it

0028-646X/87/05S003+19 $03.00/0

is sufficient to remember that, from its Greek derivation, ecology is the study of households (*oikos*, house; *logos*, study). There are therefore two components, each of which is itself divisible.

The household: it is obvious that this has a number of components, (1) the aggregation of living organisms and the way these live together and interact; (2) their background surroundings and what these provide; (3) the interactions between (1) and (2).

The study: since the word was proposed by a scientist, this implies a scientific study to obtain understanding, which can involve up to four main methods: (1) description; (2) induction, in the sense of Mill; (3) experimentation, in the sense of Bacon; (4) hypothesis/deduction, as argued by Popper (Medawar, 1979). Each of these four methods has value. We can make little progress without description. But one of the last three has to be used if we wish to discover causation, which is usually the case because we are not often satisfied just to know *what* is there; we also want to know *why*. The most sophisticated method is that in which we carry out experiments to test particular hypotheses, an approach well analyzed for science in general by Medawar (1967) and for ecology by Price, Gaud & Slobodchikoff (1984). But each of the others has its place.

Because of the nature of the material with which we have to deal, there are further distinctions in approaches to ecology which we must also recognize. Firstly, we have the following long-recognized division. (1) The holistic: where we tend to examine the household (which we may call an ecosystem) as a whole and look for activities which reflect integration within it. (2) The reductionist: where we tend to examine the properties of component species and processes, expecting by such detailed knowledge of its parts to understand the whole household.

Secondly, and in parallel with the distinction between holistic and reductionist approaches, we have the division proposed by Schroter. This arises because, whether we like it or not, species do occur almost universally in assemblages: synecology – the study of communities and their properties; autecology – the study of species and their properties.

Although the latter two pairs of terms are not totally equivalent, they each reflect a similar division of interest between the ecological world taken as a whole or taken piece by piece. In its formative years, ecology was devoted to synecology. One culmination was Tansley's remarkably perceptive survey of British vegetation (Tansley, 1939). The same approach was also followed by Odum (1975). But, at the same time, others saw the importance of studying the species which make up these communities and it is important to note that Professor A. R. Clapham, who was responsible for the founding of the UCPE, made autecology the subject of his presidential address to the British Ecological Society (Clapham, 1956). This theme carried through to the *Festschrift* published in honour of his 80th birthday (Harley & Lewis, 1984).

Thirdly, the division of ecology into two has now become a division into three; into *populations* and *species* as well as into *communities*. This recognizes that the species which occur in communities are themselves made up of individuals which react together in interesting and complex ways. The range of exciting discoveries that have been made in this field is well displayed for plants by Harper (1977) and for both animals and plants by Begon & Mortimer (1986).

On Comparison

In the hierarchy of methods of study, it is clear that comparison, as a specific exercise, appears early. It is true that we can describe anything without making comparisons but once we attempt induction then comparisons must be involved. Experimentation, whether Baconian or more directed towards hypothesis testing, obviously involves comparison of the effects of a particular treatment with those of another treatment (which can be no 'treatment' at all).

All this may seem trivial, because comparison is surely something which we take for granted. Yet do we really understand what we are doing? There is a simple premise of statistics that, between two observations, only one comparison can be made and that, between n observations, n - 1 independent comparisons can be made. This we term the number of degrees of freedom.

Of course it is possible to make only one observation. We could discover, for instance, that tree lupin on china clay waste, because of its associated Rhizobia, fixes 400 g N ha^{-1} d^{-1} (about 150 kg ha^{-1} $year^{-1}$). Such a bald fact, however, is useless. It is true that many people will think otherwise, probably because they will be very interested that tree lupin can fix so much nitrogen. But this interest only arises because they are tacitly making a comparison with other evidence, gained elsewhere, that most plants either do not fix nitrogen at all or fix far less. They are simply not using the fact by itself but comparing it with other facts already accumulated. Herein of course is a substantial problem, because the other facts have been accumulated by other people in other situations and may not provide the basis for legitimate comparison. This becomes more obvious if we remember that some woodland plants which do not bear N-fixing organisms in nodules nevertheless fix appreciable quantities of nitrogen in their rhizosphere. Is then 400 g ha^{-1} d^{-1} at all remarkable? The only way to decide is to carry out a proper comparison with all the material being tested in a similar manner and with appropriate controls and assessments of error (Fig. 1). Then we can see the real contribution of the legume.

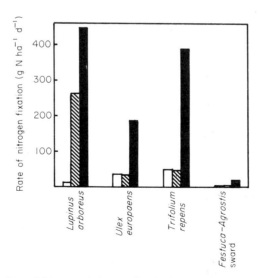

Fig. 1. A comparison of the rates of nitrogen fixation by different species of legume and by a grass sward, all growing on sand waste (rates measured successively in August 1976, October 1976 and May 1977) (from Skeffington & Bradshaw, 1980).

All of this, some will say, is totally obvious. But is it? A great deal of ecology has been, and still is, based on single observations without any proper comparisons – for example Tansley's (1939) species lists. Sometimes, it may appear otherwise because many items have gone to make up the single observation, as in the detailed description of a particular plant community or in the description of the response to density of particular animal or plant species. But few critical conclusions will be possible unless other entirely comparable evidence on other material is also available, a view supported by Strong (1980).

The response to density of a single species (Fig. 2), of course, is not entirely without interest because the experiment provides valid comparisons, and therefore evidence, about the effects of density within the single species. But we gain no idea of the significance of what we observe. There is the possibility that everything behaves in the same way. We will return to this sort of evidence and its problems later.

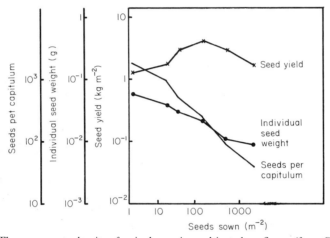

Fig. 2. The response to density of a single species, cultivated sunflower (from Clements, Weaver & Hanson, 1929 and Harper, 1977).

Finally, there is the problem of comparing things that are so unlike and unrelated that any differences we find, however interesting, are difficult to make sense of because, in comparing such very different taxa, we cannot be certain of the origin of the differences. The comparison of the temperature responses of *Urtica dioica*, an angiosperm, and *Brachythecium rutabulum*, a moss, for instance (Fig. 3), is very interesting but we must be cautious in using the differences we find to provide a full explanation of their coexistence. The differences could have historic origins and have little to do with current ecological relationships. This important point is returned to, both at the end of this paper and that by Grime & Hodgson (1987).

ON INFERENCE

Let us now assume that we have made a series of comparable observations from which we hope to make certain inferences. If we have made a number of observations, we may infer some general conclusion. But let us beware, because the process of inference has considerable limitations.

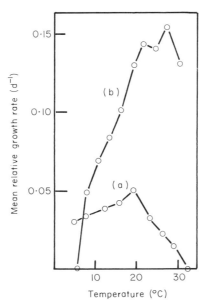

Fig. 3. The effects of temperature upon the mean relative growth rate (30 d basis) of (a) *Brachythecium rutabulum* and (b) *Urtica dioica* (from Furness, 1978).

The first and most obvious limitation is related to the number of observations and to their origin. If only two things are compared, despite great and interesting differences, being found, there is absolutely no way of determining their generality. The next two examples taken might show a completely reversed behaviour. The solution to this is, of course, well understood and is accessible to statistical analysis. The statistical significance of any comparison is related to the number of independent observations involved, the degrees of freedom. A very low number of degrees of freedom will make it difficult to achieve a significant result. The requirement is for more comparisons, which in the case of Figure 3 would mean extending this work to other species. This is the great strength of the work carried out at Sheffield. This extension must, however, involve material which, within the comparison being examined, is chosen at random and is therefore unbiasedly representative. All this is simple to do but still often not done. It does involve more work, which is not possible in all cases. This is where good Research Council or other support is invaluable.

The second limitation is the asymmetry of proof argued by Popper (1959). If we observe something which agrees with an idea, then the conclusion is that the idea is *not disproved*. But we cannot conclude that the idea is *proved*. If we observe something which disagrees with an idea then, providing we have exposed our idea to 'a cruelly critical test', what Galileo called '*Il cimento*' (Medawar, 1979), we can conclude that the idea *is disproved*. In statistical analysis, this is translated into the null hypothesis approach. Because this approach is so widely used, and because we can also make a probabilistic assessment of the error in what we conclude, I believe that the problem of asymmetry of proof has become less of a trouble for biologists. But it indicates that we must continue to produce ideas and conclusions that are testable by refutation, which means using *imagination* in constructing hypotheses.

The third and most important limitation are the troubles that can occur in attributing particular effects to particular causes because of confounding variables, recently well discussed for animals by Clutton-Brock & Harvey (1979). If we observe that a particular species, such as white clover, disappears at high altitudes, we may perhaps infer that it is more sensitive to lower temperatures than other species. But, in this case, we would be very wrong because although, for instance in North Wales, white clover generally disappears above an altitude of 500 m, it can also be found growing vigorously there even at altitudes of 1000 m on special soils which have an adequate supply of bases maintained by the geological substrata. The apparent altitudinal zonation is not due to the direct effects of climate but to associated soil effects. In statistical parlance, we must beware of silly correlations.

This sort of thing is a particular problem with observational, correlative evidence, even when a large number of comparisons is involved. Experiments designed to determine the particular effects of a particular factor, which are not very difficult to carry out in the case of plants, usually overcome this problem. But sometimes this is not so. For example, from the profound effects of the simple fertilizer treatments which emerge from the Park Grass Experiments (Brenchley, 1958; Thurston, 1969) it seems very easy to attribute appropriate causations. Yet it would, even now with so much progress in plant ecology, be a very brave person who could infer *exact* chains of causation from the differences in species composition which are evident between the different plots – because of the multifarious effects of fertilizers on an ecosystem. Pigott (1982) rightly argues that further experiments are the only answer. Tilman's work (1984, 1985), indicating the interrelations between competition for nutrients and for light, is therefore an interesting step forward.

ON ECOLOGICAL PRINCIPLES

Despite all the problems that surround our methodology, no-one can deny that we have made progress in ecology. Many will measure this progress by the number of general laws we have produced. Open any textbook and you will discover many generalizations, whether in mineral nutrition – such as the classical response of plant species to increased nutrients (Fig. 4), or in population biology – such as the

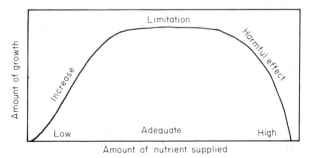

Fig. 4. The general relation between any particular nutrient or growth factor and the amount of growth made by the plant (from Russell, 1973).

$-3/2$ law of density-dependent self-thinning (Yoda *et al.*, 1963). Nearer to home, we have had the elaboration of the calcicole/calcifuge concept by Grime & Hodgson (1969) and the development of the C–S–R theory of plant strategies (Grime, 1977, 1979). All of these generalizations are very important, not least because they provide a framework of understanding.

The list of such generalizations is long. It will certainly be interesting to see (from a survey, see Bradshaw *et al.*, 1986) which ecological 'concepts' are considered important enough by members of the British Ecological Society for inclusion in their 75th Anniversary Symposium in 1988.

But is progress only to be related to the number of successful generalizations (which we will grandly call principles) which we can establish? Teachers and writers of textbooks may feel this way. The search for general principles is very important, no doubt. But the ecological world has reasons which make the search for *exceptions* equally important.

ON EVOLUTION AND SPECIES

The living material we examine is the product of evolution. This evolution is determined to a major extent by the environment. The starting point was the physical environment. As we know, there was, and is, not just one physical environment but an infinite set of them, owing to simultaneous variations in many different factors. Evolution is therefore going to set off in many different directions. Then, once this simple sort of evolution has begun, each physical environment will be complicated by the presence in it of the products of the primary evolution. Once a simple habitat offering one environment is occupied by a single species then there immediately appear two sub-environments, the space already occupied by the species and the space which is not.

Of course this is an oversimplification, because each of the spaces is itself divisible into a whole range of micro-environments. But the crucial point is that nature has offered, since life began, an immense and complex range of physical and biological environments within which evolution has occured. We may think of these environments as a two-dimensional matrix, which is complex enough. But the reality is much more complex, involving an array of niches each of which is, in the terminology of Hutchinson (1965), a multi-dimensional hyper-volume.

The great characteristic, then, of the evolutionary products of this enormous array of environments is its enormous complexity and diversity. This was well recognized by Darwin, and especially noted in the final chapter of *The Origin of Species by Means of Natural Selection* (1859), so there is nothing novel in the idea:

It is interesting to contemplate a tangled bank, clothed with many plants of many kinds, with birds singing on the bushes, with various insects flitting about, and with worms crawling through the damp earth, and to reflect that these elaborately constructed forms, so different from each other, and dependent upon each other in so complex a manner, have all been produced by laws acting around us.

But what is surprising is that, as ecologists, we have spent so much of our time looking for general principles which, for the species, means similarities in behaviour. Surely in doing this we are, wittingly or unwittingly, overlooking that great attribute of evolution, and therefore of the 'households' we study, namely the *diversity* in species which arises from the life of species within communities and therefore from evolution (Bradshaw & Mortimer, 1986).

ON SPECIES COMPARISONS

From comparisons of species, we can deduce a very simple idea: that although some ecologists may, and indeed should, look for great principles of general applicability, others must look for the exceptions, or indeed, at situations where there do not appear to be great principles at all. Some may feel that I am trying to lead them towards a sort of latter-day obscuritanism. Maybe I am, but I am in deadly earnest in warning that, if we concentrate on the generalities of ecology, we will miss some of its most elegant lessons, those which come from the inexorable principle that evolution automatically leads to diversity.

A target in ecology must, then, be understanding this diversity of species. As Clapham (1956), Grime (1984) and others have argued, this should mean looking at diversity in morphology, in physiology, in life-cycle and in any other attributes which determine the ecology of species. Perhaps the first step must be an examination of distribution, because what is this but the primary defining attribute of the ecology of a species?

Immediately, however, we realize that, for the reason already given, we have to be involved *in comparison* because we can have no idea about the significance of a particular species' distribution unless we have one for a second species with which to compare it. At the same time, this comparison cannot be based on evidence which is so general that there is no precision – '*Arrhenatherum elatius* is usually found in rough pastures and hedgerows on neutral soils' or '*Agrostis capillaris* is found on a wide range of pastures on neutral or acid soils' – because no exact comparison is possible. This leads to a requirement for precise evidence on distribution, which is so rarely available. Sheffield's *Ecological Atlas of Grassland Plants* (Grime & Lloyd, 1973) was a conspicuous exception (Fig. 5). It not only illustrated the value of precise data but also what can be gained when several species are compared on a common basis – and not just a few but 50 or 100 or, in the case of the latest compendium (Grime, Hodgson & Hunt, 1987), even more.

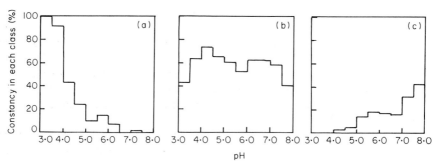

Fig. 5. The distribution with respect to soil-surface pH of three grasses within grasslands of the Sheffield region: (a) *Deschampsia flexuosa*, (b) *Festuca ovina*, (c) *Brachypodium pinnatum* (from Grime & Lloyd, 1973).

However, these sorts of data are obviously only a starting point. The pH value is a factor which affects many other characteristics of the soil, each of which is often capable of varying independently of pH – so obvious for calcium and nitrogen for instance. We have good evidence that not only do species have their own particular responses to each of these, but also that these responses do not go in parallel (Bradshaw *et al.*, 1958, 1964) as Figure 6 demonstrates. The responses of *A. elatius* and *Festuca ovina* discussed by Rorison (1987) show that there are even further complexities.

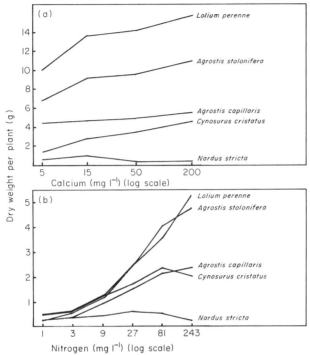

Fig. 6. The response of five grass species to calcium (a) and to nitrogen (b) (from Bradshaw *et al.*, 1958, 1964).

Of course, the ecological world is more than pH or nutrients. It is salt and temperature and water and light and a host of other physical factors. It is not difficult to set up comparative tests to see how species differ in their response to these, and remarkable differences can be found (Jarvis & Jarvis, 1963). It may be more difficult to discover how much species are affected by these factors in the field. Indeed, it is remarkable how little critical evidence there is on the relation of distributions of species to any critical physical factor. There is, for instance (with recent notable exceptions, see Rorison, 1987), very poor evidence for even such an important nutrient as nitrogen. So how can we begin to understand the factors involved in the distribution of species? Yet these species are the products of evolution and therefore of the natural selection which arises from specific environmental conditions. So we should be able to relate what we find in species to what we find in environments. But it will not necessarily be easy.

ON THE REST OF THE ENVIRONMENT

We have just argued that the environment of a species is not solely determined by the simple physical factors which occur on site. It is also profoundly shaped by how many, and what, other organisms are there too. The other plants, can, of course, be members of the same species. Indeed, these may be the most common determinant of the micro-environment of any individual. These organisms ultimately have effects which must be physical. But it is convenient, because of their origin, to distinguish them as biological.

This biological component of the environment may profoundly affect plant responses to the physical components. This can be well seen from the work of Werner & Platt (1976) on the distribution of *Solidago* species in relation to moisture (Fig. 7). The presence of other species alters the distributions considerably.

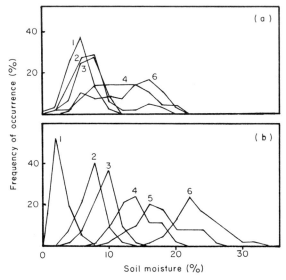

Fig. 7. Differences in the distribution of *Solidago* species in relation to moisture, in a successional old field (a) and in a mature prairie (b) (from Werner & Platt, 1976). *Solidago* species: 1, *S. nemoralis*; 2, *S. missouriensis*; 3, *S. speciosa*; 4, *S. canadensis*; 5, *S. gigantea*; 6, *S. graminifolia*.

The trouble is that these biological components of the environment are tremendously variable. Own density can, for most species, vary between very wide limits. So also can the density of other species, quite apart from their actual occurrence. If we are anxious to discover the response of a species to its own density (Fig. 8) and compare this with that of other species to *their* own density, we shall have to carry out the necessary experiments in order to demonstrate the generality of the $-3/2$ self-thinning law. However, now it is not the similarities in behaviour of the various species which become interesting, but their differences. The similarities may be no more than the results of in-built physical or physiological constraints within which the species must operate. The differences support the ecological interest. These do indeed seem to make sense (White, 1980) and make us realize that the differences in response are of ecological significance.

It is much more difficult to compare the behaviour of individual species in relation to the presence of others, if only because of the great number of these and their combinations. It is remarkable what complicated differences in response are found when only three species are grown in all possible combinations (Haizel & Harper, 1973). Even with two species, the interactions owing to 'same species' are difficult to unravel from those of 'other species' (Jolliffe, Minjas & Runeckles, 1984). The difficulty in relating such results to the distribution of the subject species is one salutary aspect of this sort of comparison.

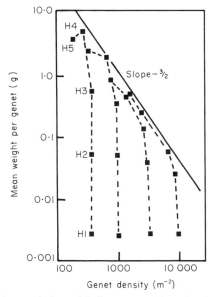

Fig. 8. Self-thinning in populations of *Lolium perenne* planted at four different densities. H1–H5 represent five successive harvests (from Kays & Harper, 1974).

Here I have put my head into the hornet's nest of competiton. What is so difficult is to find a simple way of testing the total effects of the biological environment on species. The most elegant is the reciprocal transplant, where two or more species are planted both into their own and the other's habitat(s). This was first used in a recorded manner by Bonnier (1890) but, alas, it was misused because the material was not well enough maintained and recorded so that species 'transformations' occurred (Clausen, Keck & Hiesey, 1940). Since those early days, the method has mainly been used to test intraspecific rather than interspecific differences. However, Schoen *et al.* (1986) have used it to compare *Impatiens capensis* with *I. pallida*. Interestingly, they found that differences in survivorship and fruit production were attributed primarily to features of the site of the transplants rather than to neighbours. A similar picture was obtained by other investigators (e.g. Fowler & Antonovics, 1981; Snow & Vince, 1984). But more critical work is needed, such as that where performance of transplants is examined in the presence and absence of vegetation (Wilson & Keddy, 1986). Perhaps the most elegant and penetrating use of the reciprocal transplant has been with populations of *Anthoxanthum odoratum* in the Park Grass experiment, which is discussed in the next section.

It is not possible to do justice to what is to be extracted from these sorts of comparison because of the sheer complexity of environments. At this point, it is perhaps necessary to remind ourselves that far too little has been done to define the distribution of species in terms that take into account *all* the characteristics of their environments. I personally find, for instance, that the recognition of either r–K types of adaptation (MacArthur & Wilson, 1967; Pianka, 1970) or C–S–R types of adaptation (Grime, 1977, 1984), even as parts of a continuum, are too likely to lead us to miss essential, important, attributes of differences in adaptation to be valuable tools when used on their own. We should use them for exposing anomalies and complexities and as a stepping stone to further analysis.

However, this does not excuse us from attempting a definition of the characteristics of the distributions of different species using as many attributes as we can. The first step may be in terms of the obvious overall characteristics of their habitats, as is so beautifully explored in the new Sheffield compendium (Grime *et al.*, 1987, from which Figure 9 has been abstracted). But if we are to achieve proper understanding, it is clear that a reductionist approach will be required to define the habitat preferences of species in precise physicochemical terms.

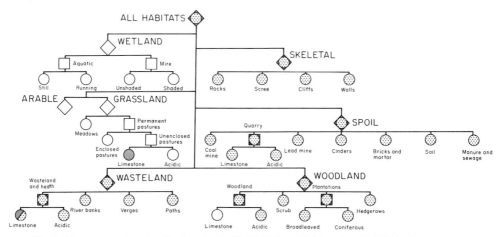

Fig. 9. Two contrasting distributions in the Sheffield flora. Habitats in which *Primula veris* occurs in 5% or more of the survey sample (dark stipple) are far fewer and more specialized than the habitats which support comparable occurrences of *Chamerion angustifolium* (pale stipple) (abstracted from Grime, Hodgson & Hunt, 1987).

ON WHAT CAN HAPPEN WITHIN SPECIES

It is rather simplistic to argue that comparative ecology shold mainly involve an examination and understanding of differences between species, because evolution has created a diversity which we must understand, if we do not also remember that these different products of evolution in different environments must have had their origin within species.

What happens *within* species can therefore be just as much part of comparative ecology as what is found *between* them. Yet, on the whole, this has been more the concern of people interested in evolution, although they may call themselves ecological geneticists. The great value of comparisons based on different populations of a species is that any differences are far more readily related to the environments in which the populations occur, and are much less confounded by changes in characteristics acquired at some time past and of little relation to the present environment in which the species occur.

The differences found within species may not be as extreme as those found between species, which could be a disadvantage. But the specificity of their relationship to the environment occupied by the populations concerned is a great advantage. At the same time, all such evidence helps to put our ecological observations into an evolutionary context. As Harper (1982) points out, all 'ultimate' biological explanation has to be in evolutionary terms.

This has been argued on a number of occasions. The recent symposium in Mexico (Dirzo & Sarukhan, 1984) provides a wide range of powerful examples. Perhaps the most eloquent and apposite case is the analysis of the Park Grass populations of *A. odoratum*. These populations indeed reveal, almost better than anything else, the subtle complexity of the environments produced by the simple fertilizer treatments (Snaydon & Davies, 1982). The reciprocal transplants, which were repeated for different pairs of plots and are therefore of general validity, then show just how fundamental those differences are to the survival of the different populations (Davies & Snaydon, 1976) (Fig. 10). Heavy metal tolerance (Baker, 1987) provides another eloquent example, for a more barbaric type of habitat, and *Phlox* (Schmidt & Levin, 1985) shows just how complex and subtle the adaptations can be.

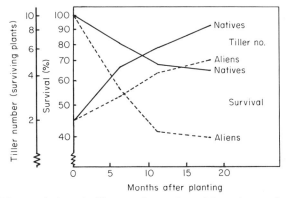

Fig. 10. The survival, and tiller members of surviving plants, of *Anthoxanthum odoratum* transplanted reciprocally into their native and into alien plots in the Park Grass experiment (from Davies & Snaydon, 1976).

Comparative ecology, to understand the diversity of environmental pressures and the characteristics of species which enable them to cope, has here a powerful tool. At the same time, population differences emphasize the evolutionary under-pinning of comparative ecology.

ON ECOLOGICAL AMPLITUDE

However, this diversity within species has another significance. It allows the species to extend its range of ecological amplitude. Nowhere is this more clear than in *F. ovina*, whose normal distribution is remarkably wide (Fig. 5). It possesses populations whose differences in behaviour are as extreme as those possessed by normal calcicole and calcifuge species (Snaydon & Bradshaw, 1961). This dependence of ecological amplitude on genetic variation is obvious in many species (Bradshaw, 1984).

To understand the ecology of a species properly, we have therefore to look at this genetic flexibility, expressed in the form of population differences, and consider, as with *Holcus lanatus* (McGrath, Rorison & Baker, 1980), what it tells us about ecology rather than evolution. It will give us more work but, without it, our analyses of species differences may well be erroneous.

I have the temerity to say that this line of enquiry can, and should, be extended downwards to the study of differences between individuals, simply because these are the components of populations and are the ultimate objects of adaptation. There is now strong evidence that variation between individuals provides a crucial contribution to the behaviour of populations and, therefore, of species (Allard & Adams, 1969; Burdon, 1980). It is, of course, also the raw material from which population and species differences evolve.

However, the ecological behaviour of species is not only affected by genetic flexibility. It is also profoundly influenced by phenotypic plasticity. Almost

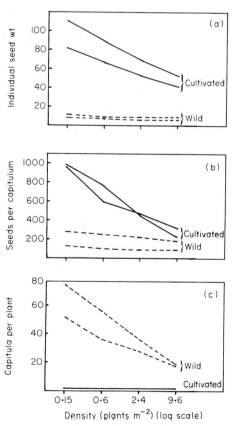

Fig. 11. The different responses to density of two populations of wild sunflower and two varieties of cultivated sunflower (compare with Fig. 2) (from Bradshaw, 1974).

no plant genotype produces only one phenotype; its phenotype, and therefore its ecological behaviour, can be altered by environment (Bradshaw, 1965). It is therefore important that, wherever possible, phenotypic response must be measured. It can give most interesting insights into differences in adaptation of different species, such as between the bryophytes *Thuidium tamariscinum* and *Lophocolea bidentata*, in response to shade (Rincon & Grime, 1985). It can also be found that the behaviour previously ascribed to a species (Fig. 2) is not typical of all of that species. In this particular case, wild sunflowers have the stability of seed size which is normal in wild plants; this is in complete contrast to cultivated material (Fig. 11). All this again means further work but, without it, the ecology of species will be improperly understood.

ON LEVELS AND TYPES OF COMPARISON

All the previous comments arise from the argument that the products of evolution are diverse and that the comparisons from which we can understand this diversity and relate it to the environment are important. This does not mean that we should not look for the generalizations that are so well argued for by Grime & Hodgson (1987). But it does mean that deviations from these laws are as important as the laws themselves and we should try to understand them.

This apparent contradiction within comparative plant ecology can indeed be understood if we bear in mind the nature of evolution. There are basically two trends which are possible: evolution which leads to divergence and evolution which leads to convergence. If we are looking for divergence, we must look for

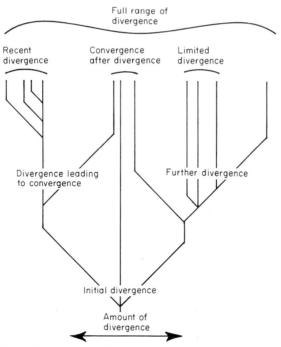

Fig. 12. Origins of the subject materials for comparative ecology: the processes of evolutionary change leading to divergence and convergence.

differences. Convergence should lead to similarities, which we may be able to recognize as general laws. Clearly, divergence is the primary step in evolution, so differences should be the primary material for study. But similarities can be equally important, not only because they are material from a second step in evolutionary adaptation, but also because parallelism in different material may well provide the replication which is needed to establish patterns in the products of primary divergence (Fig. 12).

At the same time, the occurrence either of similarities or of differences will provide primary evidence of what happens in particular situations. When two species occupy the same habitat, do we expect them to be more similar than when they do not? Is coexistence associated with convergence or with divergence? The matter is still a major problem in ecology (Begon, Harper & Townsend, 1986) and evidence is needed. The work of Fitter (1987) shows what fascinating things are to be found.

Of course, important comparisons can also be made at other quite different levels in ecology, notably at the level of community and ecosystem. Indeed, some of these urgently require attention. But these represent a different type of study which requires a separate justification.

SOME FINAL COMMENTS

In essence, the comparative ecology which I am supporting, and which is the subject of this symposium, is a reductionist and experimental autecology. As such, if it is well planned, it ought to be immune from sloppiness of evidence and argument. However, there are still further pitfalls, of which we must beware.

The first is the problem of ancestral characters. To Jacob (1977) we owe the reminder that evolution is a process of tinkering, in which the new is made out of the old. Much of what we see and examine has been evolved in the long past. At the same time, almost everything else has, inevitably, been evolved in the nearer past. Almost nothing can be said to be directly a product of the present. Harper (1982) therefore argues that we should talk about *abaptation*, rather than *adaptation*, evolution from, rather than evolution towards.

Even if we do not go to this extreme and accept that evolution *towards* existing conditions must certainly be occurring, we must remember that such evolution does not mean that the species has already arrived at a state of perfect adaptation; rather, it may simply have made some progress. We must also remember that these species must be made up mostly of ancestral characters from the past. As such, they will be only partly related to present-day conditions and the characteristics of a group of species may be determined by ancestry, as elegantly shown by Hodgson (1986) and further elaborated by Grime & Hodgson (1987). This argument has been used to question the possibility of evaluating the occurrence of adaptation at all (Gould & Lewontin, 1979).

At the same time, some non-adaptive characters may have evolved in parallel with adaptive characters by pleiotropy, or some other process leading to correlated response. All this leads to the conclusion that, if we are to assess the adaptive value of any character, we should do so by its effects rather than by its apparent causes (Clutton-Brock & Harvey, 1979), forgetting any expectation of perfection and seeking only comparative superiority. All of this is profound argument for the experimental, reductionist approaches which are so evident in this series of papers.

We must, therefore, go carefully to find out what is really adaptive – what really counts for survival and provides the *why* in ecology. However, it is only by comparison that any progress is possible. So let us make sure we employ, and analyze, comparisons properly.

ACKNOWLEDGEMENTS

I am most grateful to Alastair Fitter, Philip Grime, Roderick Hunt and, particularly, Martin Mortimer, who painstakingly read the original manuscript and suggested many valuable improvements, also in Paul Keddy and Roy Snaydon for suggesting further material.

REFERENCES

ALLARD, R. W. & ADAMS, J. (1969). Population studies in predominantly self-pollinated species. XIII. Intergenotype competition and population studies in barley and wheat. *American Naturalist*, **103**, 621–645.

BAKER, A. J. M. (1987). Metal tolerance. In: *Frontiers of Comparative Plant Ecology* (Ed. by I. H. Rorison, J. P. Grime, R. Hunt, G. A. F. Hendry & D. H. Lewis), *New Phytologist*, **106** (Suppl.), 93–111. Academic Press, New York & London.

BEGON, M. E. & MORTIMER, A. M. (1986). *Population Ecology*, 2nd Edn. Blackwell Scientific Publications, Oxford.

BEGON, M. E., HARPER, J. L. & TOWNSEND, C. I. (1986). *Ecology*. Blackwell Scientific Publications, Oxford.

BONNIER, G. (1890). Cultures expérimentales dans les Alpes et dans les Pyrénées. *Revue Général de Botanique*, **2**, 513–546.

BRADSHAW, A. D. (1965). Evolutionary significance of phenotypic plasticity in plants. *Advances in Genetics*, **13**, 115–155.

BRADSHAW, A. D. (1974). Environment and phenotypic plasticity. *Brookhaven Symposium in Biology*, **25**, 75–94.

BRADSHAW, A. D. (1984). The importance of evolutionary ideas in ecology—and vice versa. In: *Evolutionary Ecology* (Ed. by B. Sharrocks), pp. 1–25. Blackwell Scientific Publications, Oxford.

BRADSHAW, A. D. & MORTIMER, A. M. (1986). Evolution in communities. In: *Community Ecology* (Ed. by D. J. Anderson), pp. 309–341. Blackwell Scientific Publications, Oxford.

BRADSHAW, A. D., LODGE, R. W., JOWETT, D. & CHADWICK, M. J. (1958). Experimental investigations into the mineral nutrition of several grass species. I. Calcium level. *Journal of Ecology*, **46**, 749–757.

BRADSHAW, A. D., CHADWICK, M. J., JOWETT, D. & SNAYDON, R. W. (1964). Experimental investigations into the mineral nutrition of several grass species. IV. Nitrogen level. *Journal of Ecology*, **52**, 665–676.

BRADSHAW, A. D., CHERRETT, J. M., GOLDSMITH, F. B., GRUBB, P. J. & KREBS, J. R. (1986). 75th anniversary symposium: survey of the members' view. *Bulletin of the British Ecological Society*, **17**, 168–169.

BRENCHLEY, W. E. (1958). (Revised by K. Warington.) *The Park Grass Plots at Rothamsted 1856–1949*. Rothamsted Experimental Station, Harpenden, Herts.

BURDON, J. J. (1980). Intraspecific diversity in a natural population of *Trifolium repens*. *Journal of Ecology*, **68**, 717–735.

CLAPHAM, A. R. (1956). Autecological studies and the 'Biological Flora of the British Isles'. *Journal of Ecology*, **44**, 1–11.

CLAUSEN, J., KECK, D. D. & HIESEY, W. M. (1940) *Experimental Studies on the Nature of Species. I. The Effect of Varied Environments on Western North American Plants*. Carnegie Institute of Washington Publications 520.

CLEMENTS, F. E., WEAVER, J. E. & HANSON, H. C. (1929). *Competition in Cultivated Crops*. Carnegie Institute of Washington Publications 398, pp. 202–233.

CLUTTON-BROCK, T. H. & HARVEY, P. H. (1979). Comparison and adaptation. *Proceedings of the Royal Society, London, Series B*, **205**, 547–565.

DARWIN, C. (1859). *The Origin of Species by Means of Natural Selection*. Harvard Facsimile, 1st Edn 1964.

DAVIES, M. S. & SNAYDON, R. W. (1976). Rapid population differentiation in a mosaic environment. III. Measures of selection pressures. *Heredity*, **36**, 59–66.

DIRZO, R. & SARUKHAN, J. (Eds). (1984). *Perspectives on Plant Ecology*. Sinauer, Sunderland, Massachusetts.

FITTER, A. H. (1987). An architectural approach to the comparative ecology of plant root systems. In: *Frontiers of Comparative Plant Ecology* (Ed. by I. H. Rorison, J. P. Grime, R. Hunt, G. A. F. Hendry & D. H. Lewis), *New Phytologist*, **106** (Suppl.), 61–77. Academic Press, New York & London.

FOWLER, N. L. & ANTONOVICS, J. (1981). Small scale variability in the demography of transplants of two herbaceous species. *Ecology*, **62**, 1450–1457.

FURNESS, S. B. (1978). Effects of temperature upon growth. *Annual Report of the Unit of Comparative Plant Ecology (NERC), 1978*, pp. 5–6. University of Sheffield, Sheffield.

GOULD, S. J. & LEWONTIN, R. C. (1979). Spandrels of Sand Marco and the Panglossian paradigm—a critique of the adaptationist program. *Proceedings of the Royal Society, London, Series B*, **205**, 581–598.

GRIME, J. P. (1977). Evidence for the existence of three primary strategies in plants and its relevance to ecological evolutionary theory. *American Naturalist*, **111**, 1169–1194.

GRIME, J. P. (1979). *Plant Strategies and Vegetation Processes*. Wiley, Chichester.

GRIME, J. P. (1984). The ecology of species, families and communities of the contemporary British flora. *New Phytologist*, **98**, 15–33.

GRIME, J. P. & HODGSON, J. G. (1969). An investigation of the ecological significance of lime-chlorosis by means of large scale comparative experiments. In: *Ecological Aspects of the Mineral Nutrition of Plants* (Ed. by I. H. Rorison), pp. 67–99. Blackwell Scientific Publications, Oxford.

GRIME, J. P. & HODGSON, J. G. (1987). New frontiers of comparative plant ecology. In: *Frontiers of Comparative Plant Ecology* (Ed. by I. H. Rorison, J. P. Grime, R. Hunt, G. A. F. Hendry & D. H. Lewis), *New Phytologist*, **106** (Suppl.), 283–295. Academic Press, New York & London.

GRIME, J. P. & LLOYD, P. S. (1973). *An Ecological Atlas of Grassland Plants*. Edward Arnold, London.

GRIME, J. P., HODGSON, J. G. & HUNT, R. (1987). *Comparative Plant Ecology: A Functional Approach to Common British Plants*. Allen & Unwin, London. (In press.)

HAIZEL, K. A. & HARPER, J. L. (1973). The effects of density and timing or removal on interference between barley, white mustard and wild oats. *Journal of Applied Ecology*, **10**, 23–31.

HARLEY, J. L. & LEWIS, D. H. (Eds) (1984). *The Flora and Vegetation of Britain: Origins and Changes—The Facts and their Interpretation*. Academic Press, London.

HARPER, J. L. (1977). *Population Biology of Plants*. Academic Press, London.

HARPER, J. L. (1982). After description. In: *The Plant Community as a Working Mechanism* (Ed. by E. I. Newman), pp. 11–26. Blackwell Scientific Publications, Oxford.

HODGSON, J. G. (1986). Commonness and rarity in plants with special reference to the Sheffield flora. III. Taxonomic and evolutionary aspects. *Biological Conservation*, **36**, 275–296.

HUTCHINSON, G. E. (1965). *The Ecological Theater and the Evolutionary Play*. Yale University Press, Newhaven.

JACOB, F. (1977). Evolution and tinkering. *Science*, **196**, 1161–1166.

JARVIS, P. G. & JARVIS, M. S. (1963). The water relations of tree seedlings. I. Growth and water-use in relation to soil water potential. *Physiologia Plantarum*, **16**, 236–253.

JOLLIFFE, P. A., MINJAS, A. N. & RUNECKLES, V. (1984). A reinterpretation of yield relationships in replacement series experiments. *Journal of Applied Ecology*, **21**, 227, 244.

KAYS, A. & HARPER, J. L. (1974). The regulation of plant and tiller density in a grass sward. *Journal of Ecology*, **62**, 97–105.

MACARTHUR, R. H. & WILSON, E. O. (1967). *The Theory of Island Biogeography*. Princeton University Press, Princeton.

MCGRATH, S. P., RORISON, I. H. & BAKER, A. J. M. (1980). Growth and distribution of *Holcus lanatus* L. populations with reference to nitrogen source and aluminium. *Annual Report of the Unit of Comparative Plant Ecology (NERC), 1980*, pp. 14–15. University of Sheffield, Sheffield.

MCINTOSH, R. P. (1985). *The Background of Ecology*. Cambridge University Press, Cambridge.

MEDAWAR, P. B. (1967). *The Art of the Soluble*. Methuen, London.

MEDAWAR, P. B. (1979). *Advice to a Young Scientist*. Harper & Row, New York.

ODUM, E. P. (1975). *Ecology*, 2nd Edn. Holt, Rinehart & Winston, New York.

PIANKA, E. R. (1970). On *r*- and *K*-selection. *American Naturalist*, **104**, 592–597.

PIGOTT, C. D. (1982). The experimental study of vegetation. *New Phytologist*, **90**, 389–404.

POPPER, K. (1959). *The Logic of Scientific Discovery*. Hutchinson, London.

PRICE, P. W., GAUD, W. S. & SLOBODCHIKOFF, C. (1984). Introduction: Is there a new ecology? In: *A New Ecology* (Ed. by P. W. Price, C. N. Slobodchikoff & W. S. Gaud), pp. 1–11. John Wiley, New York.

RINCON, J. E. & GRIME, J. P. (1985). Plant 'foraging' for light. *Annual Report of the Unit of Comparative Plant Ecology, 1985*, pp. 21–22. University of Sheffield, Sheffield.

RORISON, I. H. (1987). Mineral nutrition in time and space. In: *Frontiers of Comparative Plant Ecology* (Ed. by I. H. Rorison, J. P. Grime, R. Hunt, G. A. F. Hendry & D. H. Lewis), *New Phytologist*, **106** (Suppl.), 79–92. Academic Press, New York & London.

RUSSELL, E. W. (1973). *Soil Conditions and Plant Growth*, 10th Edn. Longman, London.

SCHMIDT, K. P. & LEVIN, D. A. (1985). The comparative demography of reciprocally sown populations of *Phlox drummondii* Hook. I. Survivorship, fecundities and finite rates of increase. *Evolution*, **39**, 369–404.

SCHOEN, D. J., STEWART, S. C., JECHOWICZ, M. J. & BELL, G. (1986). Partitioning the transplant site effect in reciprocal transplant experiments with *Impatiens capensis* and *Impatiens pallida*. *Oecologia*, **70**, 149–154.

SKEFFINGTON, R. A. & BRADSHAW, A. D. (1980). Nitrogen fixation by plants grown on reclaimed china clay wastes. *Journal of Applied Ecology*, **17**, 469–477.

SNAYDON, R. W. & BRADSHAW, A. D. (1961). Differential responses to calcium within the species *Festuca ovina* L. *New Phytologist*, **60**, 219–234.

SNAYDON, R. W. & DAVIES, T. (1982). Rapid divergence of plant populations in response to recent changes in soil conditions. *Evolution*, **36**, 289–297.

SNOW, A. A. & VINCE, S. W. (1984). Plant zonation in an Alaska salt marsh. II. An experimental study of the role of edaphic conditions. *Journal of Ecology*, **72**, 669–684.

STRONG, D. R. (1980). Null hypotheses in ecology. *Synthese*, **43**, 271–285.

TANSLEY, A. G. (1939). *The British Islands and their Vegetation*. Cambridge University Press, Cambridge.

THURSTON, J. M. (1969). The effect of liming and fertilizers on the botanical composition of permanent grassland, and on the yield of hay. In: *Ecological Aspects of the Mineral Nutrition of Plants* (Ed. by I. H. Rorison), pp. 3–10. Blackwell Scientific Publications, Oxford.

TILMAN, D. (1984). Plant dominance along an experimental nutrient gradient. *Ecology*, **65**, 1445–1453.

TILMAN, D. (1985). The resource-ratio hypothesis of plant succession. *American Naturalist*, **125**, 827–852.

WERNER, P. A. & PLATT, W. W. (1976). Ecological relationships of co-occurring golden rods (*Solidago*: Compositae). *American Naturalist*, **110**, 959–971.

WHITE, J. (1980). Demographic factors in populations of plants. In: *Demography and Evolution in Plant Population* (Ed. by O. T. Solbrig), pp. 21–48. Blackwell Scientific Publications, Oxford.

WILSON, S. D. & KEDDY, P. A. (1986). Measuring diffuse competition along an environmental gradient: results from a shoreline plant community. *American Naturalist*, **127**, 862–869.

YODA, K., KIRA, T., OGAWA, H. & HOZOMI, K. (1963). Self thinning in overcrowded pure stands under cultivated and natural conditions. *Journal of Biology, Osaka City University*, **14**, 107–129.

New Phytol. (1987) **106** (Suppl.), 23–34

SEEDS AND SEED BANKS

By K. THOMPSON

*Department of Biological Sciences, Plymouth Polytechnic, Drake Circus,
Plymouth PL4 8AA, UK*

SUMMARY

The involvement of four factors (life history, predation, possession of a persistent seed bank and seedling establishment) in the evolution of seed size in flowering plants is discussed. Among herbacaeous plants, biennials seem to have larger seeds than annuals and perennials, a difference which may explain the capacity of some biennials to establish in closed vegetation. In many species, small seeds may have evolved at least partly as a defence against predators. Seeds < 3 mg have some immunity from vertebrate predators but much smaller seeds are consumed by invertebrates. Burial reduces or eliminates most predation by invertebrates.

Possession of a persistent seed bank is associated with small, compact, smooth seeds with exacting requirements for germination. Seeds of species which lack seed banks are larger, frequently long or flat, and often have hairs or awns and lax requirements for germination. The evolution of the former group is often intimately connected with the preferences of earthworms. Burial facilitates anchorage of seedlings.

In closed herbaceous vegetation, establishment of seedlings seems to be dependent on a low rate of exhaustion of seed reserves achieved by a large seed, low relative growth rate of seedlings or both. In many data sets, seed size and seedling growth rate are negatively correlated, but the generality of the correlation is in doubt. The large seeds of many herbs, shrubs and trees of woodland are suspected to be of critical importance in seedling emergence through tree litter.

Key words: Seed size, life history, seed predation, seed banks, seedling establishment.

INTRODUCTION

Within the angiosperms seed size varies over 10 orders of magnitude between species (Harper, Lovell & Moore, 1970). This paper considers some major selective forces and constraints which have influenced this remarkable diversity, paying particular attention to those associated with life history, predation, tendency to develop a persistent seed bank and seedling establishment. The intention is not to attempt an exhaustive review of those selective pressures which have acted on the evolution of seed size. For example, dispersal, the subject of much discussion elsewhere, will not be covered. Similarly, variation in seed size within species and within the progeny of individual plants will be ignored. Not that such variation is unimportant; indeed, its very importance demands separate and detailed treatment and therefore places it outside the scope of this paper. I intend rather to focus on critical gaps in our knowledge and to identify areas where there has perhaps been an uncritical acceptance of established dogma.

Some references to woody plants will be made but the bulk of my discussion will concern herbaceous species. Throughout, I use the term 'seed' in its loose sense of the fertilized ovule and associated structures.

LIFE HISTORY

Attempts to discover the relationships (if any) between life history and seed size are frequently confounded by the overriding effect of habitat. Thus, because

0028-646X/87/05S023+12 $03.00/0

annuals normally occupy habitats with much density-independent mortality, one might expect the trade-off between seed number and size to favour numerous small seeds, irrespective of any effect of the annual life-history *per se*. Attempts to overcome this problem have involved the comparison of seed weights of groups of plants classified with respect to life history and drawn from within the same habitat. Chalk grassland is one of the few habitats which contain enough species for this approach to be worthwhile (Silvertown, 1981). He compiled seed weights of 53 chalk grassland species and found that the seeds of annuals were significantly lighter than those of perennials and biennials, which did not differ from each other. Unfortunately, the effect of habitat may not have been eliminated by this analysis for chalk grassland is not a homogeneous one. Over half the annuals in Silvertown's study were winter annuals, a group dependent for establishment on areas of bare soil (Grubb, 1976; Grime, 1979). In contrast, at least some of the perennial species can establish in relatively small gaps in closed turf (Fenner, 1978). Annuals and perennials of chalk grassland share the same habitat but they occupy very different regeneration niches (Grubb, 1977).

Table 1. *Mean log_{10} seed weights of plants with different life histories (from Thompson, 1984)*

	Annual	Biennial	Perennial
Number of species	82	36	177
Mean log_{10} seed weight	−0·487	0·105	−0·329
Standard error of the mean	0·174	0·119	0·055

Sources: Grime *et al.* (1981) with additional data for biennials from Salisbury (1942). Species of closed woodland excluded.

Note: Mean log_{10} seed weight of biennials different from both annuals ($P < 0·001$) and perennials ($P < 0·002$). Annual and perennial seed weights do not differ significantly.

It is perhaps neither possible nor even sensible to try to separate the interactions of life history and habitat with seed size. Accordingly, Thompson (1984) largely ignored the effect of habitat in comparing seed weights of 295 herbaceous dicotyledons (Table 1). The only attempt to remove the confusing effect of interaction between habitat and life history was to exclude species of closed woodland, a habitat which in Britain is almost exclusively occupied by perennials. The surprising result of this study was that seeds of biennials are much heavier than those of annuals and perennials, which do not differ significantly in average weight. Thompson (1984) suggested that many biennials are not the fugitive species they are popularly supposed to be (Harper, 1977) but can persist in closed vegetation owing to their large seeds. This theory, admittedly speculative, depends on a positive correlation between seed size and success of establishment in closed vegetation. The question of whether such a correlation exists will be re-examined later in this paper.

PREDATION

If predators of seeds forage preferentially on particular sizes of seeds, they will tend to cause evolutionary changes in seed size. The evidence that animals do select seeds on the basis of size is overwhelming, yet the overall selective effect is complex owing to the variety of predators involved. Mammals and birds prefer large seeds but quantifying the size effect is complicated by variation in seed abundance, nutritional quality and the identity of the predator. Kelrick *et al.* (1986) found that a range of vertebrate seed predators took seeds down to about 3 mg but consumed very few below this size. Mittelbach & Gross (1984) found that *Peromyscus* foraged preferentially on seeds of *Tragopogon dubius* (6·94 mg) and largely ignored seeds of around 2 mg or less. The mouse, *Sylvaemus flavicollis*, fed on seeds of *Carex pilulifera*, which weigh only 1·3 mg, but predated whole inflorescences rather than single seeds (Kjellson, 1985). Similarly *Apodemus sylvaticus* ate *Scabiosa columbaria* seeds, which also weigh 1·3 mg, but the seeds were presented in groups of 20 (Verkaar, Schenkeveld & Huurnink, 1986). It is difficult to draw any firm conclusions from these data but it seems likely that small seeds (below about 2 to 3 mg) often escape predation by vertebrates, although the degree of immunity will be less if seeds are aggregated. Nearly all trees have seeds much larger than 3 mg and we might therefore expect the evolutionary response of trees to vertebrate seed predators to take forms other than a change in seed size. The available evidence supports this prediction: lodgepole pine responds to predation by varying the number of seeds per cone (Elliott, 1974), oaks and hickories by varying their nutritional value and shell thickness (Smith & Follmer, 1972) and many forest trees by masting (Silvertown, 1980). Probably because his experience is primarily with trees, Smith (1980) states that '... the general effect of seed predators is to select for an increased thickness of endocarp rather than to affect seed size'. This may be true for trees but, for herbs, predation by mammals is probably a potent force selecting for reduced seed size.

Seed predation by insects is likely to have more complex effects on seed size, if only because insects predate seeds in a variety of ways. Insect larvae which consume the entire contents of a seed capsule or inflorescence (e.g. Randall, 1982) may affect the 'packaging' of seeds but are unlikely to affect the size of individual seeds. Where a larva is confined to a single seed, however, there may be powerful evolutionary effects on seed size. Janzen (1969) found that a group of American legumes with large seeds (mean weight 3 g) avoid attack by bruchid beetle larvae by possessing toxic chemicals. Another group, whose seeds are attacked, has a mean seed weight of 0·26 g. The smaller seeds of the latter group do not prevent attack; their greater numbers simply ensure that some seeds escape detection. One species has very small seeds (< 3 mg) and is not attacked, presumably because it is too small to support the development of a bruchid larva.

Most work on post-dispersal insect predators has concerned ants and little is known about other insect taxa. Dispersed seeds of *C. pilulifera* are eaten by *Harpalus fuliginosus*, a carabid beetle (Kjellson, 1985) and seeds of *Lomatium* spp. are eaten by *Eleodes nigrina*, a tenebrionid beetle (Thompson, 1985). *Eleodes* seems to prefer the larger of two species of *Lomatium* but the plants differ chemically as well and, in any case, the seeds of both species are relatively large. Ants select seeds on the basis of size but selectivity varies with species of ant and abundance of seeds. The often conflicting evidence suggests that abundance of seeds may often be a critical factor. In 'cafeteria' experiments, where ants are presented with relatively large quantities of seeds, there is generally a preference for relatively large seeds (Abramsky, 1983; Mittelbach & Gross, 1984; Kelrick *et al.*, 1986). As predicted by optimal foraging theory, selectivity is greater as distance from the nest increases (Davidson, 1978). In contrast, when ants harvest lower density, natural seed populations, they often take smaller seeds. Campbell (1982), reviewing several Australian studies, noted that ants preferred small (1 to 2 mg) but not very small

($<$ 1 mg) seeds. However, very small seeds (e.g. *Filago californica*, 0·04 mg) may be the most frequent food item if they are sufficiently abundant (Inouye, Byers & Brown, 1980). The selective effect of ants on seed size is therefore likely to vary geographically, seasonally and perhaps even over short distances at the same site.

There is, however, a critical difference between foraging in rodents and in ants and beetles. Rodents can find buried seeds, while ants and beetles are strictly surface foragers (Reichmann, 1979; Abramsky, 1983). Westoby, Cousins & Grice (1982) attributed the low rate of seed harvesting by ants at Fowlers Gap, New South Wales to the cracking clay loams which allowed most seeds to become rapidly buried and therefore unavailable. Burial does not protect from predation those seeds large enough to be attractive to rodents. Small seed size itself confers some immunity from predation by mammals and small size coupled with burial may also reduce predation by invertebrates. Before exploring this further, however, we need to consider those aspects of seed size and shape which facilitate seed burial.

SEED BANKS

Considering the characteristics of seeds which form large, persistent reservoirs of buried viable seeds, Thompson & Grime (1979) concluded that: 'The majority have small seeds, and many of the more abundant species possess seeds which are exceedingly small'. Selection for possession of a persistent seed bank therefore appears also to select for small seed size. The connection between seed size and morphology, and possession of a seed bank can be examined in more detail in British Gramineae. Thirty-two common British grasses can be classified as those which possess seed banks and those which do not (Milton, 1939; Dore & Raymond, 1942; Champness & Morris, 1948; Jalloq, 1975; Thompson & Grime, 1979). Seeds of grasses with seed banks are small and compact, while those of grasses which lack seed banks are large and attenuated (Fig. 1). Furthermore, of the 20 species which lack seed banks, 16 have awns and 18 have antrorse hairs (Grime *et al.*, 1981). Of the 12 species which have seed banks, the figures are two and five, respectively. Figure 2 illustrates typical members of both groups. The ecological significance of these seed types has been investigated in Australia by Peart (1979, 1981, 1984). The small, unornamented seeds of species with seed banks rely on burial for successful seedling establishment. In Peart's studies, seedlings from surface-lying seeds of this type failed to anchor and none survived to maturity. In contrast, the larger seeds either had 'active' awns which push the seed into surface cracks or 'passive' awns which appear to influence the attitude of falling seeds, facilitating penetration of the soil surface.

In the British flora, the large seeds of species which lack seed banks tend to be short-lived in the soil (Roberts, 1986 and Fig. 3). In contrast, species with seed banks are often extremely long-lived in the soil, e.g. 100 years for *Verbascum thapsus* and 80 years for *Oenothera biennis* (Kivilaan & Bandurski, 1981), 100 years and 200 years, respectively, for *Digitalis purpurea* and *Hypericum pulchrum* (Darby & Thompson, unpublished data). These long-lived seeds are without exception very small, lending support to the idea that immunity from predation is important in the evolution of seed longevity. In plants which suffer much post-dispersal seed predation, such as large-seeded trees, seed banks are almost unknown. Indeed, masting as an anti-predation strategy (Silvertown, 1980) is clearly incompatible with extended seed longevity. The absence of seed banks among most tropical trees has been attributed to predation (Hopkins & Graham, 1983). Mittelbach & Gross

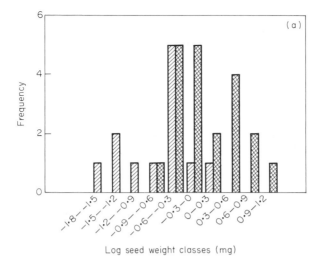

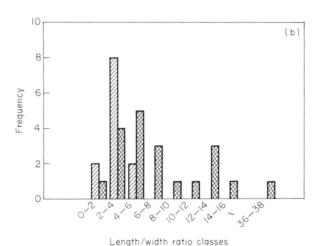

Fig. 1. Frequency of (a) seed weight and (b) length/width ratio in 32 common British grasses. ▨ Species which accumulate persistent buried seed banks. ▨ Species which do not accumulate seed banks. Data from Grime *et al.* (1981).

(1984) rejected the connection between predation and seed longevity but failed to take account of the protection from predation afforded to small seeds by burial.

There is growing evidence that, in temperate regions at least, the association between small seeds and possession of a seed bank owes much to the activities of earthworms. Earthworms ingest seeds of species with seed banks (e.g. *Poa* spp., *Agrostis* spp., *D. purpurea*, *Holcus lanatus*) far more readily than the seeds of species which lack seed banks (e.g. *Festuca* spp., *Lolium perenne*, *Cynosurus cristatus*) (McRill & Sagar, 1973; Grant, 1983). Such ingestion is important both in initial burial and in subsequent return of seeds to the surface in casts. Removal of casts from grassland reduces seed germination. In Grant's study, 70% of seedlings occurred on casts, although these made up only 24 to 28% of the surface area. It is clearly an advantage, therefore, for the seeds of species forming persistent seed banks to be ingested by earthworms and natural selection is likely to favour those

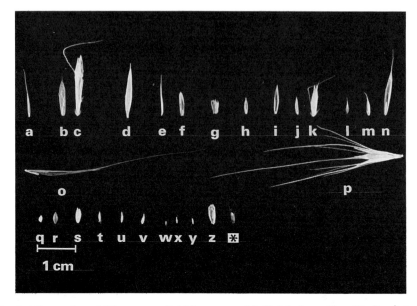

Fig. 2. Dispersules of 27 common British grasses. (a) *Vulpia bromoides*; (b) *Brachypodium pinnatum*; (c) *Avenula pratensis*; (d) *Elymus repens*; (e) *Nardus stricta*; (f) *Lolium perenne*; (g) *Briza media*; (h) *Koeleria macrantha*; (i) *Festuca rubra*; (j) *F. ovina*; (k) *Arrhenatherum elatius*; (l) *Dactylis glomerata*; (m) *Cynosurus cristatus*; (n) *Bromus erectus*; (o) *B. sterilis*; (p) *Hordeum murinum*; (q) *Phleum pratense*; (r) *Milium effusum*; (s) *Holcus lanatus*; (t) *Poa trivialis*; (u) *P. pratensis*; (v) *P. annua*; (w) *Agrostis stolonifera*; (x) *A. vineale*; (y) *A. capillaris*; (z) *Danthonia decumbens*; (*) *Anthoxanthum odoratum*. (q) to (*) accumulate persistent buried seed banks; (a) to (p) do not. Nomenclature follows Clapham, Tutin & Warburg (1981).

characteristics (small size, compact shape, lack of awns and hairs) which permit such ingestion.

It has been argued above that predation is an important ultimate cause of the link between seed longevity, small seed size and possession of a persistent seed bank. Paradoxically, however, predation may not be the proximal cause of the death of short-lived seeds in the soil. The majority of native large-seeded herbs examined in Britain have evolved physiologies which favour rapid germination under a wide range of environmental conditions (Thompson & Grime, 1979; Grime *et al.*, 1981) and consequently the main cause of death following burial is premature germination at a depth unsuitable for emergence (Schafer & Chilcote, 1970; Cook, 1980; Washitani, 1985 cf. Fenner, 1987). In contrast, many species with long-lived seeds appear to have evolved mechanisms so successful in preventing premature germination that the decay rate of the seed bank in the soil is negligible over the short term (van Baalen, 1982).

SEEDLING ESTABLISHMENT

Ever since Salisbury's (1942) demonstration of the correlation between seed size and successional maturity of a habitat, it has been assumed that a capacity for seedling establishment in closed vegetation depends, at least partly, on large seeds. Many experimental studies (e.g. Grime & Jeffrey, 1965; Gross, 1984; Keizer, van Tooren & During, 1985) support this opinion but some do not (e.g. Silvertown & Dickie, 1981). On closer examination, however, the role of seed size in seedling

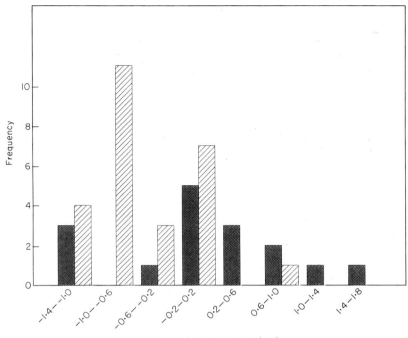

Fig. 3. Frequency classes of seed weight in 42 common British dicotyledons. ▨ Species whose seeds persist in cultivated soil for at least five years. ▧ Species whose seeds persist in cultivated soil for less than five years. Difference between mean $\log_{10}$ seed weights of two groups is significant at $P < 0.01$. Data from Grime *et al.* (1981) and Roberts (1986).

establishment is not straightforward. Fenner (1978) found that, among six ruderal and six closed-turf species, seed size was not related to 'competitive ability' of the seedling, defined as the ratio of mean yield of the seedling in tall turf to that achieved on bare soil. Seed size *was* positively correlated with mean yield in tall turf but growth was so poor that, after 13 weeks, weight of seedlings was little greater than the initial seed weight. He therefore concluded that the correlation between seed weight and seedling size in tall turf was of little significance.

Gross (1984), in a study of six biennials, also found that seed weight was correlated with seedling weight in tall turf but, unlike Fenner, concluded that this *was* evidence of the ability of large-seeded species to establish in closed vegetation. However, as in Fenner's work, the seedlings did not greatly increase their initial seed weight, suggesting again that, '... seed weight was still the main factor determining the size of the seedlings' (Fenner, 1978). Interestingly, competitive ability in tall turf (*sensu* Fenner) *was* correlated with seed weight in the species studied by Gross (1984).

In the species studied by Fenner (1978) and Gross (1984), competitive ability in tall turf was strongly and negatively correlated with potential relative growth rate (RGR) on bare soil. Species which grew fast on bare soil were disadvantaged when growing in competition with established vegetation, suggesting that establishment in closed vegetation may be more dependent on potential seedling RGR than on seed weight. There is a negative correlation between seed size and potential RGR in the species studied by Fenner (1978) and by Gross (1984), although the

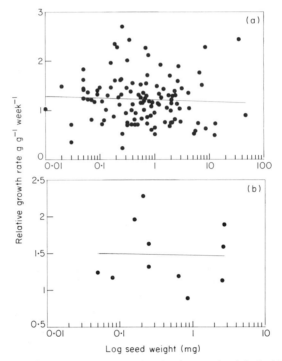

Fig. 4 Relative growth rates of seedlings plotted against $\log_{10}$ seed weight for (a) 117 species of herbs, shrubs and trees, and (b) 11 species of Compositae. Regression lines are shown: neither is significant at $P < 0.05$. Data from Grime & Hunt (1975) and Grime *et al.* (1981).

relationship is weaker in the case of Fenner's species. If such a relationship were general, and not just a chance feature of these two data sets, it might prove difficult to assess the relative importance of seed size and seedling RGR in seedling establishment. The accepted interpretation of Salisbury's (1942) results could be open to question. Several data sets exist which might be used to test the generality of the relationship. Atkinson (1973) found separate negative relationships between seedling growth rate and seed weight in 21 herbaceous dicotyledons and in 21 monocotyledons (all but one of which were grasses). Unfortunately, Atkinson's definition of growth rate (mg of plant weight/mg of seed weight/week) severely underestimated the true RGR of large-seeded species, because the embryo makes up a relatively small proportion of the mass of large seeds (Fenner, 1983). Fenner (1983) found a highly significant negative relationship between seedling RGR and seed weight in 23 species of Compositae. Undoubtedly the most extensive data set, however, is provided by Grime & Hunt (1975) and Grime *et al.* (1981), from which one can plot RGR against seed weight for 117 species [Fig. 4(a)]. These data provide no evidence of any correlation, nor is there any correlation in any substantial subset of the data in Figure 4(a), such as grasses, annual dicots or perennial dicots. There is also no relationship in the Compositae [Fig. 4(b)], even though seven of the 11 species are among those included in Fenner's (1983) study.

Why, then, did Fenner (1983) find a negative correlation between RGR and seed weight, while the data of Grime & Hunt (1975) show no correlation for a similar group of species? The answer may lie in the conditions under which the experiments were performed. Fenner's seedlings were grown in distilled water;

Grime & Hunt's were grown in complete nutrient solution. Fenner (1983) has shown that large seeds contain relatively little mineral nutrients and growth of seedlings from them may therefore be reduced, even in the first few days after germination, by absence of external nutrients. Recently, Wulff (1986) has demonstrated the same effect at an intraspecific level; the growth of seedlings from large seeds of *Desmodium paniculatum* is suppressed more by low nutrient availability than that of seedlings from small seeds. Fenner (1987) has commented on the differential rates of utilization of different elements from reserves in seeds as they germinate.

The evidence therefore suggests that seedling establishment in closed vegetation is facilitated by a low overall rate of exhaustion of seed reserves and that this end may be achieved either by large seed size or by low potential RGR of the seedling or, frequently, by both. When a sufficiently large number of species is examined, the negative correlation between seed size and potential seedling RGR evident in several studies is no longer apparent. Low growth rate and large seed size appear to have evolved as separate solutions to the problem of seedling establishment in a hostile environment.

In addition to showing that the seeds of herbaceous species increase in size with maturity of habitat, Salisbury (1942) also showed that shrubs and trees have exceptionally large seeds. He explained this finding as follows: ' ... the largest reproductive structures ... are associated with trees, for these are mostly members of the late stages of plant successions, and thus their seedlings are more liable than those of most other categories of plants to have to make considerable growth before emerging into an illumination adequate for independent growth ... ' (Salisbury, 1942; pp. 4–5). Does this explanation make ecological sense? The seedlings of plants of tall herbaceous vegetation experience steep light gradients within a few decimetres of the ground and, in such circumstances, the growth in height permitted by a large seed probably pays dividends in terms of increased interception of light. In woodlands, however, the main agents of shading are trees, which generate a light gradient near the ground which is very shallow indeed. There may, of course, be shading by the ground flora but the light gradient generated by the ground flora is certainly no steeper than that generated by herbaceous vegetation of unshaded habitats. Indeed, it is difficult to see how shade can have much influenced the evolution of large seeds in woodland herbs such as *Mercurialis perennis* and *Hyacinthoides nonscripta*. Both germinate and establish in late winter when the trees are leafless and the biomass of ground flora herbs is at a seasonal nadir. Germination and establishment of many large-seeded temperate trees also takes place during winter or early spring, e.g. *Quercus* spp. (Jones, 1959) and *Fraxinus excelsior* (Wardle, 1961). It is also clear that shade tolerance in the seedlings of temperate trees and woodland herbs owes rather more to low respiration rate (Grime, 1966; Loach, 1967; Mahmoud & Grime, 1974) than to a capacity to 'emerge into an illumination adequate for independent growth'.

It therefore seems that the evolution of large seeds in woodland plants cannot be attributed entirely to shade. A significant contributory factor, hinted at briefly by Grime (1979), may be selection for an ability to emerge from beneath tree litter. Tree litter physically inhibits the growth of mature plants (Sydes & Grime, 1981a, b) and presumably has even more drastic effects on seedlings. A large seed may be necessary to produce a seedling robust enough either to penetrate or dislodge a considerable weight of tree litter. Seedlings in grassland also have to cope with litter but the evidence suggests that herbaceous and tree litter differ qualitatively in their effect on seedlings. Gross (1984) studied the effect of simulated herbaceous litter (straw) on growth of seedlings of six biennials. Growth was often better in the litter treatment than on bare soil, regardless of seed size. Winn (1985) obtained similar results in a field study of *Prunella vulgaris*. She found that herbaceous litter promoted successful emergence but that tree litter was inhibitory, and attributed this to qualitative and quantitative differences between the two types of litter. Herbaceous litter was relatively sparse and consisted of small plant parts; its main effect was to keep surface soil moist. Tree litter consisted of, '... relatively thick layers... that could inhibit germination by blocking out light or by physically preventing emergence' (Winn, 1985).

Salisbury (1942) was probably correct to ascribe the increase in seed size from open to closed herbaceous vegetation to competition for light. However, the argument cannot be extended uncritically to woody vegetation, where ability to penetrate litter may be of equal importance. Nor, of course, are shade and litter the only selective forces tending to increase seed size in woodland plants. The fleshy fruits of many woody plants, both temperate and tropical, may need to be large to attract potential dispersers. In the wet tropics there is very little litter accumulation and, here, shade may well have been the major selective force acting to increase seed size. It must also be borne in mind that large seed size may today have acquired roles which were not necessarily involved in its evolution. For example, the demonstration that seedlings of large-seeded trees are extremely good at emerging from short, steep light gradients (Grime & Jeffrey, 1965) does not prove that selection for such emergence was instrumental in their evolution.

CONCLUSION

Numerous selective pressures have acted, and continue to act, on the evolution of seed size. The consequences of many of these pressures are poorly understood. In some cases, this is because there has been little investigation or because critical experiments have yet to be performed. These problems particularly affect our understanding of the roles of life history, post-dispersal predation of seeds other than by ants and vertebrates, and seedling RGR. Indeed, two of the most challenging areas identified in this paper are the role of predation in shaping the characteristics of seeds of herbaceous plants and the extent to which selection for large seed size in closed vegetation has been dependent on a correlation between seed size and seedling RGR. Also in future, more account will need to be taken of the interaction between these current selection pressures and phylogenetic constraints on seed size (Hodgson & Mackey, 1986). Insights into these areas will come only from careful, comparative experiments.

REFERENCES

ABRAMSKY, Z. (1983). Experiments on seed predation by rodents and ants in the Israeli desert. *Oecologia*, **57**, 328–332.
ATKINSON, D. (1973). Some general effects of phosphorus deficiency on growth and development. *New Phytologist*, **72**, 101–111.
CAMPBELL, M. H. (1982). Restricting losses of aerially sown seed due to seed-harvesting ants. In: *Ant–Plant Interactions in Australia* (Ed. by R. C. Buckley), pp. 25–30. Junk, The Hague.
CHAMPNESS, S. S. & MORRIS, K. (1948). Populations of buried viable seeds in relation to contrasting pasture and soil types. *Journal of Ecology*, **36**, 149–173.
CLAPHAM, A. R., TUTIN, T. G. & WARBURG, E. F. (1981). *Excursion Flora of the British Isles*, 3rd Edn. Cambridge University Press, Cambridge.

COOK, R. (1980). The biology of seeds in the soil. In: *Demography and Evolution in Plant Populations* (Ed. by O. T. Solbrig), pp. 107–188. Blackwell, Oxford.

DAVIDSON, D. W. (1978). Experimental tests of the optimal diet in two social insects. *Behavioural Ecology and Sociobiology*, **4**, 35–41.

DORE, W. G. & RAYMOND, L. C. (1942). Pasture studies. XXIV. Viable seeds in pasture soil and manure. *Scientific Agriculture*, **23**, 69–79.

ELLIOTT, P. F. (1974). Evolutionary responses of plants to seed-eaters: pine squirrel predation of lodgepole pine. *Evolution*, **28**, 221–231.

FENNER, M. (1978). A comparison of the abilities of colonizers and closed-turf species to establish from seed in artificial swards. *Journal of Ecology*, **66**, 953–963.

FENNER, M. (1983). Relationships between seed weight, ash content and seedling growth in twenty-four species of Compositae. *New Phytologist*, **95**, 697–706.

FENNER, M. (1987). Seedlings. In: *Frontiers of Comparative Plant Ecology* (Ed. by I. H. Rorison, J. P. Grime, R. Hunt, G. A. F. Hendry & D. H. Lewis), *New Phytologist*, **106** (Suppl.), 35–47. Academic Press, London.

GRANT, J. D. (1983). The activities of earthworms and the fate of seeds. In: *Earthworm Ecology* (Ed. by J. E. Satchell), pp. 107–122. Chapman & Hall, London.

GRIME, J. P. (1966). Shade avoidance and tolerance in flowering plants. In: *Light as an Ecological Factor* (Ed. by R. Bainbridge, G. C. Evans & O. Rackham), pp. 281–301. Blackwell, Oxford.

GRIME, J. P. (1979). *Plant Strategies and Vegetation Processes*. Wiley, Chichester

GRIME, J. P. & HUNT, R. (1975). Relative growth rate: its range and adaptive significance in a local flora. *Journal of Ecology*, **63**, 393–422.

GRIME, J. P. & JEFFREY, D. W. (1965). Seedling establishment in vertical gradients of sunlight. *Journal of Ecology*, **53**, 621–642.

GRIME, J. P., MASON, G., CURTIS, A. V., RODMAN, J., BAND, S. R., MOWFORTH, M. A. G., NEAL, A. M. & SHAW, S. (1981). A comparative study of germination charcteristics in a local flora. *Journal of Ecology*, **69**, 1017–1059.

GROSS, K. L. (1984). Effects of seed size and growth form on seedling establishment of six monocarpic perennial plants. *Journal of Ecology*, **72**, 369–387.

GRUBB, P. J. (1976). A theoretical background to the conservation of ecologically distinct groups of annuals and biennials in the chalk grassland ecosystem. *Biological Conservation*, **10**, 53–76.

GRUBB, P. J. (1977). The maintenance of species-richness in plant communities: the importance of the regeneration niche. *Biological Reviews*, **52**, 107–145.

HARPER, J. L. (1977). *The Population Biology of Plants*. Academic Press, London.

HARPER, J. L., LOVELL, P. H. & MOORE, K. G. (1970). The shapes and sizes of seeds. *Annual Review of Ecology and Systematics*, **1**, 327–356.

HODGSON, J. G. & MACKEY, J. M. L. (1986). The ecological specialization of dicotyledonous families within a local flora: some factors constraining optimization of seed size and their possible evolutionary significance. *New Phytologist*, **104**, 497–515.

HOPKINS, M. S. & GRAHAM, A. W. (1983). The species composition of soil seed banks beneath lowland tropical rainforests in north Queensland, Australia. *Biotropica*, **15**, 90–99.

INOUYE, R. S., BYERS, G. S. & BROWN, J. H. (1980). Effects of predation and competition on survivorship, fecundity, and community structure of desert annuals. *Ecology*. **61**, 1344–1351.

JALLOQ, M. C. (1975). The invasion of molehills by weeds as a possible factor in the degeneration of reseeded pasture. I. The buried viable seed population of molehills from four reseeded pastures in West Wales. *Journal of Applied Ecology*, **12**, 643–657.

JANZEN, D. H. (1969). Seed-eaters *vs* seed size, number, toxicity, and dispersal. *Evolution*, **23**, 1–27.

JONES, E. W. (1959). Biological flora of the British Isles: *Quercus* L. *Journal of Ecology*, **47**, 169–222.

KEIZER, P. J., VAN TOOREN, B. F. & DURING, H. J. (1985). Effects of bryophytes on seedling emergence and establishment of short-lived forbs in chalk grassland. *Journal of Ecology*, **73**, 493–504.

KELRICK, M. I., MACMAHON, J. A., PARMENTER, R. R. & SISSON, D. V. (1986). Native seed preferences of shrub-steppe rodents, birds and ants: the relationships of seed attributes and seed use. *Oecologia*, **68**, 327–337.

KILVILAAN, A. & BANDURSKI, R. S. (1981). The one hundred-year period for Dr Beal's seed viability experiment. *American Journal of Botany*, **68**, 1290–1292.

KJELLSON, G. (1985). Seed fate in a population of *Carex pilulifera* L. II. Seed predation and its consequences for dispersal and seed bank. *Oecologia*, **67**, 424–429.

LOACH, K. (1967). Shade tolerance in tree seedlings. I. Leaf photosynthesis and respiration in plants raised under artificial shade. *New Phytologist*, **66**, 607–621.

MAHMOUD, A. & GRIME, J. P. (1974). A comparison of negative relative growth rates in shaded seedlings. *New Phytologist*, **73**, 1215–1219.

MCRILL, M. & SAGAR, G. R. (1973). Earthworms and seeds. *Nature*, **243**, 482.

MILTON, W. E. J. (1939). The occurrence of buried viable seeds in soils at different elevations and in a salt marsh. *Journal of Ecology*, **27**, 149–159.

MITTELBACH, G. G. & GROSS, K. L. (1984). Experimental studies of seed predation in old fields. *Oecologia*, **65**, 7–13.

PEART, M. H. (1979). Experiments on the biological significance of the morphology of seed dispersal units in grasses. *Journal of Ecology*, **67**, 843–863.

PEART, M. H. (1981). Further experiments on the biological significance of the morphology of seed dispersal units in grasses. *Journal of Ecology*, **69**, 425–436.

PEART, M. H. (1984). The effects of morphology, orientation and position of grass diaspores on seedling survival. *Journal of Ecology*, **72**, 437–453.

RANDALL, M. G. (1982). The dynamics of an insect population throughout its altitudinal distribution: *Coleophora alticolella* (Lepidoptera) in Northern England. *Journal of Animal Ecology*, **51**, 933–1016.

REICHMAN, O. J. (1979). Desert granivore foraging and its impact on seed densities and distributions. *Ecology*, **60**, 1085–1092.

ROBERTS, H. A. (1986). Seed persistence in soil and seasonal emergence in plant species from different habitats. *Journal of Applied Ecology*, **23**, 639–656.

SALISBURY, E. J. (1942). *The Reproductive Capacity of Plants*. Bell, London.

SCHAFER, D. E. & CHILCOTE, D. O. (1970). Factors influencing persistence and depletion in buried seed populations. II. The effects of soil temperature and moisture. *Crop Science*, **10**, 342–345.

SILVERTOWN, J. W. (1980). The evolutionary ecology of mast-seeding in trees. *Biological Journal of the Linnean Society*, **14**, 235–250.

SILVERTOWN, J. W. (1981). Seed size, life span and germination date as co-adapted features of plant life history. *American Naturalist*, **118**, 860–864.

SILVERTOWN, J. W. & DICKIE, J. B. (1981). Seedling survivorship in natural populations of nine perennial chalk grassland plants. *New Phytologist*, **88**, 555–558.

SMITH, C. C. (1980). The coevolution of plants and seed predators. In: *Coevolution of Animals and Plants* (Ed. by P. H. Raven), pp. 53–57. University of Texas, Austin.

SMITH, C. C. & FOLLMER, D. (1972). Food preferences of squirrels. *Ecology*, **53**, 82–91.

SYDES, C. & GRIME, J. P. (1981a). Effects of tree leaf litter on herbaceous vegetation in deciduous woodland. I. Field investigations. *Journal of Ecology*, **69**, 237–248.

SYDES, C. & GRIME, J. P. (1981b). Effects of tree leaf litter on herbaceous vegetation in deciduous woodland. II. An experimental investigation. *Journal of Ecology*, **69**, 249–262.

THOMPSON, J. N. (1985). Postdispersal seed predation in *Lomatium* spp. (Umbelliferae): variation among individuals and species. *Ecology*, **66**, 1608–1616.

THOMPSON, K. (1984). Why biennials are not as few as they ought to be. *American Naturalist*, **123**, 854–861.

THOMPSON, K. & GRIME, J. P. (1979). Seasonal variation in the seed banks of herbaceous species in ten contrasting habitats. *Journal of Ecology*, **67**, 893–921.

VAN BAALEN, J. (1982). Germination ecology and seed population dynamics of *Digitalis purpurea*. *Oecologia*, **53**, 61–67.

VERKAAR, H. J., SCHENKEVELD, A. J. & HUURNIK, C. L. (1986). The fate of *Scabiosa columbaria* (Dipsacaceae) seeds in a chalk grassland. *Oikos*, **46**, 159–162.

WARDLE, P. (1961). Biological flora of the British Isles: *Fraxinus excelsior* L. *Journal of Ecology*, **49**, 739–751.

WASHITANI, I. (1985). Field fate of *Amaranthus patulus* seeds subjected to leaf-canopy inhibition of germination. *Oecologia*, **66**, 338–342.

WESTOBY, M., COUSINS, J. M. & GRICE, A. C. (1982). Rate of decline of some soil seed populations during drought in western New South Wales. In: *Ant–Plant Interactions in Australia* (Ed. by R. C. Buckley), pp. 7–10. Junk, The Hague.

WINN, A. A. (1985). Effects of seed size and microsite on seedling emergence of *Prunella vulgaris* in four habitats. *Journal of Ecology*, **73**, 831–840.

WULFF, R. D. (1986). Seed size variation in *Desmodium paniculatum*. II. Effects on seedling growth and physiological performance. *Journal of Ecology*, **74**, 99–114.

New Phytol. (1987) **106** (Suppl.), 35–47

SEEDLINGS

By M. FENNER

Biology Department, University of Southampton, Southampton SO9 5NH, UK

SUMMARY

The factors which cause the death of seedlings were examined and an attempt was made to determine the selective forces to which they are subjected in the field. After briefly dealing with the difficulties of defining a seedling, I examine a few cases from the recent literature in which the causes of seedling mortality are identified and quantified either by monitoring populations under natural conditions or by manipulating the environment experimentally. The special vulnerability of the immediate post-germination phase is illustrated and the consistent underestimation of seedling loss in many studies is suggested. Some interesting cases of the effects of neighbours on seedling mortality are examined and it is shown that nearest neighbours may affect a seedling's chance of being grazed. It is concluded that since the causes of seedling mortality are so varied and unpredictable in most habitats, individual plasticity may provide a better strategy for survival than highly specialized adaptations to specific mortality factors.

Key words: Seedling, establishment, mortality factors, survival, neighbour effects.

DEFINITION

First of all, what is a seedling? It is not easy to arrive at a precise definition. A seedling comes into being when a seed germinates but the point at which it ceases to be a seedling is much less clear. A physiologist might say that a young plant continues to be a seedling up to the point at which it becomes independent of its seed's nutrient reserves. But this is an unsatisfactory definition for two reasons. Firstly, the transference of dependence from internal to external sources of nutrients occurs gradually. A seedling can undoubtedly absorb minerals from the soil whilst still mobilizing its stored reserves. Secondly, the transference of dependence occurs at a different rate for each mineral element. This was shown in a very neat experiment by Krigel (1967) on *Trifolium subterraneum**. Newly germinated seedlings of this species were grown in nutrient solutions deficient in single essential elements for various periods and then transferred to full nutrient solutions. After 7 d, the usable seed reserves of calcium had been exploited, whereas the usable internal sources of potassium lasted 21 d. External supplies of the five elements tested were required in the following chronological order: calcium, phosphorus, nitrogen, magnesium and potassium. Krigel's data for *T. subterraneum* have been largely confirmed in experiments on *Senecio vulgaris* (Fenner, 1986). In these, seedlings were again deprived of individual essential elements. The growth attained in 3 weeks by seedlings depending only on their internal supplies of each of the major nutrient elements was taken as a measure of their dependence on external supplies of each element for early growth. Seedlings deprived of calcium grew to only 4·3 times their original (embryo) weight, whereas those deprived of iron were able to attain 28·8 times the embryo weight. Table 1 records the results for all the elements tested. Clearly, each element is limiting to a different degree and dependence on external supplies occurs sooner for some elements than for others.

* Except where authorities are given, nomenclature follows Clapham, Tutin & Warburg (1981).

0028-646X/87/05S035 + 13 $03.00/0

Table 1. *Mean dry weights of seedlings of* Senecio vulgaris *after 3 weeks growth in Hoagland's nutrient solutions with one element missing*

Nutrient solution	Multiple of embryo weight
−Ca	2·99
−N	5·06
−P	6·36
−Mg	6·44
−K	7·87
−Fe	21·8
−S	25·3
Control I (distilled water)	1·72
Control II (full nutrients)	42·4

Data are expressed as multiples of the mean embryo weight (0·138 mg). The parent plants had been grown on Hoagland's full nutrient solution. (After Fenner, 1986; data reproduced with kind permission of Blackwell Scientific Publications Ltd.)

It follows from this that, if we define a seedling in terms of the cessation of its dependence on seed nutrient reserves, we really should specify whether we mean dependence on the most limiting or the least limiting element. If we use the former, the seedling stage would probably not last more than a few days in the majority of species. If we use the latter, the plant might have to be regarded as a permanent 'seedling' because dependence on external supplies of certain trace elements may never occur in the plant's lifetime. The seed may contain more than is required for the adult. Woodbridge (1969) showed this for boron in peas (*Pisum sativum* L.), and Meagher, Johnson & Stout (1952) for molybdenum in peas (*P. sativum*) and beans (*Phaseolus vulgaris* L.).

In defining the point at which a seedling becomes an established plant, it would probably be best to avoid the question of dependence altogether and simply say that a plant is a seedling so long as it continues to mobilize any of its seed reserves, regardless of whether or not it 'needs' to. This broader definition allows us to ignore the fact that the translocation of seed reserves may continue long after the plant has any real dependence on them. For example, jays sometimes remove the cotyledons of oak seedlings without any apparent ill-effects on establishment (Bossema, 1979), though the stored nutrients continue to be absorbed if the cotyledons are left *in situ*. It is possible that, for many species, internal reserves can be dispensed with at an early stage if adequate external supplies are available. Seedlings of some tropical trees normally drop their large cotyledons soon after germination without using much of their stored nutrients, e.g. *Durio zibethinus* Murr. (Ng, 1978).

Unfortunately, even this loose definition of a seedling is rather impractical for use by ecologists as it is clearly not possible to determine whether or not the young plants observed in the field are still mobilizing their internal reserves, whether dependent on them or not. In practice, most ecological studies which use the term do not define it and we are often left with a rather vague picture of a young plant with perhaps two or three leaves (in addition to the cotyledons). In view of the diverse morphology of seedlings, it would indeed be difficult to find a uniformly applicable definition for use in the field. The best we can hope for is a definition in the context of each individual study. In this paper, I give the age of the seedlings involved when dealing with experiments of my own. When quoting other people's work, I use the word in the same sense as they do.

VULNERABILITY OF THE EARLY PHASE

How does the vulnerability of seedlings compare with that of other stages in the life-cycle? Graphs in which the log of number of survivors of a cohort of seedlings is plotted against time show how the risk of death changes throughout the life-cycle. These so-called 'Deevey curves' (after Deevey, 1947) may be convex (type I) indicating low mortality in the seedling stage followed by higher risk as the plant ages. A straight line (Deevey type II curve) represents a constant risk of death, while a concave curve (type III) indicates heavy seedling mortality followed by a safer adulthood. Inspection of the Deevey curves of a wide range of species shows that there is a general tendency for short-lived plants to have type I curves (low seedling mortality) and for long-lived plants to have type III curves (high seedling mortality). (There are of course exceptions, e.g. Sharitz & McCormick, 1973.) From a theoretical point of view, high early mortality (characteristic of trees and shrubs) would be expected to select for iteroparity and longevity as a more stable evolutionary strategy (Bell, 1976).

Most published Deevey curves give an underestimation of the mortality risk in the very earliest post-germination phase because recording usually starts with the emerged seedling. Even if the number of seeds present in the soil is known (or can be estimated), the proportion which germinates but fails to emerge is extremely difficult to determine, especially for small seeds under field conditions. In general, the number of seedlings appearing in the field is only a small fraction of the seeds present in the soil. Most losses occur between seed dissemination and seedling emergence (e.g. 99 % loss in the case of Scots pine; Guittet & Laberche, 1974) but how the death risk is distributed between the pre-germination and post-germination (but unemerged) phases is unknown.

Some indication of the possible extent of pre-emergence losses, at least in some cases, is given by an experiment (Fenner, 1985; pp. 113–114) in which seeds of two tropical weed species (*Bidens pilosa* L. and *Achyranthes aspera* L.) were buried at various depths in the field, to find out what proportion of those germinating would fail to emerge. The position of the seeds at each level was marked by sowing them within horizontally placed plastic rings. Figure 1 shows the number which germinated at each depth and the proportion of these which failed to emerge. Although germination itself is reduced at the deeper levels, virtually all seeds below 8 cm which did germinate failed to emerge and the overall losses owing to this are considerable. Clearly, deaths of seedlings in the immediate post-germination, pre-emergence phase would repay further study. If they were included in all population censuses, most Deevey curves would probably show an initial steep drop, regardless of their subsequent shape.

Another feature of demographic studies which tends to underestimate seedling mortality is the length of the period between monitoring points. Any seedlings which emerge and die between monitoring points are unrecorded. This period is often as much as 15 d to one month (e.g. Sarukhan & Harper, 1973) so large numbers of seedlings may be missed, as the authors themselves recognize.

It is probable that the susceptibility of seedlings to all the common hazards (e.g. drought, competition and grazing) is at its greatest in the earliest stages. Certainly,

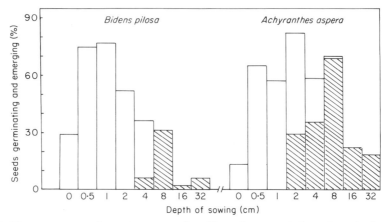

Fig. 1. The emergence and non-emergence of seedlings of two tropical weed species from seeds sown at various depths in the field. Unshaded area, emergents; shaded area, germinated but non-emergent. (From Fenner, 1985.)

there is some evidence that small (i.e. young) seedlings are more likely to be killed by grazing gastropods, whereas larger (i.e. older) seedlings may escape with being merely nibbled. This was found to be the case in a recent experiment in which uneven-aged populations of *Tagetes patula* L. seedlings were subjected to grazing by snails (*Helix aspersa* Müller). The seedlings were planted in trays over a period of 9 d, giving a randomly ordered array of seedlings which were 1, 2, 4, 6 and 9 d old. Each tray contained equal numbers of each age group. Five snails were added to each tray for 24 h, and then the numbers of seedlings in each age category which were (1) ungrazed, (2) grazed but not killed and (3) killed were recorded. The results (Fig. 2) indicate that vulnerability to grazing is greatest in the youngest seedlings and declines markedly with age. The probability of survival nearly doubled between day 1 and day 2; and by day 6 this had increased by a factor of eight. Not only is a young seedling more likely to be attacked by the snail but the attack is more likely to result in the death of the seedling. None of the surviving 1 d old seedlings had been grazed sublethally, whereas 37 % of the 9 d old ones had. This may be the result of (1) the young seedlings being more palatable, either chemically or physically (cf. Coley, 1987); or (2) the snails tending to eat not more than a fixed amount from any one seedling before moving on. Seedlings with a volume less than the critical one may be completely devoured, while larger seedlings may survive the loss of the same quantity of tissue.

Although the absolute rates of mortality quoted here relate only to the conditions of this rather artificial experiment, the principle that the youngest seedlings are most at risk (not only from grazing) may be widely applicable. The results indicate the importance of making seedling counts at no longer than daily intervals in demographic studies. Again, if this were done, a steep initial tail would be added to the Deevey curves.

CAUSES OF SEEDLING MORTALITY

What do seedlings die of in the field? In recent years, many excellent demographic studies have been carried out on a great range of species (e.g. Dolan & Sharitz, 1984; Waite, 1984; Kachi & Hirose, 1985; Burdon, Marshall & Brown, 1983;

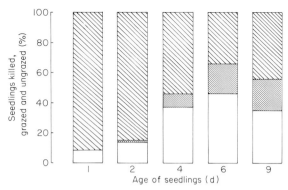

Fig. 2. The fate of seedlings of *Tagetes patula* of different ages subjected to grazing by snails (*Helix aspersa*). ☐ % seedlings ungrazed; ▨ % seedlings grazed, but not killed; ⧄ % seedlings killed. (Fenner, unpublished data.)

Klemow & Raynal, 1985). In most of these investigations, the emphasis has been largely focussed on the numbers of survivors at each stage in the life-cycle, with little attention to quantifying the causes of death in those which did not survive. In practice, determining the causes of death of seedlings presents formidable problems. A large proportion simply disappear in the intervals between monitoring. They may have been eaten, washed away or buried by earthworms. Without 24 h observation, this will remain a major problem, though possibly solvable by the use of video. Even if a dead body remains, it is often impossible to assign a cause of death. Shrivelled remains could indicate drought, grazed roots, disease or even genetic defect. The causes themselves may be complex. The death of a crowded diseased seedling could be ascribed to competition or to the pathogen. The immediate causes of death may not always be clearly distinguishable from a myriad of possible contributory factors.

In spite of these difficulties, a number of authors have been brave enough to try to specify the causes of seedling death and even to quantify them. In these attempts, three approaches have been used: (1) monitoring at frequent intervals and making reasonable deductions from the available evidence on the probable cause of death, assigning impossible cases to an 'unknown' category; (2) establishing correlations between mortality and possibly relevant factors, such as seedling density, distance from parent, density of surrounding vegetation, etc. or (3) experimental manipulation of mortality by altering the environment, e.g. by the addition of water or nutrients or exclusion of grazing.

One of the most meticulous of the monitoring type of studies is that of Mack & Pyke (1984) on *Bromus tectorum*, an annual grass with continuous germination. Most deaths occur in the first two months. Mack & Pyke categorized deaths owing to various causes in populations of this species at three sites (moist, mesic and dry) for three years (1977–78, 78–79 and 79–80). Their data (Table 2) show that, with a single species, the mortality rate and the relative importance of different causes of death vary markedly from place to place and that, even within a single population at the same site, the causes of mortality change dramatically from year to year. For example, desiccation accounted for as much as 58 % of deaths in one site in 1978–79 but for only 3 % the following year. Smut killed 31 % in 1977–78 but only 2 % in the next season. Large fluctuations were also recorded for winter deaths (caused by frost heaving) and for grazing (mainly by small mammals). The

Table 2. *Percentage of deaths due to various agents in populations of* Bromus tectorum *at three sites in eastern Washington, USA, over three years. (After Mack & Pyke, 1984; data reproduced with kind permission of Blackwell Scientific Publications Ltd.)*

	Desiccation	Winter death	Smut	Grazing	Unknown	Total numbers
Moist site						
1977–78	44·2	13·5	30·8	—	11·5	n = 104
1978–79	57·8	32·6	2·0	2·1	5·6	n = 2331
1979–80	3·0	40·6	16·8	2·0	37·6	n = 197
Mesic site						
1977–78	38·0	2·3	42·6	—	17·0	n = 176
1978–79	10·7	40·4	17·7	—	31·2	n = 936
1979–80	44·8	8·1	25·3	—	21·7	n = 442
Dry site						
1977–78	25·3	14·7	2·1	—	57·9	n = 190
1978–79	40·5	35·9	7·8	0·21	15·6	n = 1920
1979–80	—	5·3	36·6	6·2	51·9	n = 243

'unknown' fraction is often quite high, averaging 28 % in this study and accounting (or, rather, not accounting) for 58 % in one instance.

This method of monitoring is probably the most satisfactory for assigning causes for deaths but useful inferences can be drawn from cases where a correlation can be established between mortality rates and some environmental factor or population characteristic. For example, Augspurger (1983) found that mortality was positively correlated with seedling density in populations of *Platypodium elegans* J. Vogel, a wind-dispersed canopy tree in the forests of Barro Colorado Island, Panama. The immediate cause of death was attack by pathogenic fungi causing damping off. Hett (1971) also established a positive correlation between mortality and seedling density in another wind-dispersed tree species, the North American maple *Acer saccharum* Marsh. Although correlations between any two factors do not imply that one is the cause of the other, they can provide useful pointers for further possible lines of investigation.

An interesting approach to determining causes of seedling death in field populations is to monitor the effect of experimentally eliminating one possible cause of mortality. This can be done by excluding grazing, eliminating competition or ameliorating some adverse condition like drought or nutrient shortage. For example, Wellington & Noble (1985) were able to quantify the rate of mortality of seedlings of *Eucalyptus incrassata* Labill. owing to drought by noting the increased survival which resulted when water was applied regularly to the site. The addition of nutrients also reduced fatalities (probably by improving root growth, so preventing deaths from drought). In another experiment of this type, Grime & Curtis (1976) found that seedling mortality in *Arrhenatherum elatius* in a limestone grassland was much reduced by applications of water or phosphate. The size of the reduction in the death rate when these treatments are applied gives at least a rough quantitative indication of the importance of the relevant factors as causes of seedling death.

Experiments have also been done in which competition from other plants is eliminated by removal of vegetation from the vicinity of the seedlings. Gross (1980) increased the survivorship of seedlings of *Verbascum thapsus* in an old field site in Michigan by selective removal of different categories of vegetation (annuals, biennials and perennial dicots). The more categories removed, the lower was the rate of mortality. The importance of competition as a cause of seedling death was also well demonstrated by Newell, Solbrig & Kincaid (1981) for *Viola* species in New England forest sites. When established vegetation was removed from around the *Viola* seedlings their survival was approximately doubled. Figure 3 shows the effect of removal of competition in *Viola blanda* Willd. (Note that mortality is only reduced and that other factors such as drought, disease and trampling continue to operate.)

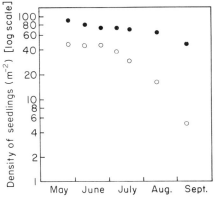

Fig. 3. Seedling mortality for two cohorts of *Viola blanda* at Concord Field Station, Bedford Massachusetts, USA, in adjacent 1 m² plots, one undisturbed (open circles), the other with all adult vegetation, including violets, removed (closed circles). (From Newell *et al.*, 1981; figure reproduced with kind permission of Blackwell Scientific Publications Ltd.)

The experimental elimination of grazing in several communities reduces seedling deaths dramatically. Christensen & Muller (1975) excluded herbivorous mammals from a chaparral understorey community and found that, between December and May, seedling density increased from 63 to 89 in protected plots but declined from 39 to zero in unprotected ones. In another grazing exclusion experiment, Parker & Salzman (1985) protected a population of *Gutierrezia microcephala* (DC) Gray (a plant of arid grasslands in New Mexico) from grasshoppers. The annual mortality of second year plants in intact vegetation was 8 % when protected compared with 47 % when exposed to natural levels of grazing.

Another possible cause of seedling death, particularly on nutrient-poor soils, is the failure to form mycorrhizas. The marked increase in growth which results from mycorrhizal associations is well established (e.g. Hayman & Mosse, 1971; Daughtridge *et al.*, 1986), and their formation by seedlings in the field is probably the rule rather than the exception (Cooke, 1977). For example, Gay, Grubb & Hudson (1982) found that, in a chalk grassland soil, seedlings of the majority of species tested (which included annuals, biennials and perennials) formed mycorrhizas within 2 weeks of germination. However, tests on a range of phosphorus-deficient soils have shown that the extent of infection induced in seedlings varies greatly from place to place, possibly because some soils contain rather few of the relevant fungi or strains which are ineffective (Mosse & Hayman, 1971). The failure to form mycorrhizas, or even a delay in doing so, would at very least put a seedling at a competitive disadvantage. The amount of mortality owing to this cause in the field is unknown for any habitat but it could be a major factor in low-nutrient soils.

Clearly, seedlings of different species die from a wide variety of causes and few generalizations can be made since the causes vary so greatly for the same species from place to place and from season to season. It seems probable, however, that in harsh habitats mortality tends to be due to abiotic factors, such as washout in salt marshes (Jefferies, Davy & Rudmik, 1981) and surface disturbance in sand dunes (Mack, 1976). In contrast, in more mesic habitats, biotic factors such as competition and grazing may account for relatively more seedling deaths. This point is neatly illustrated on a very small scale by *Sedum smallii* (Britt.) Ahles growing on granite outcrops in the southeastern USA (Sharitz & McCormick, 1973). On the shallowest soils, seedling mortality is due mainly to washout and drought. With increasing soil depth, these severe abiotic conditions are ameliorated but competition from other plants now becomes the main cause of seedling mortality. Here, the cause of death changes within a few metres. Another case in which the cause of seedling mortality changes over a short distance is given by Jerling (1982). In the lower zones of a salt marsh, seedlings of *Plantago maritima* died mainly from flooding. In the upper marsh, competition from established vegetation was the major cause of death. A temporal change in mortality factors occurs in some species with an extended period of germination. In a study of a population of *Echium plantagineum* L., Burdon *et al.* (1983) found that the main cause of death in early cohorts was drought but that seedlings of later cohorts were killed by competition.

THE EFFECTS OF NEIGHBOURS

Except in the most open habitats, seedlings will have to contend with other plants in their vicinity. In most cases, neighbouring plants, whether of the same or a different species, are likely to have a detrimental effect on the seedling by competing for resources as has been shown experimentally (e.g. Fenner, 1978; Gross & Werner, 1982; Gross, 1984). Competition amongst seedlings is often most severe from members of their own species and the importance of pre-empting the available resources by early germination and establishment has been amply demonstrated in a variety of studies (Symonides, 1977; Weaver & Cavers, 1979; Cook, 1980; Zimmerman & Weis, 1984). Autopathic effects amongst seedlings are another possible cause of mortality (Lodhi, 1979).

However, in some cases, the presence of established plants in its vicinity may increase the chances of survival of the seedling. This is especially likely to occur in harsh environments such as deserts and semi-deserts, where neighbouring vegetation may ameliorate the conditions locally and create favourable sites for establishment of seedlings in its vicinity. Turner *et al.* (1966) showed this to be the case for the saguaro cactus (*Carnegiea gigantea*) (Engelm.) Britt. & Rose in the Sonoran desert, where shade cast by the existing perennial vegetation is essential for seedling establishment. Dependence may last for five to 10 years. This 'nurse effect' is by no means confined to such extreme environments. In chalk grassland in the Netherlands, *Carlina vulgaris* and *Linum catharticum* seedlings had lower rates of mortality where bryophyte cover was highest (Keizer, van Tooren & During, 1985) and the establishment of *Carduus nutans* L. in old fields appears to be facilitated by the mulch formed by litter of *Bromus japonicus* Thumb. (Lee & Hamrick, 1983). Hillier (1984) found that the establishment of seedlings of several grassland species in 40 cm gaps in turf was facilitated by the presence of vegetative regrowth which seemed to ameliorate conditions in the gaps. The nurse effect is of course frequently used in the practice of forestry where slow growing, shade-tolerant species are protected in the initial stages by fast growing species which are later removed.

The effect of neighbouring plants in reducing seedling mortality is perhaps most interesting from an evolutionary point of view in those cases where the seedlings are nursed by contemporary seedlings of the same species. The forest floor annual, *Floerkea proserpinacoides* Willd., has seedlings which stand a better chance of successful emergence if they are clumped in a dense population rather than a sparse one. The seedlings emerge simultaneously, and collectively push aside the surface litter, the presence of which is a major cause of death for these seedlings (Smith, 1983). Seedlings which are members of such clumps are likely to be closely related and so their apparent mutual altruism does not conflict with the idea that each individual is promoting the survival of its own 'selfish' genes (Dawkins, 1976). Closely related individuals are also likely to germinate and emerge simultaneously, thereby making the co-operation more effective. The mutual promotion of *germination* by clumped seeds of the same species is known in a number of species (Linhart & Pickett, 1973) but experiments in the field have not indicated that the phenomenon is of any benefit under natural conditions (Waite & Hutchings, 1979).

It is also possible that proximity of plants, both conspecific or of other species, may aid survival not only by providing an inoculum of mycorrhizal fungi (see above) but also by providing a direct source of nutrients via mycorrhizal connections (Read, Francis & Finlay, 1985).

THE EFFECTS OF NEIGHBOURS ON THE RISK OF BEING GRAZED

An interesting factor which influences a seedling's risk of mortality is the effect that its position in relation to its neighbours has upon the likelihood of its being grazed. Dirzo & Harper (1980) suggest that the feeding behaviour of herbivores (such as slugs) may be influenced by the choice available. Experiments by Cottam (1985) indicate that the palatability of a plant to a grasshopper may be influenced as much by the presence of a neighbouring plant as by the characteristics of the first plant. With a view to investigating the effects which nearest neighbours have on the susceptibility of seedlings to grazing by snails, I set up a pilot experiment in which snails were presented with a series of two-species, 50:50, mixtures of seedlings which differed only in the pattern in which the seedlings were planted. The patterns were chosen so that each of the two species would have zero, one, two or three individuals of its own species (and consequently four, three, two or one individuals, respectively, of the other species) as its nearest neighbours (Fig. 4). The aim was to see if the 'neighbour environment' of each seedling influenced the probability of its being grazed by snails.

The two species chosen were *Taraxacum officiniale* and *Senecio jacobaea*, a highly palatable and unpalatable species, respectively (Dirzo, 1980). Newly germinated seedlings were planted in a regular 2 cm square grid at equal density in all trays, on John Innes No. 2 compost. Each of the five patterns were represented by one tray. In addition, monocultures of each of the two species were planted out. When the seedlings were 3 weeks old, three snails (*Helix aspersa*) were added to each tray. The snails were confined by covering the trays with clear plastic propagator lids. After 4 d exposure to grazing, the seedlings were scored as (1) ungrazed, (2) grazed but not killed or (3) killed (Table 3). Comparison between patterns was

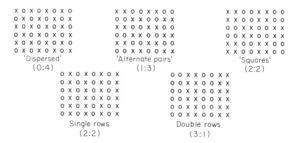

Fig. 4. The five sowing patterns used for subjecting 50:50 seedlings mixture of *Senecio jacobaea* and *Taxaracum officinale* to grazing by the snail *Helix aspersa*. The figures in brackets are the 'nearest-neighbour ratios', that is, the ratio of the number of seedlings of the same species to the number of seedlings of the other species in the four nearest neighbours. The seedlings were spaced at 2 cm intervals, 144 per tray excluding a guard row.

Table 3. *The effect of sowing pattern on the probability of a seedling being grazed by snails* (Helix aspersa) *in 50:50 mixture of* Taraxacum officinale *and* Senecio jacobaea

Pattern	Nearest-neighbour ratio (own species: other species)	% seedlings attacked (grazed or killed)		% attacked in mixture / % attacked in monoculture	
		Senecio	*Taraxacum*	*Senecio*	*Taraxacum*
Dispersed	0:4	36·0	87·5	2·07	1·40
Alternate pairs	1:3	37·5	55·5	2·16	0·89
Squares	2:2	43·8	77·5	2·52	1·24
Singles rows	2:2	27·0	10·9	1·55	0·17
Double rows	3:1	31·3	95·3	1·80	1·52
Senecio monoculture	—	17·4	—	—	—
Taraxacum monoculture	—	—	62·5	—	—

standardized by expressing the probability of being attacked (either just grazed or totally eaten) in a particular pattern as a proportion of the probability of being attacked in the corresponding monoculture (see the last two columns in Table 3). This ratio represents the factor by which the chances of being attacked change when the 'neighbour environment' changes from that in the monoculture to that in each of the various mixtures. A factor of less than 1·0 indicates that the seedling is less likely to be attacked in the mixture. The neighbours appear in such cases to protect the seedling from the herbivore. A factor of more than 1·0 indicates that the presence of neighbours of the opposite species renders the seedling more likely to be grazed.

The results show that the likelihood of a seedling being attacked varies markedly between one pattern and another. Overall, the less palatable species, *S. jacobaea*, is much more vulnerable in the mixtures than the monoculture. On average the probability of attack is increased by a factor of 2·02 (with a range of 1·55 to 2·52). The effect of the sowing pattern in mixtures is much more variable for *Taraxacum*. In some patterns (e.g. alternate pairs and single rows), the *Taraxacum* seedlings appear to be protected from grazing by their neighbours; in the other patterns they are rendered more vulnerable. There is no simple relationship between vulnerability and the 'nearest neighbour ratio' (i.e. own species:other species) for either species.

The detailed explanation of these results clearly lies in the feeding behaviour of the snails. However, the implications for the survival of an individual self-sown seedling in the field are interesting. Its survival may be determined not only by the identity of the seedlings which appear in its immediate vicinity but even by the particular conformation of those neighbours in relation to the seedling.

CONCLUSION

The survival of seedlings of any species does not depend entirely on the characteristics of the seedlings themselves. Various features of the parent plants can increase the chances of seedling survival. For example, effective dispersal may ensure escape from dominance by the parent plant; dormancy and gap-detection mechanisms facilitate escape through time from competition with established plants; synchronous fruiting and germination produce large seedling cohorts, possibly resulting in satiation of predators. These mechanisms can, however, do no more than ensure that the seedlings are favourably placed for establishment. Once germination has occurred, the seedling depends on its own morphological and physiological characteristics to cope with the various factors threatening its survival.

The demographic studies of Mack & Pyke (1984), Sharitz & McCormick (1973), Burdon *et al.* (1983), Lee & Hamrick (1983) and many others indicate clearly that, even within a single species, the causes of seedling mortality vary markedly from season to season, and from place to place. The surviving seedlings each year will have been subjected to a different set of environmental conditions (either biotic or abiotic) and have been selected for a different set of characters. We might expect this situation to have two consequences. One is the maintenance of a high degree of genetic diversity within populations (Hartgerink & Bazzaz, 1984). The second is a high degree of phenotypic plasticity in individual seedlings. Many seedlings certainly show great plasticity with respect to such features as hypocotyl extension when buried at different depths (Fenner, 1985) or overall growth rate when subjected to different degrees of competition (Fenner, 1978). Root/shoot ratios of seedlings are well known to be highly responsive to the growth conditions (Aung, 1974). In an environment which is constantly changing, we might not expect to find highly specialized 'adaptations' in seedlings to particular mortality factors. A high degree of plasticity might well be the most appropriate evolutionary strategy. Indeed, the mortality factors recorded in the literature are so wide ranging in character, so complex in their operation and so capricious in their occurrence that it is difficult to envisage any 'adaptation' in individual seedlings other than phenotypic plasticity which would enable even a few seedlings to survive in any habitat.

REFERENCES

AUGSPURGER, C. K. (1983). Seed dispersal of the tropical tree *Platypodium elegans*, and the escape of its seedlings from fungal pathogens. *Journal of Ecology*, **71**, 759–771.

AUNG, L. H. (1974). Root–shoot relationships. In: *The Plant Root and its Environment* (Ed. E. W. Carson), pp. 29–61. University Press, Virginia.

BELL, G. (1976). On breeding more than once. *American Naturalist*, **110**, 57–77.

BOSSEMA, I. (1979). Jays and oaks: an eco-ethological study of a symbiosis. *Behaviour*, **70**, 1–117.

BURDON, J. J., MARSHALL, D. R. & BROWN, A. H. D. (1983). Demographic and genetic changes in populations of *Echium plantagineum*. *Journal of Ecology*, **71**, 667–679.

CHRISTENSEN, N. L. & MULLER, C. H. (1975). Relative importance of factors controlling germination and seedling survival in *Adenostoma* chaparral. *American Midland Naturalist*, **93**, 71–78.

CLAPHAM, A. R., TUTIN, T. G. & WARBURG, E. F. (1981). *Excursion Flora of the British Isles*, 3rd Edn. Cambridge University Press, Cambridge.

COLEY, P. D. (1987). Interspecific variation in plant anti-herbivore properties: the role of habitat quality and rate of disturbance. In: *Frontiers of Comparative Plant Ecology* (Ed. by I. H. Rorison, J. P. Grime, R. Hunt, G. A. F. Hendry & D. H. Lewis), *New Phytologist*, **106** (Suppl.), 251–263. Academic Press, London.

COOK, R. E. (1980). Germination and size-dependent mortality in *Viola blanda*. *Oecologia*, **47**, 115–117.

COOKE, R. (1977). *The Biology of Symbiotic Fungi*. Wiley, London.

COTTAM, D. A. (1985). Frequency-dependent grazing by slugs and grasshoppers. *Journal of Ecology*, **73**, 925–933.

DAUGHTRIDGE, A. T., PALLARDY, S. G., GARRETT, H. G. & SANDER, I. L. (1986). Growth analysis of mycorrhizal and non-mycorrhizal black oak (*Quercus velutina* Lam.) seedlings. *New Phytologist*, **103**, 473–480.

DAWKINS, R. (1976). *The Selfish Gene*. Oxford University Press, Oxford.

DEEVEY, E. S. (1947). Life tables for natural populations of animals. *Quarterly Review of Biology*, **22**, 283–314.

DIRZO, R. (1980). Experimental studies on slug–plant interactions. I. The acceptability of thirty plant species to the slug *Agriolimax caruanae*. *Journal of Ecology*, **68**, 981–998.

DIRZO, R. & HARPER, J. L. (1980). Experimental studies on slug–plant interactions. II. The effect of grazing by slugs on high density monocultures of *Capsella bursa-pastoris* and *Poa annua*. *Journal of Ecology*, **68**, 999–1011.

DOLAN, R. W. & SHARITZ, R. R. (1984). Population dynamics of *Ludwigia leptocarpa* (Onagraceae) and some factors affecting size hierarchies in a natural population. *Journal of Ecology*, **72**, 1031–1041.

FENNER, M. (1978). A comparison of the abilities of colonizers and closed-turf species to establish from seed in artificial swards. *Journal of Ecology*, **66**, 953–963.

FENNER, M. (1985). *Seed Ecology*. Chapman & Hall, London.

FENNER, M. (1986). A bioassay to determine the limiting minerals for seeds from nutrient-deprived *Senecio vulgaris* plants. *Journal of Ecology*, **74**, 497–505.

GAY, P. E., GRUBB, P. J. & HUDSON, H. J. (1982). Seasonal changes in the concentrations of nitrogen, phosphorus and potassium, and in the density of mycorrhiza in biennial and matrix-forming perennial species of closed chalkland turf. *Journal of Ecology*, **70**, 571–593.

GRIME, J. P. & CURTIS, A. V. (1976). The interaction of drought and mineral nutrient stress in calcareous grassland. *Journal of Ecology*, **64**, 975–988.

GROSS, K. L. (1980). Colonization by *Verbascum thapsus* (mullein) of an old-field in Michigan: experiments on the effects of vegetation. *Journal of Ecology*, **68**, 919–927.

GROSS, K. L. (1984). Effects of seed and growth form on seedling establishment of six monocarpic perennial plants. *Journal of Ecology*, **72**, 369–387.

GROSS, K. L. & WERNER, P. A. (1982). Colonizing abilities of 'biennial' plant species in relation to ground cover: implications for their distributions in a successional sere. *Ecology*, **63**, 921–931.

GUITTET, J. & LABERCHE, J. C. (1974). L'implantation naturelle du pin sylvestre sur pelouse xérophile en forêt de Fontainebleau. II. Démographie des graines et des plantules au voisinage des vieux arbres. *Oecologia Plantarum*, **9**, 111–130.

HARTGERINK, A. P. & BAZZAZ, F. A. (1984). Seedling-scale environmental heterogeneity influences individual fitness and population structure. *Ecology*, **65**, 198–206.

HAYMAN, D. S. & MOSSE, B. (1971). Plant growth responses to vesicular–arbuscular mycorrhiza. I. Growth of *Endogone*-inoculated plants in phosphate-deficient soils. *New Phytologist*, **70**, 19–27.

HETT, J. M. (1971). A dynamic analysis of age in sugar maple seedlings. *Ecology*, **52**, 1071–1074.

HILLIER, S. H. (1984). *A quantitative study of gap recolonization in two contrasted limestone grasslands*. Ph.D. thesis, University of Sheffield, UK.

JEFFERIES, R. L., DAVY, A. J. & RUDMIK, T. (1981). Population biology of the salt marsh annual *Salicornia europaea* agg. *Journal of Ecology*, **69**, 17–31.

JERLING, L. (1982). *Population dynamic of a perennial herb* (Plantago maritima L.) *along a distributional gradient – a demographical study*. Doctorate thesis, University of Stockholm, Sweden.

KACHI, N. & HIROSE, T. (1985). Population dynamics of *Oenothera glazioviana* in a sand-dune system with special reference to the adaptive significance of size-dependent reproduction. *Journal of Ecology*, **73**, 887–901.

KEIZER, P. J., VAN TOOREN, B. F. & DURING, H. J. (1985). Effects of bryophytes on seedling emergence and establishment of short-lived forbes in chalk grassland. *Journal of Ecology*, **73**, 493–504.

KLEMOW, K. M. & RAYNAL, D. J. (1985). Demography of two facultative biennial plant species in an unproductive habitat. *Journal of Ecology*, **73**, 147–167.

KRIGEL, I. (1967). The early requirement for plant nutrients by subterranean clover seedlings (*Trifolium subterraneum*). *Australian Journal of Agricultural Research*, **18**, 879–886.

Lee, J. M. & Hamrick, J. L. (1983). Demography of two natural populations of musk thistle (*Carduus nutans*). *Journal of Ecology*, **71**, 923–936.

Linhart, Y. B. & Pickett, R. A. (1973). Physiological factors associated with density-dependent seed germination in *Boisduvallia glabella* (Ongraceae). *Zeitschrift für Pflanzenphysiologie*, **70**, 367–370.

Lodhi, M. A. K. (1979). Germination and decreased growth of *Kochia scoparia* in relation to its autoallelopathy. *Canadian Journal of Botany*, **57**, 1083–1088.

Mack, R. N. (1976). Survivorship of *Cerastium atrovirens* at Aberffraw, Anglesey. *Journal of Ecology*, **64**, 309–312.

Mack, R. N. & Pyke, D. A. (1984). The demography of *Bromus tectorum*: the role of microclimate, grazing and disease. *Journal of Ecology*, **72**, 731–748.

Meagher, W. R., Johnson, C. M. & Stout, P. R. (1952). Molybdenum requirements of leguminous plants supplied with fixed nitrogen. *Plant Physiology*, **27**, 223–230.

Mosse, B. & Hayman, D. S. (1971). Plant growth responses to vesicular–arbuscular mycorrhiza. II. In unsterilized field soils. *New Phytologist*, **70**, 29–34.

Newell, S. J., Solbrig, O. T. & Kincaid, D. T. (1981). Studies on the population biology of the genus *Viola*. III. The demography of *Viola blanda* and *Viola pallens*. *Journal of Ecology*, **69**, 997–1016.

Ng, F. S. P. (1978). Strategies of establishment in Malayan forest trees. In: *Tropical Trees as Living Systems* (Ed. by P. B. Tomlinson & M. H. Zimmermann), pp. 129–162. Cambridge University Press, Cambridge.

Parker, M. A. & Salzman, A. G. (1985). Herbivore exclosure and competitor removal: effects on juvenile survivorship and growth in the shrub *Gutierrezia microcephala*. *Journal of Ecology*, **73**, 903–913.

Read, D. J., Francis, R. & Finlay, R. D. (1985). Mycorrhizal mycelia and nutrient cycling in plant communities. In: *Ecological Interactions in Soil* (Ed. by A. H. Fitter), pp. 193–217. Blackwell, Oxford.

Sarukhan, J. & Harper, J. L. (1973). Studies on plant demography: *Ranunculus repens* L., *R. bulbosus* L. and *R. acris* L. I. Population flux and survivorship. *Journal of Ecology*, **61**, 675–716.

Sharitz, R. R. & McCormick, J. F. (1973). Population dynamics of two competing annual plant species. *Ecology*, **54**, 723–740.

Smith, B. H. (1983). Demography of *Floerkea proserpinacoides*, a forest-floor annual. III. Dynamics of seed and seedling populations. *Journal of Ecology*, **71**, 413–425.

Symonides, E. (1977). Mortality of seedlings in natural psammophyte populations. *Ekologia Polska*, **25**, 635–651.

Turner, R. M., Alcorn, S. M., Olin, G. & Booth, J. A. (1966). The influence of shade, soil and water on saguaro establishment. *Botanical Gazette*, **127**, 95–102.

Waite, S. (1984). Changes in the demography of *Plantago coronopus* at two coastal sites. *Journal of Ecology*, **72**, 809–826.

Waite, S. & Hutchings, M. J. (1979). A comparative study of establishment of *Plantago coronopus* L. from seeds sown randomly and in clumps. *New Phytologist*, **82**, 575–583.

Weaver, S. E. & Carver, P. B. (1979). The effects of date of emergence and emergence order on seedling survival rates in *Rumex crispus* and *R. obtusifolius*. *Canadian Journal of Botany*, **57**, 730–738.

Wellington, A. B. & Noble, I. R. (1985). Post-fire recruitment and mortality in a population of the mallee *Eucalyptus incrassata* in semi-arid, south-eastern Australia. *Journal of Ecology*, **73**, 645–656.

Woodbridge, C. G. (1969). Boron deficiency in pea *Pisum sativum* cv. 'Alaska'. *Journal of the American Society for Horticultural Science*, **94**, 542–544.

Zimmerman, J. K. & Weis, I. M. (1984). Factors affecting survivorship, growth, and fruit production in a beach population of *Xanthium strumarium*. *Canadian Journal of Botany*, **62**, 2122–2127.

New Phytol. (1987) **106** (Suppl.), 49–60

POPULATION DYNAMICS – THE INFLUENCE OF HERBIVORY

By H. J. VERKAAR*

Department of Population Biology, University of Leiden, PO Box 9516, 2300 RA Leiden, The Netherlands

SUMMARY

Numbers in plant populations often vary widely and, as yet, only a few studies describe adequately how they are determined. The importance of abiotic factors in establishment as well as in determining the carrying capacity of the habitat has been widely recognized. It is also accepted that intraspecific competition for limiting resources may limit numbers in a density-dependent way. In addition, both herbivores and pathogens may reduce numbers of plants through mechanisms which are either dependent or independent of population density.

Herbivory affects plants not only in the vegetative and reproductive phases but also in the pre-dispersal and post-dispersal phases of seeds. Because many studies on the population dynamics of plants have been confined to the former phases, the relative importance of herbivores is likely to have been underestimated for the latter. Furthermore, because herbivory often does not result in mortality but in reduced growth and seed production, its effects may be manifested only as reductions in the size of later generations. The effects of herbivory can therefore only be assessed properly if they are studied over several generations.

Examples are presented of plant populations in which numbers are determined by competition for limiting resources. However, other situations are recognized in which it is the activity of herbivores which appears to determine numbers. There is a continuing need for studies of population dynamics, but there is an urgent requirement for these studies to be related to a unifying conceptual base.

Key words: Herbivory, intraspecific competition, population dynamics, regulation, survivorship.

INTRODUCTION

One of the major concerns of our celebrant of this Jubilee, the Unit of Comparative Plant Ecology, has been and still is the development of general concepts in plant ecology. A major step towards a coherent set of unified concepts has been made by the recognition of primary and secondary strategies (Grime, 1979). The distinction between various strategies provides a basis upon which to classify and interpret 'life-table' studies, although to link these two approaches more securely further research and theory is needed, to recognize which factors ultimately determine plant numbers and regulate population size in various ecological situations.

Although many papers bear the word 'regulation' in their title, there is much confusion about the mechanisms underlying this phenomenon. Over many years, there has been an active scientific controversy concerning the way population sizes of plants are limited, and much heat in the debate seems to be coupled with a distinct lack of concrete evidence. Some insight into the division of opinion on this topic can be obtained by reference to the views of its main protagonists. On the

* Present address: Ministry of Transport and Public Works, Road and Hydraulic Engineering Division, Section Research on Natural Environment and Landscape, PO Box 5044, 2600 GA Delft, The Netherlands.

0028-646X/87/05S049 + 12 $03.00/0

one hand, Brues (1946) concluded that the influence of feeding by insects on the growth and reproduction of plants was considerable and acted as a major determinant in both natural vegetation and plant communities associated with agriculture, horticulture and forestry. Chew (1974) and Mattson & Addy (1975) also supported the hypothesis that consumers, especially insects, regulated whole ecosystems. In sharp contrast, Hairston, Smith & Slobodkin (1960), Slobodkin, Smith & Hairston (1967) and Jermy (1984, 1985), who compared the regulation of various trophic levels, maintained that abiotic conditions exercise the main limitations upon plant numbers and biomass. This was thought to occur by reduction of the number of suitable habitats as well as by reduction of numbers and biomass within one habitat. These abiotic conditions were said to operate either as density-independent factors determining the carrying capacity or as density-dependent factors influencing intraspecific competition for limiting resources. In addition, it was proposed that populations of herbivores would be limited by predators and unfavourable abiotic circumstances rather than food limitation (Andrewartha & Birch, 1954; Hairston et al., 1960; Andrewartha, 1970).

The argument was developed further by Nicholson (1954, 1957), Klomp (1962) and Bakker (1964), who suggested that abiotic factors cannot regulate plant populations in a density-dependent way. They expressed the view that 'weather does not regulate numbers, but determines the number of suitable habitats' (Klomp, 1962). They recognized, however, that abiotic conditions do influence the outcome of competition, which is of course a density-dependent process.

Hairston's view received much criticism with regard to its premises, consistency and methodology, and its lack of evidence from natural situations (Murdoch, 1966; Ehrlich & Birch, 1967). As recognized by Dempster & Pollard (1986), density-dependent processes are difficult to study under field conditions. Nevertheless, Schoener (1983) and Connell (1983) have surveyed existing field studies on competition and have felt able to conclude that these support Hairston's hypothesis. In studies by both Schoener (1983) and Connell (1983), the surveys which were examined included a great variety of species and habitats, but for a major part they dealt with plants in the vegetative and reproductive phase. In some cases the seedling phase was considered also, but little attention was paid to processes in the seed bank.

Obviously, investigators who study plant population dynamics have to consider the processes in all life phases during several generations to understand the factors that regulate numbers (Antonovics & Levin, 1980; Crawley, 1983). Although Harper (1977) described a wealth of studies on population dynamics, one can hardly find a single investigation of population regulation under natural conditions in his handbook. This deficiency has not been rectified by more recent publications, although the five-year study of population dynamics in the summer annual, *Erucastrum gallicum**, by Klemow & Raynal (1983) and the investigation of *Cakile edentula* (Bigel.) Hook. by Keddy (1981) are notable contributions.

So far, competitive ability and competitive exclusion have been generally regarded as prime determinants of plant numbers. In animal population ecology, the universal importance of these processes is now questioned (e.g. den Boer, 1986) and, as I shall show later for plants, the general validity of the competition-based approach is also under critical review. In its place, other interspecific relations which operate in a density-dependent way, especially plant–animal

---

* Nomenclature, except when authorities are given, follows Clapham, Tutin & Warburg (1981).

relationships, are receiving increased attention. Particularly in this latter field, much new work is under way, and it may be seen as a major frontier in population biology at the present time. In this paper, therefore, I will briefly discuss some of the pioneer studies in this field.

Two phenomena underlying the mechanism of the determination of numbers of plants in populations may be identified. Firstly, mainly density-independent factors, abiotic as well as interspecific competition, limit the maximum that is possible at a certain place at a particular moment, i.e. the carrying capacity. Secondly, density-dependent processes may contribute to the regulation of the population size.

INCREASE AND DECREASE OF NUMBERS

In natural communities, wide variation in abundance is found not only in the number of species represented but also in the number of specimens of each species. Populations of some species are characterized by the occurrence of only a few specimens, whereas the individuals of others are extremely numerous. In populations of the same species within the same micro-habitat, there is also often a large variation in numbers from season to season and from year to year. Especially in short-lived plants, numbers of reproductive plants can vary more than 100-fold when years are compared (Runge, 1968; Sagar & Mortimer, 1976; Watt, 1981; Grubb, Kelly & Mitchley, 1982; van der Meijden *et al.*, 1985). The numbers of germinable seeds in the seed bank can vary even more (Grime, 1978; Thompson & Grime, 1979; Schenkeveld & Verkaar, 1984a). In trees, especially in those with non-fleshy fruits, mast years occur in which, when weather conditions are favourable, seed production is very abundant (Larson & Schubert, 1970; Janzen, 1975; Silvertown, 1980).

In population ecology, some generalizations have been developed, partly based on 'life-table' studies; the distinction between r- and K-selected species is one of them (MacArthur & Wilson, 1967; Roughgarden, 1971; Hickman, 1979). r-Selected species show high seedling mortality and high seed output, whereas K-selected species have low seedling mortality, mainly density-independent, a high survival potential and low seed output. More or less comparable with this distinction are those derived from case studies in a particular habitat. Bartolome (1979) distinguished between a 'conservative strategy' with a low mortality in early life phases and a potentially low seed production, and of an 'opportunistic strategy' with a high seedling mortality and a high seed production. In their study of survival in various micro-habitats, Schenkeveld & Verkaar (1981) noted that some species showed a shift from 'optimal' population density where most plants of a certain life phase were found to another 'optimal' population density in the next life phase. This may result in a considerable loss of individuals during life. Such species with a low seed weight, a relatively high seed production, a short life span and a high seedling mortality were named 'spenders' (During *et al.*, 1985). Conversely, 'savers' do not show such a shift and have a higher seed weight, a lower seedling mortality, lower seed production and a longer life span.

The pool of carbohydrates invested in reproductive organs can be distributed between either a few heavy seeds or many light seeds (Smith & Fretwell, 1974). Because seedling mortality is related to seed weight in many cases (Grime & Jeffrey, 1965), there is usually a positive correlation between seedling survivorship and seed size within groups of plants of similar life history and coexisting in the same communities (e.g. Verkaar & Schenkeveld, 1984). In vascular plants, maintenance or expansion of populations depends on either vegetative multiplication or seed production. The proportion of the carbohydrate pool involved in reproductive effort differs widely between species (Harper, Lovell & Moore, 1970). Monocarpic plants generally invest more in reproductive organs than polycarpics, and annuals more than perennials. In general, annuals have lighter seeds than longer-lived species [Silvertown (1981), but see Thompson, (1987)].

DENSITY-INDEPENDENT FACTORS

Density-independent factors may determine the number of plants in populations. For example, the number of flowering plants of Senecio jacobaea in a certain year may be strongly correlated with the amount of precipitation in the preceding summer (Dempster & Lakhani, 1979; van der Meijden et al., 1985). Similarly, in Cynoglossum officinale, the percentage of individuals flowering is related to the rainfall of the previous summer and the number of days with frost in the preceding winter (de Jong, Klinkhamer & Prins, 1986). Hunt, Hope-Simpson & Snape (1985) described the population size of ramets of Pyrola rotundifolia ssp. maritima and discovered a strong correlation between increase in numbers of ramets and the occurrence of cold nights, particularly in early spring, and to a lesser extent, to low rainfall in the latter part of the preceding summer. In Dutch chalk grasslands, mortality of the seedlings of some annual and biennial plants is highly correlated with the number of days with showers, the amount of precipitation, the number of days with frost or the mean daily air humidity (Schenkeveld & Verkaar, 1984b). Thus, among other factors, climatic conditions can determine population size.

In British annual and biennial plants of chalk grassland, Kelly (1982) found little evidence for the occurrence of density-dependent regulation. For short-lived plants of Dutch chalk grasslands, Schenkeveld & Verkaar (1984b) suggested that competitive exclusion within this group of plants is not very likely, since the densities of plants were generally rather low and the role of predation and fungal disease appeared to have a more significant role in the determination of population size. Fowler (1986) reached a similar conclusion for Texan grassland populations, mentioning that the highly variable climate of this region may make the relaxation or absence of density-dependent population regulation common.

INTRASPECIFIC COMPETITION

From numerous studies, there is evidence that intraspecific competition may reduce growth and accelerate mortality (Yoda et al., 1963; White, 1979; Westoby, 1984). Much of our knowledge of intraspecific competition has been derived from agricultural systems and from forestry (van Dobben, 1967; Harper, 1977). However, in natural populations also, intraspecific competition for limiting resources (nutrients, light, water, pollinators, etc.) may contribute to the determination of numbers.

The classical study of Symonides (1979a, b, c) showed that with increasing density there was a reduction of plant size, reproductive capacity and sometimes also higher mortality in Polish inland sand-dune populations of Plantago indica L., Tragopogon heterospermus Schweigg., Androsace septentrionalis L. and Ceras-

tium holosteoides although, in some species, e.g. *P. indica*, density-dependent mortality was less important. In another sand-dune study (Watkinson & Harper, 1978), the numbers of the annual grass, *Vulpia fasciculata*, were not determined by density-dependent mortality, but plant size and reproductive output were strongly and negatively correlated with density. In ramets of the clonal perennial, *Ranunculus repens*, density-dependent mortality has been documented in both grassland and woodland sites (Lovett Doust, 1981).

A very elaborate study of the importance of density-dependent processes has been carried out by Smith (1983a, b, c) on the spring ephemeral annual, *Floerkea proserpinacoides* Willd., from North American deciduous forests. A major feature of this study is that all of the plant's life phases were considered. Vegetative plants of *F. proserpinacoides* grow in high densities. The survivorship curves were of similar shape at different densities, but plants at high density died earlier than did those of low density. At higher density, growth was lower and the production of buds and branches was also significantly smaller, although no difference in phenology could be found (Smith, 1983a). The reproductive output of *F. proserpinacoides* was affected at higher densities; the number of flowers per plant, the proportion of flowers that successfully set seed and the number of seeds set per successful fruit decreased as density increased. Remarkably, no variation in seed weight at increasing density was reported (Smith, 1983b). Up to 97% of this plant's seeds of the seed bank emerged each year and, although seed-eating mammals were common, seed predation was insignificant, possibly owing to their concentrated and diverse flavonoid content. In natural conditions, germination was not affected by density-dependent processes (Smith, 1983c). The distribution of *F. proserpinacoides* in woodland was strongly and negatively correlated with the accumulation of leaf litter on the forest floor, such that a dense litter layer resulted in high seedling mortality. At high densities, seedling mortality was even lower than at low densities, since dense seedling populations could collectively push aside litter. Local extinctions of populations may occur as a consequence of deep accumulations of litter or of disturbances by factors such as fire and tree falls which may cause destruction of the entire seed bank. Thus, density-independent factors, i.e. litter accumulation and disturbance, determine the numbers of the current generation in this woodland habitat for *F. proserpinacoides*, and intraspecific competition during vegetative growth and reproduction may affect the population size attained in the following generation. Unfortunately, this study did not extend over several generations so that, again, uncertainties exist concerning the relative importance in the longer term of the various processes implicated in population regulation.

THE IMPORTANCE OF HERBIVORES

Herbivores can drastically reduce population size. In many populations, seed-eating animals destroy almost the whole seed output, either in the pre- or post-dispersal phase (Watt, 1923; Shaw, 1968; Heithaus, 1981; Louda, 1982; Forsyth & Watson, 1985; Kjellsson, 1985) and, in some instances, removal rates of 20% per day have been found (Mittelbach & Gross, 1984; Verkaar, Schenkeveld & Huurnink, 1986b). For example, the local scarcity of *Carlina vulgaris* at some sites in Britain (Greig-Smith & Sagar, 1981) has been attributed to extensive granivory by small mammals.

The influence of herbivores is not invariably harmful since, in some cases, it has been reported that insect defloration may induce a higher reproductive output. In one study (Hendrix, 1979), large plants of *Pastinaca sativa* in which the primary umbels were damaged by the larvae of *Depressaria pastinacella* Duponchel had a higher seed production than control plants, although in general seeds were considerably lighter. Seedling establishment from the smaller seeds was, however substantially less than that from large seeds (Hendrix, 1984). As yet, the results of this study remain inconclusive in that it is unclear whether the final yield of reproductive offspring is larger from the damaged parent plants than from the control plants.

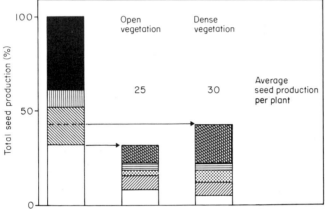

Fig. 1. The fate of seeds of *Scabiosa columbaria* in an annually mown chalk grassland in the Netherlands. y, process probably density-dependent; n, process probably density-independent. ■ Non-viable (n); ▨ pre-dispersal predation (y); ▨ post-dispersal predation before mowing (n); ▨ post-dispersal predation after mowing (n); ▨ incapable of germination (n); ▨ seedling mortality (n); ▨ rosette mortality (n); □ seed production (n).

In an investigation of the role of seed predation in *Scabiosa columbaria* in a Dutch chalk grassland site, 38·9% of the mature seeds were not fully developed and incapable of germination (Fig. 1), probably owing to unfavourable environmental conditions during fruit set. On average, 7·2% of the seeds were damaged by lepidopteran larvae in the pre-dispersal phase and their germination capacity was negligible. In the seed bank, between 20 and 60% of the seeds were lost and 86% of the remaining seeds were capable of germination. The resulting amount of seedling recruitment (20% of the seeds produced initially) agrees well with the transition values between seed production and seedling emergence found in demographic studies (Verkaar *et al.*, 1986b).

Herbivores can affect the population size not only in the reproductive phase but also in the vegetative phase. The cover of *Senecio jacobaea* correlates with the number of egg batches deposited by the cinnabar moth (*Tyria jacobaeae* L.) in the previous year (van der Meijden *et al.*, 1985). Recent studies have revealed that the effects of a single partial defoliation may have an important impact on the reproductive capacity and life span of the monocarpic herb *Verbascum thapsus*, according to calculations based on field and laboratory data (Fig. 2). When 25% of the leaf area of this species was removed once in the second year, longevity of the rosettes could be extended for up to four years and the production of fruits is increased in comparison with that of undamaged plants which survive for three

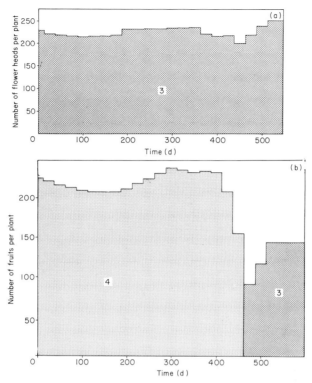

Fig. 2. Final reproductive output and life length of two monocarpic plants in a dune habitat, when 25 % of the leaf area is removed in the second or third year [calculated with a deterministic model based on field and laboratory data, according to Verkaar (in preparation)]. Day 1 is 1 April in the second year of the life span; (a) *Senecio jacobaea*; (b) *Verbascum thapsus*. Life length (years) is indicated in the figure.

years. The same rate of defoliation leads to much less drastic effects in *S. jacobaea*; seed output is little affected by one event of partial defoliation and the life span remains at three years despite the temporary loss of leaf area. This difference between species in their capacity to recover after defoliation must be attributed to differences in investment in the shoot, in differences in weight loss of the roots after defoliation and the rate at which carbohydrates are invested in new leaf tissue. Hence, the effects of herbivores on the future population size depend on the degree of damage sustained, the responses of the individual parent plant and the influence of environmental factors upon the plant both before and after defoliation (Verkaar, van der Meijden & Breebaart, 1986a).

HERBIVORES AND DENSITY DEPENDENCE

Several factors determine the probability of attack by herbivores and/or pathogens, and both population size and population density may be among them (Augspurger 1983a, b; Dirzo, 1984; Clark & Clark, 1984; Becker *et al.*, 1985; Fenner, 1987). In *S. jacobaea*, the incidence of attack by *T. jacobaeae* is correlated with plant cover (van der Meijden, 1979). Insect and predators damage more fruits of *Crataegus monogyna* if plants are large and the distance to two nearest neighbours is small (Courtney & Manzur, 1985). The likelihood of infection by stem- and fruit-boring larvae of *Epiblema scutulana* Den. & Schiff. is correlated with population size in *Cirsium vulgare* (van Leeuwen, 1983). Similarly, the probability of predation of *Aralia nudicaulis* L. by moose is considerably higher at high population densities (Edwards, 1985).

If dispersion patterns of food plants are compared, distance between individuals can be more important than density. If larval mobility is limited, e.g. in young *Pieris rapae* L. larvae, plants of *Brassica oleracea* grown in clumps were more elusive for the herbivores than plants spaced at regular intervals (Cain, Eccleston & Kareiva, 1985). In *Cornus amomum* Mill., populations of *Rhagoletis cornivora* Bush (Diptera; Tephritidae) do not eat fruits in a density-dependent way (Borowicz & Juliano, 1986). Thus, there is evidence that in some cases the influence of herbivores acts in a density-dependent way and sometimes it does not.

The occurrence of density-dependent predation of buried seeds in the soil seed bank is not very likely. However, in the few studies in which the density and pattern of surface-lying seeds was varied experimentally, rates of removal by rodents were usually linearly related to seed density but, in one experiment, no clear difference occurred in choice between clumped and scattered seed patches (Price & Heinz, 1984; Price & Waser, 1985). We must conclude that there is still uncertainty concerning the circumstances in which density-independent, post-dispersal predation of seeds occurs under field conditions.

CONCLUDING REMARKS

From the examples reviewed in this paper, it is clear that population dynamics of plants are influenced both by density-independent processes and by density-dependent processes, either owing to intraspecific competition or to density-dependent attack by herbivores or pathogens. Situations may arise in which populations are reduced to densities where density-dependent regulation by intraspecific competition does not occur, e.g. short-lived plants in a matrix of perennial plants of a chalk grassland. Of course, interspecific competition may be important here if gaps are short-lived and small. These cases partly support Hairston's view that abiotic factors and interspecific competition reduce the sizes of plant populations.

Regulation may occur both by density-dependent mortality and/or by density-dependent reproduction and vegetative multiplication. In general, the effects of herbivory which do not result in direct mortality are insufficiently assessed to conclude that herbivory is a minor factor in the determination of population size as suggested by Hairston *et al.* (1960).

Since reduction of plant numbers occurs during the whole life-cycle, all life phases must be considered in any attempt to discover how the determination of numbers takes place. It is not sufficient to consider the processes operating upon mature and reproductive individuals, since regulation often occurs during juvenile phases. Furthermore, as Hassell (1985) has pointed out, random population fluctuations and those with an element of density dependence can hardly be distinguished, and spatial heterogeneity of the sampled area may contribute to the difficulty of detecting density-dependent processes. Density dependence should be assessed by studying plant populations over several generations and in all life phases. Only when investigations are conducted with the thoroughness of that of Klemow & Raynal (1983), and when experimental approaches are applied in field studies to assess factors that regulate populations, will it be possible for studies of demography to be translated into models of population regulation.

I suspect that there is a broad gamut of populations in which, at one extreme, numbers are more or less determined by density-dependent processes whilst at the other extreme population size is primarily affected by density-independent factors. With Fowler (1986), however, one must conclude that little is known about the determination of numbers in most natural populations. This conclusion is rather disappointing, particularly in a symposium entitled *The Frontiers of Comparative Plant Ecology*. It does, however, indicate the continuing need for research on population dynamics in spite of lowering budgets for ecological research in many countries.

As with most 'frontiers' in science, active accumulation of data is taking place but there is now a great need for more generalized and unifying concepts in the field of plant population dynamics. In this respect, the comparative approach, fundamental to the work of the Sheffield unit, may once more prove to be very valuable.

ACKNOWLEDGEMENTS

I am very grateful to Dr E. van der Meijden (Leiden) and Dr H. J. During (Utrecht) for their valuable comments on an earlier draft of the manuscript. I thank Mr J. Herzberg for preparing the figures.

REFERENCES

ANDREWARTHA, H. G. (1970). *Introduction to the Study of Animal Population*, 2nd Edn. Methuen, London.

ANDREWARTHA, H. G. & BIRCH, L. C. (1954). *The Distribution and Abundance of Animals*. University of Chicago Press, Chicago.

ANTONOVICS, J. & LEVIN, D. A. (1980). The ecological and genetic consequences of density-dependent regulation in plants. *Annual Review of Ecology and Systematics*, **11**, 411–452.

AUGSPURGER, C. K. (1983a). Offspring recruitment around tropical trees: changes in cohort distance with time. *Oikos*, **40**, 189–196.

AUGSPURGER, C. K. (1983b). Seed dispersal of the tropical tree *Platypodium elegans*, and the escape of its seedlings from fungal pathogens. *Journal of Ecology*, **71**, 759–771.

BAKKER, K. (1964). Backgrounds of controversies about population theories and their terminologies. *Zeitschrift für angewandte Entomologie*, **53**, 187–208.

BARTOLOME, J. W. (1979). Germination and seedling establishment in California annual grassland. *Journal of Ecology*, **67**, 273–281.

BECKER, P., LEE, L. W., ROTHMAN, E. D. & HAMILTON, W. D. (1985). Seed predation and the coexistence of tree species: Hubbell's models revisited. *Oikos*, **44**, 382–390.

BOER, P. J. DEN (1986). The present status of the competitive exclusion principle. *Trends in Ecology and Evolution*, **1**, 25–28.

BOROWICZ, V. A. & JULIANO, S. A. (1986). Inverse density-dependent parasitism of *Cornus amomum* fruit by *Rhagoletis cornivora*. *Ecology*, **67**, 639–643.

BRUES, C. T. (1946). *Insect Dietary. An Account of the Food Habits of Insects*. Harvard University Press, Cambridge, Massachusetts.

CAIN, M. C., ECCLESTON, J. & KAREIVA, P. M. (1985). The influence of food plant dispersion on caterpillar searching success. *Ecological Entomology*, **10**, 1–7.

CHEW, R. M. (1974). Consumers as regulators of ecosystems: an alternative to energetics. *Ohio Journal of Science*, **74**, 359–370.

CLAPHAM, A. R., TUTIN, T. G. & WARBURG, E. F. (1981). *Excursion Flora of the British Isles*, 3rd Edn. Cambridge University Press, Cambridge.

CLARK, D. A. & CLARK, D. B. (1984). Spacing dynamics of a tropical rain forest tree: evaluations of the Janzen-Connell model. *American Naturalist*, **124**, 769–788.

CONNELL, J. H. (1983). On the prevalence and relative importance of interspecific competition: evidence from field experiments. *American Naturalist*, **122**, 661–696.

COURTNEY, S. P. & MANZUR, M. I. (1985). Fruiting and fitness in *Crataegus monogyna*: the effects of frugivores and seed predators. *Oikos*, **44**, 398–406.

CRAWLEY, M. J. (1983). *Herbivory – The Dynamics of Animal–Plant Interactions*. Blackwell Scientific Publications, Oxford.

DEMPSTER, J. P. & LAKHANI, K. H. (1979). A population model for cinnabar moth and its food plant, ragwort. *Journal of Animal Ecology*, **48**, 143–164.

DEMPSTER, J. P. & POLLARD, E. (1986). Spatial heterogeneity, stochasticity and the detection of density dependence in animal populations. *Oikos*, **46**, 413–416.

DIRZO, R. (1984). Herbivory: a phytocentric overview. In: *Perspectives on Plant Population Ecology* (Ed. by R. Dirzo & J. Sarukhán), pp. 141–165. Sinauer, Sunderland, Massachusetts.

DOBBEN, W. H. VAN (1967). Individu en populatie bij plant en dier. In: *Populatiebiologie* (Ed. by D. J. Kuenen), pp. 80–107. Pudoc, Wageningen.

DURING, H. J., SCHENKEVELD, A. J., VERKAAR, H. J. & WILLEMS, J. H. (1985). Demography of short-lived forbs in chalk grassland in relation to vegetation structure. In: *The Population Structure of Vegetation* (Ed. by J. White), pp. 341–370. Junk, Dordrecht.

EDWARDS, J. (1985). Effects of herbivory by moose on flower and fruit production of *Aralia nudicaulis*. *Journal of Ecology*, **73**, 861–868.

EHRLICH, P. R. & BIRCH, L. C. (1967). The 'balance of nature' and 'population control'. *The American Naturalist*, **101**, 97–107.

FENNER, M. (1987). Seedlings. In: *Frontiers of Comparative Plant Ecology* (Ed. by I. H. Rorison, J. P. Grime, R. Hunt, G. A. F. Hendry & D. H. Lewis), *New Phytologist*, **106** (Suppl.), 35–47. Academic Press, New York & London.

FORSYTH, S. F. & WATSON, A. K. (1985). Predispersal seed predation of Canada thistle. *Canadian Entomologist*, **117**, 1075–1081.

FOWLER, N. L. (1986). Density-dependent population regulation in a Texas grassland. *Ecology*, **67**, 545–554.

GREIG-SMITH, J. & SAGAR, G. R. (1981). Biological causes of local rarity of *Carlina vulgaris*. In: *The Biology of Rare Plant Conservation* (Ed. by H. Synge), ppp. 389–400. Wiley, Chichester.

GRIME, J. P. (1978). Interpretation of small-scale patterns in the distribution of plant species in space and time. In: *Structure and Functioning of Plant Populations* (Ed. by A. H. J. Freijsen & J. W. Woldendorp), pp. 101–121. North-Holland Publishing Company, Amsterdam.

GRIME, J. P. (1979). *Plant Strategies and Vegetation Processes*. Wiley, Chichester.

GRIME, J. P. & JEFFREY, D. W. (1965). Seedling establishment in vertical gradients of sunlight. *Journal of Ecology*, **53**, 621–642.

GRUBB, P. J., KELLY, D. & MITCHLEY, J. (1982). The control of relative abundance in communities of herbaceous plants. In: *The Plant Community as a Working Mechanism* (Ed. by E. I. Newman), pp. 79–97. Blackwell, Oxford.

HAIRSTON, N. G., SMITH, F. C. & SLOBODKIN, L. B. (1960). Community structure, population control, and competition. *American Naturalist*, **94**, 421–425.

HARPER, J. L. (1977). *Population Biology of Plants*. Academic Press, London.

HARPER, J. L., LOVELL, P. H. & MOORE, K. G. (1970). The shapes and sizes of seeds. *Annual Review of Ecology and Systematics*, **1**, 327–356.

HASSELL, M. P. (1985). Insect natural enemies as regulating factors. *Journal of Animal Ecology*, **54**, 323–334.

HEITHAUS, E. R. (1981). Seed predation by rodents on three ant-dispersed plants. *Ecology*, **62**, 136–145.

HENDRIX, S. D. (1979). Compensatory reproduction in a biennial herb following insect defloration. *Oecologia*, **42**, 107–118.

HENDRIX, S. D. (1984). Variation in seed weight and its effects on germination in *Pastinaca sativa* L. (Umbelliferae). *American Journal of Botany*, **71**, 795–802.

HICKMAN, J. C. (1979). The basic biology of plant numbers. In: *Topics in Plant Population Biology* (Ed. by O. T. Solbrig, S. Jain, G. B. Johnson & P. H. Raven), pp. 232–263. Columbia University Press, Columbia, Missouri.

HUNT, R., HOPE-SIMPSON, J. F. & SNAPE, J. B. (1985). Growth of the dune wintergreen (*Pyrola rotundifolia* ssp. *maritima*) at Braunton Burrows in relation to weather factors. *International Journal of Biometeorology*, **29**, 323–334.

JANZEN, D. H. (1975). Behaviour of *Hymenaea courbaril* when its predispersal seed predator is absent. *Science*, **189**, 145–147.

JERMY, T. (1984). Evolution on insect/host plant relationships. *American Naturalist*, **124**, 609–630.

JERMY, T. (1985). Is there competition between phytophagous insects? *Zeitschrift für zoologische Systematik und Evolutionsforschung*, **23**, 275–285.

JONG, T. J. DE, KLINKHAMER, P. G. L. & PRINS, A. H. (1986). Flowering behaviour of the monocarpic perennial *Cynoglossum officinale* L. *New Phytologist*, **103**, 219–229.

KEDDY, P. A. (1981). Experimental demography of the sand dune annual *Cakile edentula* growing along an environmental gradient in Nova Scotia. *Journal of Ecology*, **69**, 615–630.

KELLY, D. (1982). *Demography, population control and stability in short-lived plants of chalk grassland.* Ph.D. thesis, Cambridge.

KJELLSSON, G. (1985). Seed fate in a population of *Carex pilulifera* L. I. Seed dispersal and ant–seed mutualism. *Oecologia*, **67**, 416–423.

KLEMOW, K. M. & RAYNAL, D. J. (1983). Population biology of an annual plant in a temporally variable habitat. *Journal of Ecology*, **71**, 691–703.

KLOMP, H. (1962). The influence of climate and weather on the mean density level, the fluctuations and the regulation of animal populations. *Archives Néerlandaises de Zoologie*, **15**, 68–109.

LARSON, M. M. & SCHUBERT, G. H. (1970). *Cone Crops of Ponderosa Pine in Central Arizona, Including the Influence of Abert Squirrels.* USDA Forest Service Research Paper R.M. 58.

LEEUWEN, B. H. VAN (1983). The consequences of predation in the population biology of the monocarpic species *Cirsium palustre* and *Cirsium vulgare*. *Oecologia*, **58**, 178–187.

LOUDA, S. M. (1982). Limitation of the recruitment of the shrub *Haplopappus squarrosus* (Astraceae) by flower- and seed-feeding insects. *Journal of Ecology*, **70**, 43–53.

LOVETT DOUST, L. (1981). Population dynamics and local specialization in a clonal perennial (*Ranunculus repens*). I. The dynamics of ramets in contrasting habitats. *Journal of Ecology*, **69**, 743–755.

MACARTHUR, R. H. & WILSON, E. O. (1967). *The Theory of Island Biogeography.* Princeton University Press, Princeton.

MATTSON, W. J. & ADDY, N. D. (1975). Phytophagous insects as regulators of forest primary production. *Science*, **190**, 515–522.

MEIJDEN, E. VAN DER (1979). Herbivore exploitation of a fugitive plant species: local survival and extinction of the cinnabar moth and ragwort in a heterogeneous environment. *Oecologia*, **42**, 307–323.

MEIJDEN, E. VAN DER, JONG, T. J. DE, KLINKHAMER, P. G. L. & KOOI, R. E. (1985). Temporal and spatial dynamics in populations of biennial plants. In: *Structure and Functioning of Plant Populations*, vol. II (Ed. by J. Haeck & J. W. Woldendorp), pp. 91–103. North-Holland, Publishing Company, Amsterdam.

MITTELBACH, G. G. & GROSS, K. L. (1984). Experimental studies of seed predation in old fields. *Oecologia*, **65**, 7–13.

MURDOCH, W. W. (1966). Community structure, population control, and competition – a critique. *American Naturalist*, **100**, 219–226.

NICHOLSON, A. J. (1954). An outline of the dynamics of animal populations. *Australian Journal of Zoology*, **2**, 9–65.

NICHOLSON, A. J. (1957). The self-adjustment of populations to change. In: *Population Studies: Animal Ecology and Demography* (Ed. by K. B. Warren), pp. 153–173. Cold Spring Harbor Symposium on Quantitative Biology 22. Cold Spring Harbor, New York.

PRICE, M. V. & HEINZ, K. M. (1984). Effects of body size, seed density and soil characteristics on rates of seed harvest by heteromyid rodents. *Oecologia*, **61**, 420–425.

PRICE, M. V. & WASER, N. M. (1985). Microhabitat use of heteromyid rodents: effects on artificial seed patches. *Ecology*, **66**, 211–219.

ROUGHGARDEN, J. (1971). Density-dependent natural selection. *Ecology*, **52**, 453–468.

RUNGE, F. (1968). Die Artmächtigkeitsschwankungen in einem nordwest-deutschen Enzian-Zwenkenrasen. *Vegetatio*, **15**, 124–128.

SAGAR, G. R. & MORTIMER, A. M. (1976). An approach to the study of the population dynamics of plants with special reference to weeds. *Applied Biology*, **1**, 1–47.

SCHENKEVELD, A. J. & VERKAAR, H. J. (1981). Hapaxanthic species in limestone grasslands: conscious savers or big spenders? *Acta Botanica Neerlandica*, **30**, 134.

SCHENKEVELD, A. J. & VERKAAR, H. J. (1984a). On the ecology of short-lived forbs in chalk grasslands: distribution of germinative seeds and its significance for seedling emergence. *Journal of Biogeography*, **11**, 251–260.

SCHENKEVELD, A. J. & VERKAAR, H. J. (1984b). *On the ecology of short-lived forbs in chalk grasslands.* Thesis, Utrecht.

SCHOENER, T. W. (1983). Field experiments on interspecific competition. *American Naturalist*, **122**, 240–285.

SHAW, M. W. (1968). Factors affecting the natural regeneration of sessile oak (*Quercus petraea*) in North Wales. II. Acorn losses and germination under field conditions. *Journal of Ecology*, **56**, 647–660.

SILVERTOWN, J. W. (1980). The evolutionary ecology of mast seeding in trees. *Biological Journal of the Linnean Society*, **14**, 235–250.

SILVERTOWN, J. W. (1981). Seed size, life span, and germination date as co-adapted features of plant life history. *American Naturalist*, **118**, 860–864.

SLOBODKIN, L. B., SMITH, F. E. & HAIRSTON, N. G. (1967). Regulation in terrestrial ecosystems, and the implied balance of nature. *American Naturalist*, **101**, 109–124.

SMITH, B. H. (1983a). Demography of *Floerkea proserpinacoides*, a forest-floor anual. I. Density-dependent growth and mortality. *Journal of Ecology*, **71**, 391–404.

SMITH, B. H. (1983b). Demography of *Floerkea proserpinacoides*, a forest-floor annual. II. Density-dependent reproduction. *Journal of Ecology*, **71**, 405–412.

SMITH, B. H. (1983c). Demography of *Floerkea proserpinacoides*, a forest-floor annual. III. Dynamics of seed and seedling populations. *Journal of Ecology*, **71**, 413–425.

SMITH, C. C. & FRETWELL, S. D. (1974). The optimal balance between size and number of offspring. *American Naturalist*, **108**, 499–506.

SYMONIDES, E. (1979a). The structure and population dynamics of psammophytes on inland dunes. II. Loose-sod populations. *Ekológia Polska*, **27**, 191–234.

SYMONIDES, E. (1979b). The structure and population dynamics of psammophytes on inland dunes. III. Populations of compact psammophyte communities. *Ekológia Polska*, **27**, 235–257.

SYMONIDES, E. (1979c). The structure and population dynamics of psammophytes on inland dunes. IV. Population phenomena as a phytocenose-forming factor (a summing-up discussion). *Ekológia Polska*, **27**, 259–281.

THOMPSON, K. (1987). Seeds and seed banks. In: *Frontiers of Comparative Plant Ecology* (Ed. by I. H. Rorison, J. P. Grime, R. Hunt, G. A. F. Hendry & D. H. Lewis), *New Phytologist*, **106** (Suppl.), 23–34. Academic Press, New York & London.

THOMPSON, K. & GRIME, J. P. (1979). Seasonal variation in the seed banks of herbaceous species in ten contrasting habitats. *Journal of Ecology*, **67**, 893–921.

VERKAAR, H. J. & SCHENKEVELD, A. J. (1984). On the ecology of short-lived forbs in chalk grasslands: development under low photon flux density conditions. *Flora*, **175**, 135–141.

VERKAAR, H. J., MEIJDEN, E. VAN DER & BREEBAART, L. (1986a). The response of *Cynoglossum officinale* L. and *Verbascum thapsus* L. to defoliation in relation to nitrogen supply. *New Phytologist*, **104**, 121–129.

VERKAAR, H. J., SCHENKEVELD, A. J. & HUURNINK, C. (1968b). The fate of *Scabiosa columbaria* L. (Dipsacaceae) seeds in a chalk grassland. *Oikos*, **46**, 159–162.

WATKINSON, A. R. & HARPER, J. L. (1978). The demography of a sand dune annual: *Vulpia fasciculata*. I. The natural regulation of populations. *Journal of Ecology*, **66**, 15–33.

WATT, A. S. (1923). On the ecology of British beechwoods with special reference to their regeneration. *Journal of Ecology*, **11**, 1–48.

WATT, A. S. (1981). Further observations on the effects of excluding rabbits from grassland A in East Anglian Breckland: the pattern of change and factors affecting it (1936–73). *Journal of Ecology*, **69**, 509–536.

WESTOBY, M. (1984). The self-thinning rule. *Advances in Ecological Research*, **14**, 167–225.

WHITE, J. (1979). Demographic factors in populations of plants. In: *Demography and Evolution in Plant Populations* (Ed. by O. T. Solbrig), pp. 21–48. Botanical Monographs 15. University of California Press, Berkeley.

YODA, K., KIRA, T., OGAWA, H. & HOZUMI, K. (1963). Intraspecific competition among higher plants. XI. Self-thinning in overcrowded pure stands under cultivated and natural conditions. *Journal of Biology of Osaka City University*, **14**, 107–129.

New Phytol. (1987) **106** (Suppl.), 61–77

AN ARCHITECTURAL APPROACH TO THE COMPARATIVE ECOLOGY OF PLANT ROOT SYSTEMS

By A. H. FITTER

Department of Biology, University of York, York YO1 5DD, UK

SUMMARY

Although plants devote a large proportion of their resources to roots, we have a poor understanding of the constraints under which root systems function. Roots are much less variable morphologically than leaves and it is likely that root systems rather than individual roots are the focus of natural selection. In other words, architecture is more important than morphology. Existing classifications of root systems, based on the developmental model, have failed to provide much insight into their functioning and an alternative, topological model is outlined, in which the link is the basic unit of classification. Other components of the architecture of root systems, including link lengths, branching angles and diameters, are considered and the ecological implications of variation in each is discussed.

Simulation models of transport and space exploration are discussed and it is shown that resource cost, transport efficiency and exploration efficiency cannot be simultaneously minimized and that optimum form may vary with the mobility of the resource. In general, a 'herringbone' structure seems to be the most efficient at exploration of space but the least transport-efficient and the most expensive.

Key words: Root systems, architecture, topology, geometry, transport, exploitation.

INTRODUCTION

Roots and their functions

Roots account for between 40 and 85 % of net primary production in a wide range of ecosystems from grassland to forest (Fogel, 1985). Typically, plant growth in non-agricultural conditions is limited more by soil-derived resources than by CO_2 or solar radiation (Fitter, 1986a). It seems axiomatic, therefore, that an understanding of the functioning of plants within natural communities must demand an equal understanding of the behaviour of roots and root systems. The function and activity of root systems is closely linked to their normal environment – soil. The two main functions of the above-ground parts of plants – capture of photons and production of dispersal propagules – can only be performed above ground. Equally, the primary root function, the capture of water and ions, is manifestly a below-ground activity. Another commonly identified root function, anchorage, is clearly a secondary consequence of competition for photons above ground, while a third, the storage of resources and the production of persistent propagules (for dispersal in time rather than space), could in principle be equally well carried out above ground. That persistence has tended to become a below-ground phenomenon, exemplified by bulbs, corms, tubers and rhizomes as well as actual root storage, is presumably due to the greater safety afforded by a predominantly solid medium, which has, for example, precluded the evolution of large subterranean herbivores.

0028-646X/87/05S061 + 17 $03.00/0

Types of roots

It is perhaps at first sight surprising that roots, which clearly perform at least as many functions as shoots, should show so little morphological variation. Because of their uniformity, the identification of plants by their underground parts is an arcane skill and this uniformity reflects the relatively equable environment in which roots grow, free from problems such as overheating and the need to attract pollinators. The major morphological features in which differences can be discerned are radius, development of root hairs and colour. The first two have considerable importance and will be discussed later, while the latter may be connected with defences against pathogens and grazers.

The internal anatomy of roots, however, varies to a much greater extent, in particular in the number of vascular poles in the primary root, which ranges from two upwards. The significance of this lies in the usual restriction of formation of lateral roots either to these poles or to the regions between them, which in principle could lead to a defined redial pattern of laterals, much as in the phyllotaxis of leaves. The number of vascular poles in a root, though varying between species and higher taxa, is by no means species-specific. Fine roots tend to have fewer poles than coarse roots (Wardlaw, 1928), proximal laterals more than distal ones (Torrey, 1976) and, in excised apices, those treated with IAA more than those without (Torrey, 1957).

Types of root systems

An ecological perspective of the functions of root systems must, therefore, concentrate more on the morphology of root *systems* – root system architecture – than on that of individual roots, although the architecture may be determined by anatomical features involving branch generation. There have been a number of attempts to produce architectural classifications of root systems (Freidenfeldt, 1902; Cannon, 1949; Weaver, 1958; Krasilnikov, 1968) but none has been particularly successful, partly because of the great plasticity of root systems. Nevertheless, there are a number of architectural features that play a large role in determining the overall form of root systems.

(i) Balance of primary and adventitious roots: species with extensive production of adventitious roots are particularly common in the monocotyledons; their root systems generally have many more or less similar parts all joined to the stem base and lack the single dominant axis that can be achieved by primary root systems.

(ii) Degree of branching: both primary and adventitious root systems may be unbranched or heavily branched. The Orchidaceae, Alliaceae and Liliaceae, for example, all have many, little-branched adventitious roots, while *Plantago media** is an example of a species with a little-branched primary root. In contrast, the primary roots of most dicotyledons and adventitious roots of grasses have many laterals. At the extreme are many annuals, such as *Senecio vulgaris* or *Stellaria media*, where the primary root soon dies and growth is continued entirely by laterals.

(iii) Plasticity of branching: Nobbe (1862) and Höveler (1892) were perhaps the first to demonstrate the propensity of lateral branches to proliferate in more fertile soil zones, an observation extended in natural soils by Sprague (1933), in fertilizer

* Nomenclature of British plants follows Clapham, Tutin & Warburg (1981); that of other species is as given in papers cited.

bands by Duncan & Ohlrogge (1958) and in controlled conditions by Hackett (1972) and by Drew and his co-workers [Drew, Saker & Ashley (1973) and later papers].

It is surprising that so little attention has been given to the ecological significance of this response and to the possibility that there may be interspecific variation in plasticity, which could be adaptive. In an important series of experiments, Grime, Crick & Rincon (1986) grew two contrasting species, the nutrient-responsive grass, *Agrostis stolonifera*, and the stress-tolerant sedge, *Scirpus sylvaticus*, in complex root 'choice chambers' and offered different spatial and temporal patterns of nutrients. The grass was highly plastic and developed root systems with many branches in permanently high-nutrient chambers, where root growth rate was nearly six times that in permanently low nutrient chambers. The sedge, in contrast, responded only slightly to this spatial pattern. Where nutrients were supplied to individual chambers in short pulses (24 h), the sedge was better able to obtain nitrogen than the grass, compared with a treatment where the pulses lasted 9 d. This aspect of morphological plasticity urgently needs further attention.

This plasticity, however, renders most classification systems unworkable and vitiates the value of much root research based on labour-intensive excavations of small samples of root systems growing in the field (e.g. Kutschera, 1960). Nevertheless, it is clear that wide variation in root architecture occurs and that it contains both genotypic and phenotypic components. As such, it is presumably the outcome of some 400 million years of natural selection, based on the different abilities of various architectures to carry out important root functions under diverse environmental circumstances. The identification of the link between architecture and function is therefore critical and I propose to examine that link for one root function, the exploration of soil for resources (water and ions). Other functions (storage and anchorage) may have an important role to play in determining the fitness of a particular architecture, but that role awaits elucidation.

ROOT SYSTEM ARCHITECTURE

Measurable variables

The architecture of a root system can be dissected into a number of measurable variables, of which the most important are topology, link length (i.e. inter-branch distances and root–tip lengths), branching angles (including both the angle of origin and the radial direction) and link radius. Other important variables include the duration or longevity of individual parts and the presence of symbiotic associations which may alter the functioning of any given architectural structure. Of all these, the most fundamental in determining the overall structure is topology, which underlies for example, the major distinctions in root system form identified in most classifications of root systems.

Topology

Topology encompasses the non-metric aspects of branching structure. I have discussed elsewhere the theory underlying a topological classification of root systems (Fitter, 1985a, 1986b). In brief, it relies on a basipetal ordering system, in which external links (root segments between a meristem and a branch junction) have magnitude 1 and the magnitude of any internal link (a segment between two junctions) has a magnitude equal to the sum of the magnitudes of the links that it developmentally gave rise to. The magnitude of a system is the same as that of the final link and equal to the number of external links (Fig. 1). For a given system magnitude, numerous topologies are possible, which can be grouped into topologically distinct networks (TDNs). Members of a TDN set can be interconverted by plastic deformation but cannot be transformed into another

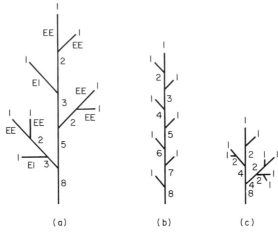

Fig. 1. Three diagrammatic root systems [(a) to (c)] to show the magnitude ordering scheme adopted for a topological classification. External links (those terminating in a meristem) have magnitude 1 and all internal links have a magnitude equal to the sum of their two daughter links. The distinction between external–external (EE) and external–internal (EI) links (see p. 66) is illustrated in (a); (b) and (c) represent the topological extremes at magnitude 8.

TDN (Fig. 2). A further classification into ambilateral classes, ignoring left–right switches, is possible. Members of a TDN set can be characterized by their external path length (p_e), the sum of the number of links in all paths from all external links to the base link, and by their diameter (d) or altitude (a), the longest individual path length in the system (Fig. 3). Although I have previously used the term diameter to describe this, I propose here to revert to the less confusing term, altitude, which in any case has priority (Cayley, 1857).

Both p_e and a vary between set limits, their minimum being associated with a dichotomous branching system and maximum with a herringbone pattern; formulas for these are in Fitter (1986b). These minima and maxima vary with magnitude and so the slope of a plot of p_e or a against magnitude is a valuable topological index. Unsurprisingly, this index is not a fixed characteristic for a species. In *Trifolium pratense*, at least, it can be altered by watering rate (Fitter, 1986b). At a watering rate of 2·5 ml d^{-1}, which severely restricted growth, the index for altitude was 0·88, not significantly different from the theoretical maximum of 1·0; at 20 ml d^{-1} it was as low as 0·46 (Fig. 4). These figures show that quite different topologies were adopted by the plants, which had many more herringbone-like root systems when deprived of water but were more diffusely branched in moist soil.

In contrast, nutrient concentrations seem to exert a much smaller effect on topology. In a factorial experiment involving three concentrations of N and three of P in sand culture, I detected no changes in this topological index for *T. pratense* (Fitter, in preparation).

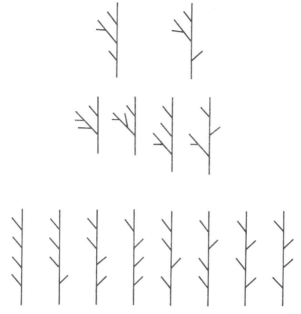

Fig. 2. The 14 topologically distinct networks (TDNs) for magnitude 5. Each TDN is unique and cannot be transformed into any other by simple plastic deformation. Those in each row can be interconverted by left–right switches and fall into a single ambilateral class (Smart, 1978). Members of an ambilateral class have the same topological characteristics.

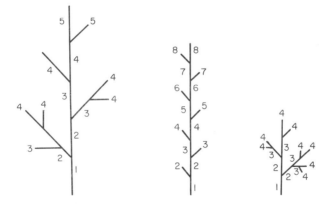

Fig. 3. The same networks as in Figure 1 to show path lengths, counted from the origin. The altitude (*a*) is the longest individual path length (5, 8 and 4, respectively) and the external path length (p_e) is the sum of all paths to external links (33, 43 and 32, respectively).

These findings raise two questions. How general are these patterns and what is their functional significance? The generality has not yet been tested but I have shown, using a different but analogous technique (Strahler ordering; Fitter, 1982), that root systems of some plant families appear to have higher branching ratios (R_b) than others. This is equivalent to showing that they have a high topological index or, in other words, that they are more herringbone-like. In particular, the Polygonaceae (four species sampled), Chenopodiaceae (four spp.), Plantaginaceae (three spp.) and Gramineae (six spp.) had consistently higher values than did the Caryophyllaceae (eight spp.), Cruciferae (six spp.) and Leguminosae (nine spp.).

Geometry

Lengths. Topologically identical root systems can still take on very different appearances if they vary in metric aspects of their geometry, particularly the lengths of individual links. In a topological classification, links can be regarded as

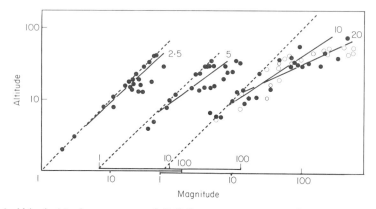

Fig. 4. Altitude (*a*) of root systems of *Trifolium pratense* grown at four watering rates (2·5, 5, 10 and 20 ml d⁻¹). Data for 10 and 20 ml d⁻¹ plants are drawn on a common base line and the latter are distinguished by solid symbols. The dashed line represents the maximum possible value of *a* (*a* = *n*) and the solid line the fitted regression. [Redrawn from Fitter (1986b).]

external (terminating in a meristem) or internal (Fig. 3), and external links are usefully further classified into those that join other external links [external–external (EE) links (Smart, 1978)] and those that join internal links [external–internal (EI) links]. All networks have *n*-1 internal links and *n* external links, and herringbone networks have a high proportion of EI links.

The length of external links therefore represents the distance behind the apex at which branching occurs. Observations suggest that this figure varies widely among root systems. In *Quercus rubra*, it may be around 10 cm (Lyford, 1980) and, in cereals, can be as high as 15 cm (Hackett, 1968) but in *T. pratense* it averages 1 to 2 cm (Fitter, in preparation) and is very sensitive to soil chemical status (Fig. 5). Both Al^{3+} and NO_3^- ions reduce external link length, although the mechanisms of action are likely to be very different.

Internal link length represents the frequency of branch production once initiated and has an important influence on the form of root systems (Lungley, 1973; Fitter, 1982) and is influenced by treatment. Hackett (1972) found up to 10 laterals cm⁻¹ of main axis in NO_3^--supplemented zones of barley compared to five to six in NO_3^--free zones. These represent internal link lengths of 1 to 2 mm, while Crossett, Campbell, & Stewart (1975) found 2 to 3 mm to be typical for barley. Similarly, Dittmer (1938) recorded values from 2 to 20 mm in a range of field crop plants and Lyford (1975) found 2 mm to be the average for both yellow birch and red oak, with a range from 0·5 to 10 mm in field samples, as compared to 2 to 3 mm for a number of trees grown in a rhizotron. In *T. pratense*, mean lengths of internal

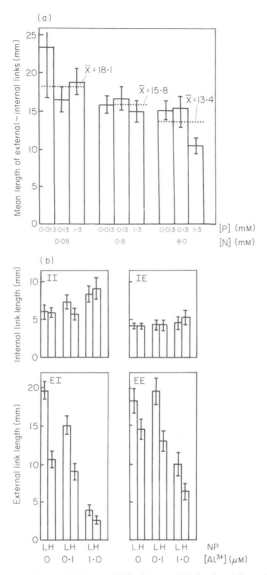

Fig. 5. (a) Mean length of external–internal links [external links that join an internal link; Smart (1978)] as a function of N and P supply for *Trifolium pratense* in sand culture. Nitrogen does not affect link length but the effect of P is significant ($F_{2, 116} = 4.52$, $P = 0.013$). (b) Mean length of internal and external links as a function of Al^{3+} and N and P supply rates for *T. pratense* in sand culture. Low NP is 0.4 mM NH_4^+ and 0.1 mM $H_2PO_4^-$, high NP 10 times that. F values for individual effects are (** $P < 0.01$; *** $P < 0.001$):

	Al	NP	Al × NP
Degrees of freedom	2.84	1.84	2.84
Link type			
Internal–internal	4.74***	0.18	0.79
Internal–external	2.08	0.36	0.55
External–internal	86.68***	49.06***	8.83***
External–external	12.14***	8.66**	0.39.

links vary from about 3 to 9 mm but are little altered by watering rate (Fitter, 1986b) or ionic environment [Fig. 5(b)], although Al^{3+} seems to increase them.

Angles. Branching angles are clearly a key component of root system architecture but almost no information is available on them, beyond the general observation that primary roots tend to be positively geotropic, their daughters to

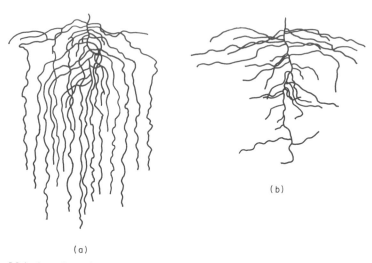

(b)

(a)

Fig. 6. Main branches of root systems of (a) *Viola arvensis* and (b) *Gnaphalium* (*Filaginella*) *uliginosum* redrawn from Kutschera (1960) to show change in geotropism of primary laterals of *V. arvensis*.

be diageotropic and further branches to be ageotropic. However, many roots show a change from a diageotropic to a positively geotropic response after a period of growth, which results in an increase in the volume of soil exploited (Fig. 6).

There are two separate angles to be considered (Fig. 7), which I shall refer to as the branch angle and radial angle, respectively. Inspection of root diagrams such as those in Weaver (1919) or Kutschera (1960) suggests that the branch angle tends to be large, often between 60 and 90°. Even in narrowly conical root systems, the initial branch angle is often high but later growth produces an effective branch angle of 45° or less. Since laterals arise in the pericycle and must penetrate the cortex before emerging, it seems reasonable to expect them to take the shortest path, which would mean they would normally emerge at a 90° branch angle and then adjust direction according to environmental stimuli.

The optimum angle for a lateral root after emerging from the epidermis of the parent root should be determined by the distance to the outer shell of the depletion zone around the root for some limiting resource. This is determined by the age of the parent root at the point of initiation and the growth rate of the parent on the one hand, and the diffusivity of the resource in question on the other. Although it is possible to derive predictions of optimum branch angle on this basis, which generally indicate that high values ($> 75°$) are most probable, no experimental data are as yet available to test these predictions.

Just as the branching angle is constrained by the way lateral roots emerge, the radial angle depends upon the internal anatomy of the root. Laterals normally arise opposite the protoxylem poles or, in some cases (Gramineae, Cyperaceae and Juncaceae), opposite the intervening protophloem poles. Since roots have a defined but variable number of such poles, the lateral roots should emerge in similarly defined ranks. The question that arises, therefore, is whether successive laterals

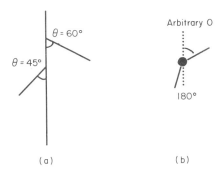

Fig. 7. The distinction between the branch angle (a) (drawn in longitudinal section) and the radial angle (b) (drawn in transverse section).

tend to arise at different poles, giving a pattern of 'rhizotaxis' analogous to phyllotaxis? Riopel (1966, 1969) in *Musa acuminata*, Mallory *et al.* (1970) in *Cucurbita maxima* and Charlton (1975) in *Pontaderia cordata* all suggested a greater or lesser tendency for successive laterals to arise in different ranks, particularly in *Cucurbita* where around 95% of all laterals apparently emerged in a distinct sequence. Charlton (1983), however, from more detailed studies of *Pontaderia*, *Pistia stratiotes* and *Potentilla palustris*, believes that this sequence may be the result of a spacing effect within ranks, such that a lateral tends not to arise at the same pole as an existing lateral for a certain distance, combined with a lack of correlation between ranks. Under such a mechanism, no well-defined rhizotactic patterns would emerge. In terms of the exploitation of soil, the distribution of laterals is clearly important and this is another area where information required to make sensible statements about comparative ecology is sadly lacking.

Root radius and root hairs. There are two sources of variation in root radius – that between links of the same magnitude and that between links of different magnitude. Radius is important both because of its effects on root function and the volume of soil that is available for exploitation (Nye, 1973; Silberbush & Barber, 1983), and because of its influence on root costs. Since the radius of the depletion zone for any ion is approximately

$$a + 2\sqrt{(Dt)} \quad \text{(Nye \& Tinker, 1977)}, \tag{1}$$

where a is root radius (cm), D the diffusion coefficient of the ion in soil (cm² s⁻¹) and t time in seconds, the ratio of the volume of soil exploited to cost (measured as root volume) is proportional to:

$$\frac{[a + 2\sqrt{(Dt)}]^2}{a^2}. \tag{2}$$

On this basis, the finer the root, the greater the return for unit investment. The counterbalancing effect to this is that fine roots have limited growth potential and transport capacity. Lyford (1975) speculates that minimum diameter and maximum length are correlated, such that transport of assimilates from the stem may limit extension growth. Equally, if a root is to be anything more than an ephemeral absorptive organ, and it is of course possible that many fine branchlets

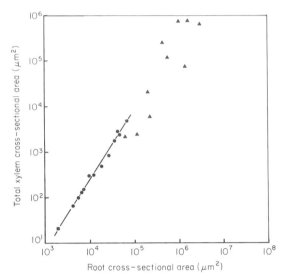

Fig. 8. Total xylem cross-sectional area (μm^2) as a function of root cross-sectional area for *Zea mays* [●; Miller (1981)] and various xerophytes [▲; Dittmer (1969)]. The line is the fitted regression to the *Zea* data and the square the intersection for minimum xylem area.

do act in this way (Reynolds, 1975), it must develop a central stele and this involves a minimum root radius. There is good correlation between xylem cross-sectional area and root cross-sectional area, both within a species [*Zea mays*; Miller (1981), Fig. 8] and between species [various xerophytes; Dittmer (1969), Fig. 8]. The smallest xylem area for a single vessel given by Miller is 20.5 μm^2 and, on the relationship of Figure 8, this would be found in a root of 25 μm radius, which is indeed approximately the radius of the finest roots reported in many studies.

Such fine roots are obviously extremely vulnerable to physical damage (e.g. desiccation) and to grazing and pathogens. Increase in diameter can therefore be viewed as an insurance against such damage, increasing the potential life span of the link and so its chances of becoming a link of higher magnitude giving rise to further exterior links. The rate at which link radius (and therefore cost) changes with link magnitude determines the cost of different architectures. At a given magnitude, a herringbone structure has the most high magnitude links and therefore potentially is the most expensive architecture. Variation in link radius of a given magnitude, for example of exterior links (magnitude 1), occurs both within and between species. It is well known that some plants have characteristically thick roots [so-called 'magnolioid' root systems, Baylis (1975)] and these may be more dependent on infection by mycorrhizal fungi for adequate uptake of phosphate. Hyphae of *Glomus* spp. have characteristic radii of to 5 μm, suggesting a large increase in cost-efficiency over lateral root growth. A similar effect can be achieved by root hairs, which are of comparable radius (2 to 10 μm; Dittmer, 1949) to vesicular–arbuscular mycorrhizal hyphae and which extend the radius of the depletion zone at a low cost (Lewis & Quirk, 1967). There is great variation in the length and density of root hairs between plant species. One curious feature is that no filamentous or branched root hairs appear ever to have evolved.

The radius of external links varies with their developmental order and is also affected by environmental factors, such as soil bulk density (Goss, 1977). It could

Table 1. *Mean radius of external links of root systems of plants in two distinct spatiotemporal guilds (defined by root activity) in alluvial grassland at Wheldrake Ings, Yorkshire*

	Species	Radius (μm)
Guild 1	*Agrostis capillaris*	44
	Anthoxanthum odoratum	60
	Lolium perenne	74
	Festuca rubra	55
	Phleum pratense	49
	$\bar{x}$	56
Guild 2	*Bromus hordeaceus*	81
	Deschampsia cespitosa	73
	Filipendula ulmaria	80
	Plantago lanceolata	125
	Ranunculus acris	98
	$\bar{x}$	91

therefore have an influence on the coexistence of plants by determining pore sizes that roots of different species are able to occupy (Sheikh & Rutter, 1969). Fitter (1987) examines this possibility and reports experiments by Martin (1979), in which coexistence of species with roots of different radius was promoted by growth in heterogeneous rather than homogeneous sand.

To test this further, I have calculated niche overlaps (Pianka, 1973) for the data in Fitter (1986c) of root activity at different soil depths and times for a group of species in an alluvial grassland. Cluster analysis of the data on niche overlap indicated two main groups of species with similar root activity in space and time (Table 1). Group I contained six grasses with a mean external link radius of 56 μm, whereas group II, containing two grasses and three forbs, had a mean radius of 91 μm; the difference was highly significant. Far from indicating coexistence brought about by occupation of distinct pores, therefore, this analysis implies that external link radius is a common factor to groups of coexisting species, perhaps in some way determining their fundamental niche.

Dynamics

I have so far discussed root system architecture largely from a static viewpoint, whereas any realistic assessment of the architecture–function relationship will certainly have to take the dynamics of the system into account. For example, the volume of soil exploited by a link is a function of time [see Eqn. (1)] and so the amount of resource obtained increases with time until adjacent roots' depletion zones make contact. On the other hand, root longevity is certainly related to radius (Reynolds, 1975) and therefore to costs of construction, although the relationship is not determined. Equally, there are recurrent costs such as respiration, repair and exudation which will increase with the longevity of the link. It is not apparently known whether, if branches are actually shed by root systems, this is a controlled senescence process with resources being withdrawn during senescence as in leaves, or whether it involves the loss of the entire investment. If the former is the case then a longer life span increases the probability that resources will be lost to grazers or pathogens and so not recovered. Where mycorrhizal infection occurs, longevity may determine the balance between profit and loss (Fitter, 1985b), since infection is unlikely to be of immediate benefit.

What is missing from all this speculation is any extensive information on root longevity. This is still a major lacuna in our knowledge (Caldwell, 1979). In contrast, root extension rates have been well documented and, in many models of root function (e.g. Baldwin, 1976), this rate is one of the critical parameters. Such models, however, rarely take into account the costs of elongation.

INTEGRATION

Approach

To integrate these various aspects of root system architecture is difficult. My attempt here is based on simulation modelling, with all its inevitable drawbacks. My excuse is simply that an analytical solution to the problems is probably not yet possible. I shall examine below the two principal functions of root systems that I identified at the outset – transport and resource capture.

Transport

I have already shown (Fitter, 1985a) that topologically herringbone-like structures are less transport-efficient, simply because their total external path length is greater than in more dichotomous systems. More importantly, *T. pratense* plants grown at low rates of water supply, which adopted herringbone structures of high topological index, proved to be less efficient in transport simulations than those grown with ample water (Fitter, 1986b). If the root systems of the well-watered plants had adopted a transport-efficient architecture, it is worth asking what benefits the others had gained by sacrificing this efficiency.

Resource capture

Computerized model of root growth. To simulate this aspect of the architecture–function relationship, I have devised a simple, two-dimensional model of root growth which can be implemented on an Acorn BBC microcomputer. Details of the model are in Appendix I, but in essence it is a node-based model, in which all existing links grow a defined amount in each growth period (approximating 1 d) and each node (the terminus of a day's growth) may generate a daughter link as well, in a randomly determined direction (left or right). Growth rates and probabilities of branch generation depend upon the developmental order of the node. The input variables are the branch angle, growth rate of nodes of developmental order 1 (i.e. the axis) and a probability coefficient for branch generation. Resource capture is simulated by allowing a coloured depletion zone to spread out from each link following Eqn (1), and the total extent of soil explored is assessed by scanning the screen for the number of pixels in the depletion zone colour. Resource diffusion coefficient is also therefore input as a variable. The programme calculates the magnitude (μ) of each link, assigns it a radius:

$$\text{radius} = (0{\cdot}2 + 0{\cdot}01\ \mu)\ \text{cm,} \tag{3}$$

and calculates the altitude and magnitude of the whole system and the slope of the regression of log altitude on log magnitude in order to estimate the topological index. Resource capture is expressed in terms of exploitation efficiency:

$$\text{mm}^2 \text{ of 'soil' exploited/mm}^3 \text{ of root produced.} \qquad (4)$$

All runs simulated 10 days' growth and, because of the stochastic elements in the model, a number of runs were made for each set of parameter values and the results plotted graphically [Fig. 9(a) to (c)].

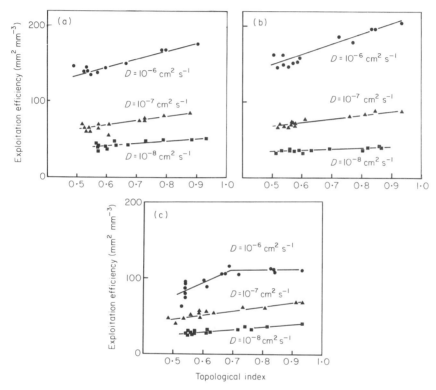

Fig. 9. Simulation of exploitation efficiency as a function of topological index at three resource diffusion coefficients and at three axial growth rates of (a) 3, (b) 5 and (c) 1 cm d^{-1}, and branch angle of 75°. Low values of the index represent many-branched root systems whereas high values tend to herringbone structures (see text).

Effect of topology and growth rate. Topological index can be altered in the model by the branch probability coefficient. If this is zero, branching is equiprobable at all nodes. With increasing values, branching is restricted to the developmental axis. The effect of topological index depends on the resource diffusion coefficient (Fig. 9). Where $D \geqslant 10^{-8}$ cm^2 s^{-1} and at a branch angle of 75° and an axial growth rate of 3 cm d^{-1}, exploitation efficiency increases with the index, but with $D < 10^{-8}$ cm^2 s^{-1}, architecture is ineffective [Fig. 9(a)]. Increasing the axial growth rate to 5 cm d^{-1} [Fig. 9(b)] makes architecture less significant at low values of D, presumably because depletion zones are now unlikely to overlap but, at 1 cm d^{-1}, the slope of the plot of exploitation efficiency against topological index at high D reaches a plateau, above which no further change in architecture offers any benefit [Fig. 9(c)]. In other words, topology becomes less important. These simulations suggest that the herringbone patterns of high topological index are more efficient at resource capture for mobile resources such as water and nitrate ions, at least where axial growth rates can be maintained. This is clearly because the wide depletion zones that develop for such resources rapidly overlap where intricate branching patterns of low topological index are produced.

The model does not take into account variations in resource availability in different soil regions and continued herringbone growth might quickly take the root into deeper, less fertile horizons. However, within the constraints of a uniform soil, a herringbone architecture is favoured.

CONCLUSIONS

That the geometry of root systems is fundamental to their function was recognized by Bray (1954). Quantitative analyses of root morphology have, however, concentrated on features of individual roots rather than root systems (Barley, 1970; Baldwin, 1976; Cushman, 1979; Silberbush & Barber, 1983). Robinson & Rorison (1983) demonstrate that root weight per unit soil volume is much more influential in the determination of NH_4^+ uptake than of the more mobile NO_3^- ion and it is generally accepted that root morphology is only significant for relatively immobile ions, particularly $H_2PO_4^-$. This is tantamount to saying that root thickness (length per unit weight or specific root length, SRL) is more important for immobile ions and Fitter (1985a) has shown that SRL is generally higher (roots are finer) at low rates of nutrient supply.

The analysis presented in this paper, however, is concerned with the disposition of the various members of a root system with respect to each other. It demonstrates that architecture is more important for mobile resources, because it also takes into account the root system function. As presented, these simulations apply only to young root systems in isolation and at low root density but the approach adopted can be used to investigate more realistic conditions. It is of course possible that the increase in exploitation efficiency achieved by young root systems of high topological index declines as root densities increase and this would be consonant with the observation that there is often a switch from a high to a low value of the index at around magnitudes 20 to 50 (Fig. 4). These conclusions therefore apply most strongly to young root systems.

It appears that there is a geometrical conflict in root system architecture. The most efficient systems for exploitation of soil, those of high topological index (herringbones), are the least efficient at transporting materials to the shoot system and, in any case, the most expensive. This may explain why root systems of red clover adopted a high topological index when deprived of water but a low value when water was plentiful (Fitter, 1986b; Fig. 4). It may also give some insight into the root systems found in many common weeds of disturbed ground, in which the primary root soon dies, and several laterals develop and branch extensively. These systems have low topological indices and typically develop in soils where resources are abundant and exploitation efficiency may not be of prime importance. Similarly, the contrast obvious in the diagrams of Cannon (1911) and Weaver (1919) of the root systems of arid zone plants is between shallow rooting, highly branched systems of plants active in the wet season and deeper-rooting, herringbone-like patterns of more xerophytic species. This also conforms to the conclusions derived from these simulations.

There are many further adaptive features of root systems to be investigated than those here but I believe that the trade-off between maximum exploitation efficiency (particularly for water), on the one hand, and maximum transport efficiency and minimum cost, on the other, has played a fundamental role in the evolution of root systems architecture.

ACKNOWLEDGEMENTS

I am grateful to Dr Michael Woldenberg for stimulating discussions on the functional significance of topology and geometry in which many of the ideas here have had their genesis. Ruth Nichols gathered much of the data presented here and drew the figures.

REFERENCES

BALDWIN, J. P. (1976). Competition for plant nutrients in soil: a theoretical approach. *Journal of Agricultural Science*, **87**, 341–356.

BARLEY, K. P. (1970). The configuration of the root system in relation to nutrient uptake. *Advances in Agronomy*, **22**, 159–201.

BAYLIS, G. T. S. (1975). The magnolioid mycorrhiza and mycotrophy in root systems derived from it. In: *Endomycorrhizas* (Ed. by F. E. Sanders, B. Mosse & P. B. Tinker), pp. 373–389. Academic Press, London.

BRAY, R. H. (1954). A nutrient mobility concept of soil–plant relationships. *Soil Science*, **78**, 9–22.

CALDWELL, M. M. (1979). Root structure: the considerable cost of below-ground function. In: *Topics in Plant Population Biology* (Ed. by O. T. Solbrig), pp. 408–427. MacMillan, London.

CANNON, W. A. (1911). *Root Habits of Desert Plants*. Carnegie Institute of Washington Publication No. 131.

CANNON, W. A. (1949). A tentative classification of root systems. *Ecology*, **30**, 542–548.

CAYLEY, A. (1857). On the analytical forms called trees. *Philosophical Magazine*, **18**, 374–378.

CHARLTON, W. A. (1975). Distribution of lateral roots and pattern of lateral initiation in *Pontaderia cordata* L. *Botanical Gazette*, **136**, 225–235.

CHARLTON, W. A. (1983). Patterns of distribution of lateral root primordia. *Annals of Botany*, **51**, 417–427.

CLAPHAM, A. R., TUTIN, T. G. & WARBURG, E. F. (1981). *Excursion Flora of the British Isles*, 3rd Edn. Cambridge University Press, Cambridge.

CROSSETT, R. N., CAMPBELL, D. J. & STEWART, H. E. (1975). Compensating growth in cereal root systems. *Plant and Soil*, **42**, 673–683.

CUSHMAN, J. H. (1979). Analytical solution to solute transport near root surfaces for low initial concentration. I. Equations and development. *Soil Science Society of American Journal*, **43**, 1087–1090.

DITTMER, H. J. (1938). A comparative study of the subterranean members of three field grasses. *Science*, **88**, 482.

DITTMER, H. J. (1949). Root hair variation in plant species. *American Journal of Botany*, **36**, 152–155.

DITTMER, H. J. (1969). Characteristics of the roots of some xerophytes. In: *Physiological Systems in Semiarid–Arid Environments* (Ed. by C. C. Hoff & M. L. Riedesel), pp. 231–238. University of New Mexico Press, Albuquerque.

DREW, M. C., SAKER, L. R. & ASHLEY, T. W. (1973). Nutrient supply and the growth of the seminal root system in barley. I. *Journal of Experimental Botany*, **24**, 1189–1202.

DUNCAN, W. G. & OHLROGGE, A. J. (1958). Principles of nutrient uptake from fertilizer bands. II. *Agronomy Journal*, **50**, 605–608.

FITTER, A. H. (1982). Morphometric analysis of root systems: application of the technique and influence of soil fertility on root system development in two herbaceous species. *Plant, Cell and Environment*, **5**, 313–322.

FITTER, A. H. (1985a). Functional significance of root morphology and root system architecture. In: *Ecological Interactions in Soil* (Special Publication of the British Ecological Society, No. 4; Ed. by A. H. Fitter, D. Atkinson, D. J. Read & M. B. Usher), pp. 87–106. Blackwell, Oxford.

FITTER, A. H. (1985b). Functioning of vesicular–arbuscular mycorrhizas under field conditions. *New Phytologist*, **99**, 257–265.

FITTER, A. H. (1986a). Acquisition and utilisation of resources. In: *Plant Ecology* (Ed. by M. J. Crawley), pp. 375–405. Blackwell, Oxford.

FITTER, A. H. (1986b). The topology and geometry of plant root systems: influence of watering rate on root system topology in *Trifolium pratense*. *Annals of Botany*, **57**, 91–101.

FITTER, A. H. (1986c). Spatial and temporal patterns of root activity in a species-rich alluvial grassland. *Oecologia*, **69**, 594–599.

FITTER, A. H. (1987). Spatial and temporal separation of activity in plant communities: prerequisite or consequence of coexistence? In: *Organisation of Communities: Past and Present* (Ed. by P. S. Giller & J. Gee). British Ecological Society Symposium, No. 26, Blackwell, Oxford. (In press.)

FOGEL, R. (1985). Roots as primary producers in below-ground ecosystems. In: *Ecological Interactions in Soil* (Special Publication of the British Ecological Society, No. 4; Ed. by A. H. Fitter, D. Atkinson, D. J. Read & M. B. Usher), pp. 23–36. Blackwell, Oxford.

FREIDENFELDT, T. (1902). Studien über die Wurzeln kräutiger Pflanzen. I. *Flora*, **91**, 115–128.

Goss, M. J. (1977). Effect of mechanical inpedence on growth of seedlings. *Journal of Experimental Botany*, **28**, 96–111.

GRIME, J. P., CRICK, J. C. & RINCON, J. A. (1986). The ecological significance of plasticity. In: *Plasticity in Plants* (Ed. by D. H. Jennings & A. Trewavas), pp. 7–29. Society for Experimental Biology Symposium No. 40. Cambridge University Press, Cambridge.

HACKETT, C. (1968). A study of the root system of barley. I. Effects of nutrition in two varieties. *New Phytologist*, **67**, 287–298.

HACKETT, C. (1972). A method of applying nutrients locally to roots under controlled conditions and some morphological effects of locally applied nitrate on the branching of wheat roots. *Australian Journal of Biological Science*, **25**, 1169–1180.

HÖVELER, W. (1892). Über die Verwerthung des Humus bei der Ernährung der chlorophyllführenden Pflanzen. *Jahrbuch der Wissenschaftlichen Botanik*, **24**, 283–315.

KRASILNIKOV, P. K. (1968). On the classification of the root systems of trees and shrubs. In: *Methods of Productivity Studies in Root Systems and Rhizosphere Organisms* (USSR Academy of Sciences), pp. 106–114. NAUKA, Leningrad.

KUTSCHERA, L. (1960). *Wurzelatlas mitteleuropäischer Ackerunkräuter und Kulturpflanzen*. DLG Verlag, Frankfurt-am-Main.

LEWIS, D. G. & QUIRK, J. P. (1967). Phosphate diffusion in soil and uptake by plants. IV. Computed uptake by model roots as a result of diffusive flow. *Plant and Soil*, **26**, 454–468.

LUNGLEY, D. R. (1973). The growth of root systems – a numerical computer simulation model. *Plant and Soil*, **38**, 145–159.

LYFORD, W. H. (1975). Rhizography of non-woody roots of trees in the forest floor. In: *The Development and Function of Roots* (Ed. by J. G. Torrey & D. T. Clarkson), pp. 179–196. Academic Press, New York.

LYFORD, W. H. (1980). *Development of the Root System of Northern Red Oak* (Quercus rubra *L.*). Harvard Forest Paper, No. 21. Harvard University, Petersham.

MALLORY, T. E., CHIANG, S., CUTTER, E. G. & GIFFORD, E. M. (1970). Sequence and pattern of lateral root formation in five selected species. *American Journal of Botany*, **57**, 800–809.

MARTIN, S. (1979). *The effects of pore sizes on root diameters and plant interference*. Unpublished B.Sc. thesis, University of York.

MAY, L.H., CHAPMAN, F. H. & ASPINALL, D. (1965). Quantitative studies of root development. I. The influence of nutrient concentration. *Australian Journal of Biological Sciences*, **18**, 25–35.

MILLER, D. M. (1981). Studies of root function in *Zea mays*. II. Dimensions of the root system. *Canadian Journal of Botany*, **59**, 811–818.

NOBBE, F. (1862). Über die feinere Verastelung der Pflanzenwurzeln. *Landwirtschaft Versuchstationen*, **4**, 212–224.

NYE, P. H. (1973). The relation between the radius of a root and its nutrient absorbing power. *Journal of Experimental Botany*, **24**, 783–786.

NYE, P. H. & TINKER, P. B. (1977). *Solute Movement in the Soil–Root System*. Blackwell, Oxford.

PIANKA, E. R. (1973). The structure of lizard communities. *Annual Review of Ecology and Systematics*, **4**, 53–74.

REYNOLDS, E. R. C. (1975). Tree rootlets and their distribution. In: *The Development and Function of Roots* (Ed. by J. G. Torrey & D. T. Clarkson), pp. 163–177. Academic Press, New York.

RIOPEL, J. L. (1966). The distribution of lateral roots in *Musa acuminata* 'Gros Michel'. *American Journal of Botany*, **53**, 403–407.

RIOPEL, J. L. (1969). Regulation of root positions. *Botanical Gazette*, **130**, 80–83.

ROBINSON, D. & RORISON, I. H. (1983). Relationships between root morphology and nitrogen availability in a recent theoretical model describing nitrogen uptake from soil. *Plant, Cell and Environment*, **6**, 641–647.

SCHUURMAN, J. J. & DE BOER, J. J. H. (1970). The developmental pattern of roots and shoots of oats under favourable conditions. *Netherlands Journal of Agricultural Science*, **18**, 168–181.

SHEIKH, K. H. & RUTTER, A. J. (1969). The responses of *Molinia caerulea* and *Erica tetralix* to soil aeration and related factors. I. Root distribution in relation to soil porosity. *Journal of Ecology*, **57**, 713–726.

SILBERBUSH, M. & BARBER, S. A. (1983). Sensitivity of simulated phosphorus uptake to parameters used by a mechanistic mathematical model. *Plant and Soil*, **74**, 93–100.

SMART, J. S. (1978). The analysis of drainage network composition. *Earth Surface Processes*, **3**, 129–170.

SPRAGUE, H. B. (1933). Root development of perennial grasses and its relation to soil conditions. *Soil Science*, **36**, 189–209.

TORREY, J. G. (1957). Auxin control of vascular pattern formation in regenerating pea root meristems grown in vitro. *American Journal of Botany*, **44**, 859–870.

TORREY, J. G. (1976). Root hormones and plant growth. *Annual Review of Plant Physiology*, **27**, 435–459.

WARDLAW, C. W. (1928). Size in relation to internal morphology. 3. The vascular system of roots. *Transactions of the Royal Society of Edinburgh*, **56**, 19–55.

WEAVER, J. E. (1919). *The Ecological Relations of Roots*. Carnegie Institute of Washington, Publication No. 286.

WEAVER, J. E. (1958). Classification of root systems of forbs of grassland and a consideration of their significance. *Ecology*, **39**, 393–401.

APPENDIX I

The flow-chart for the root growth model is as follows.

(1) Set up initial axis of variable growth rate (RATE), by defining x–y co-ordinates of two nodes. These nodes have developmental order 1.

(2) Grow new link at each external node, in direction of and at same rate as in previous time period. Link terminates in grown daughter node.

(3) Generate branches by allowing one branched daughter link to propagate at each unbranched node. Probability of branch generation is given by

$$p_\omega = \frac{e^{-\text{PROB} * \omega}}{\sum\limits_{i=1}^{\omega_{max}} (e^{-\text{PROB} * i})}$$

where ω is the developmental order of the node, and PROB is a variable such that, as PROB $\to 0$, branching becomes equiprobable at all nodes and, as PROB becomes large (> 1), branching is increasingly confined to nodes of $\omega = 1$.

(4) Grow branches at nodes selected in 3.

Angle of branching is given by a variable ANGLE (0–90). It remains constant for all developmental orders.

Direction of branching (left or right) is random but negative geotropism is not allowed.

Growth rate of branches depends upon ω, with the rate for branches arising at order 1 nodes being half the initial rate, and the rate for higher order nodes being;

$$\text{RATE} (\omega) = \frac{\text{RATE} (1)}{(\omega-1)^2}.$$

These rates are comparable to those published by May, Chapman & Aspinall (1965), Schuurman & de Boer (1970) and others.

(5) Calculate the radius (x) of the depletion zone around each link as:

$$x = 2 (Dt)^{\frac{1}{2}} \quad \text{(Nye \& Tinker, 1977)}.$$

This is drawn in colour on the screen. The total area of screen 'depleted' is assessed by scanning a 1 % sample of the pixels in the root area to estimate the proportion occupied by depletion zones. This is used as an index of the volume of soil that would be exploited by the root system.

(6) Determine the magnitude of each node and so that of the link between it and its parent node.

Radius (r_L, mm) of each link is determined.

$$r_L = 0{\cdot}2 + 0{\cdot}01 \, \mu.$$

These figures were determined by observations of field-grown root systems. The volume (= 'cost') of each link and so of the whole system can then be determined.

(7) Return to stage 2 unless a given size or age is reached.

New Phytol. (1987) **106** (Suppl.), 79–92

MINERAL NUTRITION IN TIME AND SPACE

By I. H. RORISON

*Unit of Comparative Plant Ecology (NERC), Department of Botany,
The University, Sheffield S10 2TN, UK*

SUMMARY

Interest in mineral nutrition during the last 20 years has centred increasingly on its dynamic aspects. This has led to a consideration of temporal variation, both diurnal and seasonal, and of spatial patterns as they affect nutrient supply. The response of plants to such variations are reviewed in the light of distribution and survival. Although all plants require the same small number of mineral elements, there is evidence of fine tuning in relation to soil heterogeneity. This allows a large number of nutritional options which in turn may influence coexistence and change. Selected examples for nitrogen and phosphorus are given and attention drawn to the need for further study, not only of nutrient interactions, of the energy costs of different nutrient pathways and of messenger functions of key nutrients (e.g. N and P linking activities of both root and shoot), but also of activities around the root surface; the micro-environment produced and the involvement of microflora and fauna.

Key words: Efficiency, heterogeneity, nitrogen source, nutrient and temperature gradients, root-surface activity.

INTRODUCTION

The study of mineral nutrition has had mixed support over the years but has tended to flourish in times of stress. It thrived during and for some time after World War II (Hewitt, 1966) and it is experiencing a resurgence in the wake of the more recent oil crisis.

As economic pressures lead to further limitation in the use of fertilizers, so the call goes out for 'nutrient-efficient' genotypes (Epstein, 1983). Other disciplines are looking to ecologists for the identification of characteristics shown by uncultivated plants that could be the basis of new breeds of crop. Emphasis has shifted from productivity to tolerance. The release of land from intensive agricultural use offers a fresh challenge to those attempting to create species-rich amenity and conservation areas.

The need for widespread comparative studies is highlighted by the latest review devoted to the mineral nutrition of wild plants (Chapin, 1980) which still perforce had to draw much of its experimental evidence from crop plants.

It may be thought that because the growth of all plants requires the same few mineral elements, there is little scope for survival and species diversity to be influenced by the chemical nature of the soil. The theoretical possibilities of such an outcome have been considered in broad context by, *inter alia*, Grubb (1977), Grime (1979) and Tilman (1982). A comparison of plants' tolerance of acidity, alkalinity and salinity, and the interactive effects which lead to fine tuning of response all indicate possible reasons for spatial separation. There is also a temporal component relating to the different phenologies and morphologies displayed by plants in both above- and below-ground parts.

The large number of nutritional options thus created must not be ignored in our search for an understanding of coexistence and change. As Slobodkin (1986)

0028-646X/87/05S079+14 $03.00/0

says in his essay on minimalism in art and science, 'much of the polemic in current ecology and evolution seems to arise from attempting to fill verbally the gap between narrow, formal theory and the richness of nature'. And again, 'when a science is not well formulated it may help to reconsider extremely simple situations in order to dispel the rhetorical clouds'.

It is proposed to make such an approach by describing a few key experiments aimed at relating form and function and considering the ecological implications of the results.

We are concerned to find evidence of how plants cope with heterogeneity in nutrient supply across the landscape, down the soil profile and even along the length of an individual root. To be meaningful, our study also requires consideration of conditions and of plant response during unpredictable as well as diurnal and seasonal periods of activity on both macro- and micro-scales. In probing less glamorous subterranean parts, we are also reminded of the need to study the regulatory role of root processes on the functioning of the plant as a whole.

WHOLE PLANT RESPONSE

Introduction

The influence of mineral nutrition may be seen across a broad landscape of, say, chalk grassland or greensand heath (Hope-Simpson, 1938; Kinzel, 1983) and, in more detail, at the Park Grass Plots of Rothamsted (Thurston, 1969) where increased fertility has led to a decrease in diversity.

Wide-scale surveys indicate the pH range of individual species (Grime & Lloyd, 1973; Ellenberg, 1982). These in turn can be related to the chemical components of the soil reaction complex (Rorison & Robinson, 1984) with soils of intermediate pH tending to be more fertile and more balanced in nutrient composition than soils at either end of the pH scale.

Current studies of the ecological aspects of plant mineral nutrition rely upon both laboratory and field investigation and range in level of biological organization from enzymes to the nutrient dynamics of ecosystems (Rorison & Robinson, 1986). In the space available, it is only possible to consider the fine tuning of plant response to a limited number of conditons and I have chosen some responses to the supply of nitrogen and phosphorus, partly because of the differences in their availability and release from soils (Nye & Tinker, 1977) and partly because of their key roles in plant metabolism (Bieleski & Ferguson, 1983; Pate, 1983).

Effects of pH

The form of both phosphorus and nitrogen in the soil is influenced by pH and, in the case of nitrogen, by temperature as well. For example, a plot of the nitrogen content of grassland soils in the Sheffield area shows a predominance of available NH_4^+ in acidic soils and a predominance of NO_3^- increasing as pH rises towards neutrality (McGrath & Rorison, 1982). The solubility of P, if not the absolute amount of P available, tends to follow the NH_4^+ curve.

At one extreme, tolerance of NH_4^+ ensures an adequate supply of N for plants surviving in acidic soils, although NH_4^+ is one of the major deterrents to the establishment of plants from other soil types (Gigon & Rorison, 1972). Under less extreme conditions of soil reaction, plants may survive on a mixture of nitrogen sources whose proportions of NH_4^+ to NO_3^- and absolute supply varies with the season.

Effects of temperature

Plants respond differently to such differences in resources, as may be deduced from a simple experiment of Peterkin (1981). He started from the basic tenet that NH_4^+ predominates in cold soils because low temperatures do not favour nitrification. There is also evidence (Rorison, Peterkin & Clarkson, 1983b) that plants absorb NH_4^+ in preference to NO_3^- at low temperatures. He grew three species in solution culture containing N as NH_4NO_3 at pH 4·5 and exposed them to one of two temperature regimes: a warm (15/20 °C diurnal cycle) and a cool (5/10 °C).

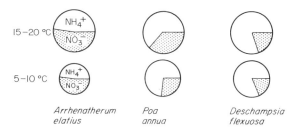

Fig. 1. The effect of species on the proportions of NO_3^- and HN_4^+ absorbed at warm or cool temperatures. Areas of the circles represent total uptake of N from NH_4NO_3 after 24 h. The area of the sectors represents proportions of total uptake of NO_3^- (◎) and NH_4^+ (◯) (Peterkin, 1981).

Arrhenatherum elatius*, a calcareous grassland population of a potentially fast-growing species, was non-selective in its uptake of $^{15}NH_4^+$ and $^{15}NO_3^-$ but total N uptake was much higher at the higher temperatures. This is in keeping with both its habitat and its habit of a marked peak of growth in summer (Fig. 1). *Poa annua*, which is advantaged by early spring growth when temperatures may be too low for maximum nitrification, showed selection in favour of NH_4^+ at low temperatures. *Deschampsia flexuosa*, as befitted its slow growth pattern, showed only a small temperature response for total uptake but was highly selective for NH_4^+ at both temperatures. This is in keeping with its distribution on highly acidic soils with a predominance of NH_4^+. It could also reflect its 'efficiency' in obtaining nitrogen at a relatively lower cost (Raven, 1985) than species, often potentially faster-growing, that utilize relatively more NO_3^- but tend not to grow on into the winter (Rorison *et al.*, 1983b).

Effect of localized supply

Since supplies of nutrients are distributed irregularly throughout the root profile, growth and nutrient uptake in response to localized supply have been studied in a number of crop plants. A typical pattern of compensation for restricted availability includes increases in specific absorption rate (SAR, nutrient uptake rates per unit of root), increases in root:shoot dry weight ratios, reductions in relative growth rates and proliferations of lateral roots within the nutrient-rich zones.

Robinson & Rorison (1983) confirmed such patterns of response for *Lolium perenne* and another fast-growing grass, *Holcus lanatus*, but revealed in the

* Details of ecological and physiological characteristics of species mentioned in this paper may be found in Grime, Hodgson & Hunt (1987). Nomenclature follows Clapham, Tutin & Warburg (1981).

slow-growing *Deschampsia flexuosa* a further facet of its ability to cope with restricted supplies of nitrogen. Figure 2 shows that when, after 14 d growth in full nutrient supply, nitrogen is withdrawn from 90 % of its root system, the growth (and total N content) of *D. flexuosa* is not significantly affected, whereas that of the two faster-growing species is. *D. flexuosa* maintained normal growth not only because it has a low potential relative growth rate linked to an innately low absolute demand for N (conservation) but also because it achieves greater proportional increases in specific N absorption rate (efficiency). All three species showed changes in root morphology but, whereas the changes (more fine roots and root hairs) in *D. flexuosa* were confined to the roots receiving nitrogen, in *H. lanatus* and *L. perenne* there was a response throughout their root system. This was relatively costly in terms of N and photosynthate transported to the −N roots.

In *D. flexuosa*, the amount of root material needed to absorb the necessary N was small and no adjustment in morphology of the −N roots occurred, thus removing the need to transport a greater proportion of N into those roots. This allowed increased N transport to the shoot which satisfied the demand for N there so that its normal growth was maintained. This in turn stabilized root:shoot growth and helped maintain maximum relative growth rate.

Again, the preferential movement of N absorbed as NH_4^+ to the shoot of *D. flexuosa* under conditions of limited supply may also reduce its energy costs relative to those of the other species (Fig. 3). However, we still need experimental clarification of the interactions of N and C metabolism in plants growing under optimal and suboptimal conditions in order to produce a precise explanation.

Temporal and spatial variation (laboratory experiments)

It is easier to control temporal variability in nutrient supply than spatial variability (Clement, Jones & Hopper, 1979; Rorison & Robinson, 1986) and a comparative study of plant response could be very rewarding. Figure 4 shows changes with time of root:shoot ratios of *Rumex acetosa* in response to four orders of magnitude changes in supply of P and compared with a steady state supply.

This immediate response to external change is typical of several relatively fast-growing herbaceous plants although some, e.g. *Urtica dioica*, only respond at the higher concentrations. In contrast, slower-growing species (e.g. *D. flexuosa*) have a slower response, accompanied in some cases (*Scabiosa columbaria*) by accumulation of the element which is in increased supply (Rorison & Gupta, 1974).

A combination of both temporal and spatial fluctuation in nutrient supply has been achieved experimentally by Grime, Crick & Rincon (1986). Using root systems distributed in nine separate growth 'arenas', they showed that *Agrostis stolonifera*, a species of fertile soils, differed from *Scirpus sylvaticus*, a plant restricted to relatively unproductive soils, in both morphological plasticity and N capture.

Greater absolute growth and nutrient uptake were achieved by *A. stolonifera* than *S. sylvaticus* under all conditions except when short, localized and spatially and temporally unpredictable pulses of N were given. Here (and also in terms of excess N taken up as a percentage increase over response to the basic treatment), *S. sylvaticus* proved relatively more efficient than *A. stolonifera*. It appeared that the relatively greater root mass (and root:shoot ratio) of the former allowed a more immediate response to increased nutrient supply. But when higher nutrient concentrations were maintained, *A. stolonifera* grew faster. As the authors con-

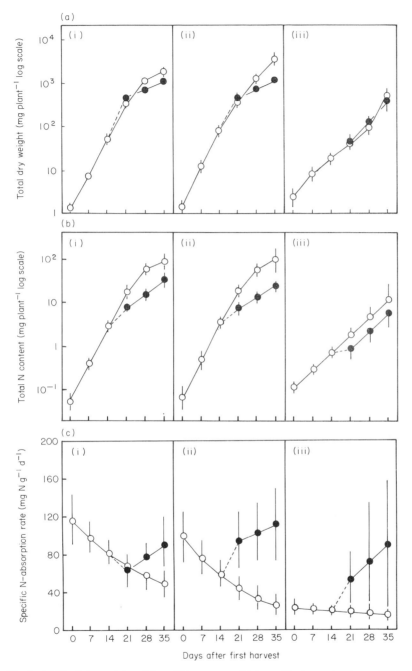

Fig. 2. (a) Total dry weights, (b) total N content and (c) specific absorption rates of N in (i) *Lolium perenne*, (ii) *Holcus lanatus* and (iii) *Deschampsia flexuosa*. ○, controls; ●, split-root treatments. Each symbol is a fitted mean (±95% confidence limits). Observed means lie within the confidence limits in all cases (Robinson & Rorison, 1983).

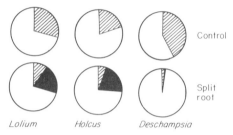

Fig. 3. Proportions of ammonium-N absorbed during 24 h in *Lolium perenne*, *Holcus lanatus* and *Deschampsia flexuosa*. ○, Shoot; ⊘, root + N; ●, root − N (Robinson & Rorison, 1983).

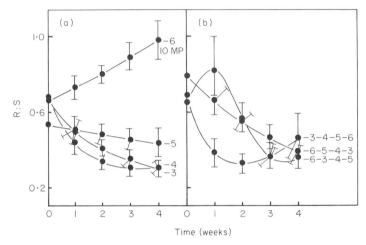

Fig. 4. Root:shoot (R:S) ratios of *Rumex acetosa* duing a 4-week experimental period of (a) steady supply and (b) changing supply of phosphorus. Molarities in full nutrient solution and 95 % confidence limits are indicated (Rorison & Gupta, 1974).

clude, 'a more appropriate test of the relative abilities of *S. sylvaticus* and *A. stolonifera* to capture and retain nitrogen from an infertile soil would of necessity involve a longer-term study under natural climatic conditions'. The question arises – how long is long enough?

Temporal and spatial variation (field experiments)

Fitter (1986) reported an experiment involving the activity of roots of a number of coexisting grassland plants. He made simultaneous injections of three chemical tracers, Li, Rb and Sr, at three different soil depths and measured the concentration of each tracer in leaf tissue. From harvests taken during the growing season April to June, two main periods of activity were recognized, one in spring (April to May) and one in summer (June). Species that were active early in the season tended to be less productive and shallower-rooted and it is suggested that this combination of characters allowed them to escape from competition with more productive species by being active at a time when deeper soil layers were less hospitable (cf. Tilman, 1986). Certainly in spring, surface soils tend to be warmer than those below (Rorison, Sutton & Hunt, 1986).

In sown plots, Veresoglou & Fitter (1984) recorded an earlier peak in uptake of P and K by *H. lanatus* than in *Arrhenatherum elatius*. This work was carried out over two years and showed the importance of following growth and nutrient uptake beyond the normal growing season.

A major problem in such work is to measure the growth and activity of individual root systems. One way is to grow plants individually in equal volumes of soil in separate containers. It removes the component of coexistence but allows for the recovery of both shoot and root systems.

An experiment of this type, carried out in Sheffield during 1981 to 1983, also emphasized the differences in response of individual species. Two species which coexist on shallow rendzinas on slopes in Derbyshire were used. *A. elatius* is potentially fast-growing, deep-rooted and tends to die back in winter; *Festuca ovina* has a slower growth rate and its leaves tend to remain green over winter. Although their native soil is as rich as a good agricultural soil in total N and P, very little of this is available ($NaHCO_3^-$-soluble) to the plants growing in it.

In order to produce precise and replicated measurements on whole plants the experiment ran for two successive years under natural climatic conditions. Undue heterogeneity and predation in the field was avoided by sowing plants in replicated individual tubes 20 cm high and containing 1 l of fresh soil. This represented the maximum soil depth and volume likely to be available to each plant in the field.

A series of harvests revealed two distinct patterns resulting in approximately the same mean dry weight per plant after one year. *A. elatius* showed rapid summer growth, die-back and weight loss of root and shoot in winter and rapid recovery in spring. *F. ovina* showed slower more continuous growth throughout and no net loss of weight over winter. Peterkin (1981) has called this aptly 'the tortoise and hare syndrome'. Controlled environment experiments revealed that die-back in *A. elatius* was not related to short days or low temperatures.

A further factorial experiment with initial additions of N and/or P indicated a significant response to P (Fig. 5) but not to N. Growth and nutrient uptake (including N) was significantly increased in plants receiving additional P. Even *A. elatius* continued to gain weight over winter in response to an enhancement of P supply (Rorison, Gupta & Spencer, 1983a), i.e. the limitation on growth was nutritional rather than climatic (as in the case of white clover in North Wales cited by Bradshaw (1987).

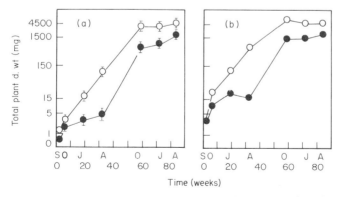

Fig. 5. The growth of (a) *Festuca ovina* and (b) *Arrhenatherum elatius* from September 1981 until May 1983. Fitted curves of total dry weight per plant are presented. Unless indicated by a vertical line, ±95% confidence limits lie within the symbols, as did observed means. There was no significant difference in the control and the +N treatments (●) or in the +P and +NP treatments (○), so only one set of data is plotted in each case (Rorison *et al.*, 1983a).

The distribution of nitrogen between September 1981 and October 1982 reveals three major points of interest. Firstly, there is no response to added N, probably because the seedlings were too small initially to utilize the N before it was leached out or fixed into the soil profile. Secondly, there is a significant increase in N removed by +P plants (Table 1) owing to greater exploitation by roots and triggered by the small addition of P and, thirdly, a very small proportion of the total N (largely organic N) is involved in turnover.

Table 1. *Nitrogen content of 1 l columns of rendzina soil showing original additions, rainwater input and losses due to leaching and uptake by plants (Rorison, Gupta & Spencer, 1983a)*

Initially (September 1981)			
Total pot (*c.* 70% N_o)	5000 mg		
NH_4NO_3 added	25 mg		
By October 1982			
+ in rain	*c.* 4 mg NO_3-N		
+	*c.* 3 mg NH_4-N		
− net loss	*c.* 60 mg (control)		
− total N plant mg		*Festuca ovina*	*Arrhenatherum elatius*
Control		15·4	19·0*
+ N		13·8	17·5
+ P		31·0	37·8
+ NP		29·9	40·1

N_o, organic nitrogen.
* Figures not linked by a vertical line are significantly different at $P < 0.05$.

Thus, there is strong evidence of phenological differences under conditions of low temperature and low nutrient (P) availability which can lead to the coexistence of *A. elatius* and *F. ovina*.

Such results underline the importance of measuring growth throughout the year under natural climatic conditions. They also illustrate a case where, despite high total soil N and P, an initial growth response was only obtained from an addition of P (Gupta & Rorison, 1986).

ACTIVITY AT THE ROOT SURFACE

Introduction

Having discussed the general pattern of plant distribution/survival in relation to N source and pH down to the level of the individual plant, it is appropriate to focus attention on the root systems because the form of N taken up by the plant is known to affect the pH of the root surface. Here again, it is proposed to describe a general pattern and then to examine the fine tuning of variation within a root system and even along the same root.

Uptake of N

In order to maintain electrochemical neutrality within the plant, there must be a balanced uptake/efflux of cations and anions. There is an efflux of hydrogen ions during the uptake of NH_4^+ which leads to the acidification of the moisture film at the root surface. The stoichiometric ratio of NH_4^+ uptake to H^+ efflux is usually quoted as $1:1$ (Raven & Smith, 1976; Raven, 1986). With the uptake of NO_3^-, the picture is less simple (Clarkson, 1986). Bicarbonate ions (HCO_3^-) are lost as a result of the uptake of NO_3^- but the ratio of uptake : efflux is rarely $1:1$. This is because nitrate reduction (NR) may occur to varying degrees in the root and shoot of plants. In some, NR is difficult to detect in the root [e.g. *Brassica napus* (rape); Moorby, Nye & White, 1985]. In others, the root is a major site for NR. In most vascular land plants, at least half of the OH^- assimilation is neutralized by the synthesis of organic acids and the cost can vary (Raven, 1986).

If NO_3^- is photoreduced in the shoot, the resulting OH^- may be disposed of (1) by precipitation of $CaCO_3$; (2) by synthesis of oxalic acid and storage of oxalate; (3) by synthesis of malic acid in shoots, transport of malate in phloem and consumption of malate in roots; or (4) by reduction of NO_3^- in the root. Both (3) and (4) lead to efflux of $OH^- + CO_2$ from roots.

The relative costs of growth, of N assimilation and of pH regulation using these forms of assimilation of NO_3^- (and comparing them with those of NH_4^+ and N_2) are discussed by Gutschick (1981) and by Raven (1985), who also points out the possible ecological significance of trends in the energy costs incurred.

These comparisons refer to plants receiving an adequate supply of nitrogen. Recently, Moorby *et al.* (1985) and Moorby (1986) have recorded similar changes of pH at the root surface owing to nitrogen (and phosphate) deficiency.

N deprivation

In experiments with roots of rape (*B. napus* cv. Emerald) supplied with nitrate nitrogen, as expected, the pH of the surrounding medium increased. Within 10 minutes, roots deprived of nitrogen reduced the pH of the medium, a condition which was rapidly reversible and independent of the nitrogen status of the plant. A split-root experiment demonstrated that although the acidification takes place over the entire root surface deprived of NO_3^-, other parts of the root system still receiving nitrate continued to increase the pH of the medium.

P deprivation

When phosphate was removed from the nutrient solution, a shift in pH (of the medium) also occurs but is delayed for several days and, when acidification does eventually occur, the efflux of H^+ is restricted to the terminal 2 to 3 cm of the root. This efflux is apparently dependent upon the P status of the plant. As this falls, there is a fall in uptake of NO_3^- which coincides with the production of H^+. No effect on NR was detected until several days later.

Localized changes in the cation : anion balance of the plant were cited as a reason for the efflux of either H^+ or HCO_3^-. A deficiency of anions externally leads to a relatively high uptake of cations near the root tip, especially Ca^{2+}, leading to a need for an efflux of H^+ (Ferguson & Clarkson, 1976). The pH continued to increase around the older portion of the root where there is relatively less uptake of Ca^{2+} and Mg^{2+}.

As usual, the effect of prevailing nutrient status on the response of the plant is less simple than the results of single factor experiments might imply.

Moorby *et al.* (1985) suggest that the ability of different parts of the root system (of rape) to secrete H^+ or HCO_3^- according to the nitrate supply available has an adaptive significance for plants growing on poor soil. Some nutrients, notably phosphate, occur in soils in forms unavailable to plants at neutral and higher pH, but become increasingly available as the pH is lowered.

This is a point worth taking further while bearing in mind the possible chemical minefield that is exposed when pH of soil falls below $c.$ $5\cdot0$. It must also be taken into account that what is good for the non-mycorrhizal rape may not be of such significance for the majority of other herbaceous species which have mycorrhizal associations, or for the rest of the rhizosphere's microbial population (Francis, 1986). Uptake from a mixture of $NH_4^+ -$ and NO_3^- would also tend to reduce the likelihood of a pH shift because of the efflux of both H^+ and HCO_3^-.

N_2 fixation

The accumulation of H^+ lost from the roots of some leguminous crops causes considerable acidity at the root surface (Jarvis & Hatch, 1985) but not to the extent reported for *Alnus glutinosa* which will grow normally down to a pH of $2\cdot8$ in solution culture (Troelstra, van Dijk & Blacqueire, 1985). It is important to remember that inter- and intraspecific differences in the rate of acidification will also vary with the ontogeny of the plant.

However, these results (and those described earlier for NO_3^- and NH_4^+) indicate that localized effects in response to heterogeneous supplies of nutrients in the soil can recur and whether the plant's response is likely to be instantaneous or delayed. Moorby's experiments included an ingenious combination of nutrient film culture and agar film techniques to identify efflux zones.

Marschner & Römheld (1983) had already demonstrated interspecific differences in response to N source (NO_3^-, NH_4^+ and N_2 fixation), as indicated by infiltration of the root systems by agar containing bromocresol purple. The pH was also measured with micro-electrodes. They included two monocotyledons, *Zea mays* (maize) and *Triticum aestivum* (wheat), and two dicotyledonous N_2 fixers, *Trifolium repens* (clover) and *Cicer arietinum* (chickpea). An adequate supply of NO_3^- to roots of maize resulted in a net efflux of HCO_3^- but, with less available N, there was acidification, particularly in the apical sections of the root. With a proliferation of young lateral roots near the top of the soil profile acidification was more rapid, but whether this was due to excess uptake of cations or depletion of N was not determined. When maize and chickpea (not N_2-fixing at the time!) were grown together in an initially adequate supply of NO_3^-, the former continued to produce a net efflux of HCO_3^- but the roots of the latter rapidly became acidified.

It is possible that roots proliferating near the surface were competing for NO_3^- and, in its absence, lost H^+. It is also possible that chickpea, with a higher requirement for Ca^{2+} and Mg^{2+} than maize, produces a net efflux of H^+ even with adequate supplies of NO_3^-.

DISCUSSION

These results indicate that plant roots do not only act as a sink for mineral nutrients delivered by mass flow and diffusion but that they can also influence the solubility of mineral nutrients by (root-induced) pH changes. The importance of these changes in the dynamics of mineral nutrient release from the soil and uptake by the plants needs further study in terms of plant response and soil type. For example, for a crop with high demand for NPK growing on an acid to neutral soil, the addition of NH_4^+ could lead to an efflux of H^+ and possibly to an increase in the uptake of P, Mn, Zn and Fe. However, unless nitrification of $NH_4^+ - N$ is prevented, the uptake of resulting NO_3^- will again lead to a rise in surface pH until a state of N deficiency is reached.

Also, in unbuffered cultured solution uptake of aluminium into susceptible varieties of wheat can be reduced by an excess uptake of NO_3^- leading to an efflux of HCO_3^-, a rise in pH and subsequent precipitation of aluminium salts (Foy & Fleming, 1978). However, it must be borne in mind that, under natural conditions of soil pH, i.e. < 5.0, where Al^{3+} is likely to be toxic to susceptible species, NH_4^+ and not NO_3^- is likely to be the most common form of N and its uptake is liable to lead to further acidification and increased solubility and toxicity of Al.

In a timely review, Loneragan (1979) concluded: 'To understand the effect of the chemical environment on root growth and function, more information is needed on the activities of particular regions of the root and of the micro-environment produced at the soil interface by those activities'. Marschner and others have now shown the way to providing the information. But there are still many questions unanswered.

More information is also needed on those factors which affect the formation, development and function of mycorrhizas.

For example, we have still to identify the role and significance of vesicular–arbuscular infection on the nutrition of plants *in the field*. What are the *consequences* of infection for the survival of plants in natural communities such as grassland? Does the timing and extent of infection vary? How does uptake by mycorrhizal associations relate to uptake from uninfected sections of the root? Does infection merely represent an extension of the surface area available? Is it always beneficial for the uptake of water, P, N and other nutrients (Smith *et al.*, 1986)? There are conditions where it may not be so (Fitter, 1977, 1985), as when parents may be feeding potential competitors of other species as well as their own offspring!

According to Harley & Smith (1983), the relationship of different kinds of infection to different environments is still to be defined but this has been explored by Read (1984). A cost–benefit analysis is still needed.

CONCLUSIONS

There has been no time to consider variations due to interactions between nutrients, mycorrhizal involvement, microbial responses in general, cost–benefits of the pathways used by different species, the possible role of N as messenger material controlling metabolic processes (such as photosynthesis), and linking activities in both root and shoot. All provide fascinating avenues for research in the future. What has emerged is evidence for pulses in the supply of nutrients from soil in spring and autumn and for heterogeneity of placement and movement in the profile.

Plants respond to this variability in different ways. Some slow-growing species are not only better husbanders of resources but are also more efficient in their utilization and redistribution than faster-growing species. Efficiency is manifest morphologically in the growth of fine roots 'where it matters', physiologically in greater flexibility in specific absorption rates and phenologically in extending the season of growth.

Changes in soil pHs at root surfaces not only provide evidence of specific differences on a relatively small scale but have given further insight into the role of cation:anion balance.

There is therefore plenty of scope for further understanding of the role of mineral nutrition in the functioning of the whole plant and its interactions with the environment. For the future we still need to know more about functions in general and particularly about (1) activities at root surfaces – the micro-environment produced, (2) nutrient interactions, (3) involvement of microflora and fauna in the field (they are *there*, so what is their function *vis-à-vis* natural communities); (4) energy costs of pathways in different species; (5) messenger activities of mineral nutrients [the role of N (Cooper *et al.*, 1986) and P (Herold, 1980) as messenger material linking activities in both root and shoot].

These are fascinating avenues for future research, revealing what goes on 'under the blankets' of the soil and at the same time enabling us to understand more fully how the more familiar superstructure above is functioning.

ACKNOWLEDGEMENTS

I would like to thank all my colleagues for stimulating discussions over the years. Their names appear among the references; in particular recent research students, Steve McGrath, John Peterkin and David Robinson, and Lal Gupta, Rita Spencer and Frank Sutton who have given generous support throughout.

REFERENCES

BIELESKI, R. L. & FERGUSON, I. B. (1983). Physiology and metabolism of phosphate and its compounds. In: *Inorganic Plant Nutrition. Encyclopedia of Plant Physiology. New Series, 15A* (Ed. by A. Läuchli & R. L. Bieleski), pp. 422–449. Springer-Verlag, Berlin.

BRADSHAW, A. D. (1987). Comparison – its scope and limits. In: *Frontiers of Comparative Plant Ecology* (Ed. by I. H. Rorison, J. P. Grime, G. A. F. Hendry, R. Hunt & D. H. Lewis), *New Phytologist*, **106** (Suppl.), 3–21. Academic Press, New York & London.

CHAPIN, F. S. (1980). The mineral nutrition of wild plants. *Annual Review of Ecology and Systematics*, **11**, 233–260.

CLAPHAM, A. R., TUTIN, T. G. & WARBURG, E. F. (1981). *Excursion Flora of the British Isles*, 3rd Edn. Cambridge University Press, Cambridge.

CLARKSON, D. T. (1986). Regulation of the absorption and release of nitrate by plant cells: a review of current ideas and methodology. In: *Fundamental, Ecological and Agricultural Aspects of Nitrogen Metabolism in Higher Plants* (Ed. by H. Lambers, J. J. Neeteson & I. Stulen), pp. 3–27. Martinus Nijhoff, Dordrecht.

CLEMENT, C. R., JONES, L. H. P. & HOPPER, M. J. (1979). Uptake of nitrogen from flowing nutrient solution: effect of terminated and intermittent nitrate supplies. In: *Nitrogen Assimilation of Plants* (Ed. by E. J. Hewitt & C. V. Cutting), pp. 123–133. Academic Press, London.

COOPER, H. D., CLARKSON, D. T., JOHNSON, M. G., WHITEWAY, J. N. & LOUGHMAN, B. C. (1986). Cycling of amino nitrogen between shoots and roots in wheat seedlings. *Plant and Soil*, **91**, 319–322.

ELLENBERG, H. (1982). *Vegetation Mitteleuropas mit den Alpen in ökologischer Sicht*. Verlag Eugen Ulmer, Stuttgart.

EPSTEIN, E. (1983). Foreword. In: *Inorganic Plant Nutrition. Encyclopedia of Plant Physiology. New Series, 15* (Ed. by A. Läuchli & R. L. Bieleski), pp. v–ix. Springer-Verlag, Berlin.

FERGUSON, I. B. & CLARKSON, D. T. (1976). Simultaneous uptake and translocation of magnesium and calcium in barley (*Hordeum vulgare*) roots. *Planta*, **128**, 267–269.

FITTER, A. H. (1977). Influence of mycorrhizal infection on competition for phosphorus and potassium by two grasses. *New Phytologist*, **79**, 119–125.

FITTER, A. H. (1985). Functioning of vesicular–arbuscular mycorrhizas under field conditions. *New Phytologist*, **99**, 257–267.

FITTER, A. H. (1986). Spatial and temporal patterns of root activity in a species-rich alluvial grassland. *Oecologia (Berlin)*, **69**, 594–599.

FOY, C. & FLEMING, A. L. (1978). The physiology of plant tolerance to excess available aluminium and manganese in acid soils. In: *Crop Tolerance to Suboptimal Land Conditions* (Ed. by C. A. Jung), pp. 301–328. American Society of Agronomy, Madison.

FRANCIS, A. J. (1986). Acid rain effects on soil and aquatic microbial processes. *Experientia*, **42**, 455–465.

GIGON, A. & RORISON, I. H. (1972). The response of some ecologically distinct plant species to nitrate- and to ammonium-nitrogen. *Journal of Ecology*, **60**, 93–102.

GRIME, J. P. (1979). *Plant Strategies and Vegetation Processes*. Wiley, Chichester.

GRIME, J. P. & LLOYD, P. S. (1973). *An Ecological Atlas of Grassland Plants*. Edward Arnold, London.

GRIME, J. P., CRICK, J. C. & RINCON, J. E. (1986). The ecological significance of plasticity. In: *Plasticity in Plants* (Ed. by D. H. Jennings & A. J. Trewevas), pp. 5–29. Blackwell Scientific Publications, Oxford.

GRIME, J. P., HODGSON, J. G. & HUNT, R. (1987). *Comparative Plant Ecology: A Functional Approach to Common British Plants and Communities*. George Allen & Unwin, London. (In press.)

GRUBB, P. J. (1977). The maintenance of species-richness in plant communities: the importance of the regeneration niche. *Biological Reviews*, **52**, 107–145.

GUPTA, P. L. & RORISON, I. H. (1986). Interspecific response by grasses to P in infertile and uncultivated soils. *Journal of the Science of Food and Agriculture*, **37**, 14–15.

GUTSCHICK, V. P. (1981). Evolved strategies in nitrogen acquisition by plants. *American Naturalist*, **118**, 607–637.

HARLEY, J. L. & SMITH, S. E. (1983). *Mycorrhizal Symbiosis*. Academic Press, London.

HEROLD, A. (1980). Regulation of photosynthesis by sink activity—the missing link. *New Phytologist*, **86**, 131–144.

HEWITT, E. J. (1966). *Sand and Water Culture Methods Used in the Study of Plant Nutrition*, 2nd Edn. Technical Communication No. 22, Commonwealth Agricultural Bureaux, Farnham Royal.

HOPE-SIMPSON, J. F. (1938). A chalk flora of the Lower Greensand; its use in interpreting the calcicole habit. *Journal of Ecology*, **26**, 218–235.

KINZEL, H. (1983). Influence of limestone, silicates and soil pH on vegetation. In: *Physiological Plant Ecology. III. Responses to the Chemical and Biological Environment. Encyclopedia of Plant Physiology, New Series, 12C* (Ed. by O. L. Lange, P. S. Nobel, C. B. Osmond & H. Ziegler), pp. 201–244. Springer-Verlag, Berlin.

JARVIS, S. C. & HATCH, D. J. (1985). Rates of hydrogen efflux by nodulated legumes grown in flowing culture solution with continuous pH monitoring and adjustment. *Annals of Botany*, **55**, 41–51.

LONERAGAN, J. F. (1979). The interface in relation to root function and growth. In: *The Soil–Root Interface* (Ed. by J. L. Harley & R. Scott-Russell), pp. 351–367. Academic Press, London.

MARSCHNER, H. & RÖMHELD, V. (1983). *In vivo* measurements of root induced pH changes at the soil–root interface: effect of plant species and nitrogen source. *Zeitschrift für Pflanzenphysiologie*, **111**, 241–251.

McGRATH, S. P. & RORISON, I. H. (1982). The influence of nitrogen source on the tolerance of *Holcus lanatus* L. and *Bromus erectus* Huds. to manganese. *New Phytologist*, **91**, 443–452.

MOORBY, H. (1986). *Environmental conditions affecting acid-base changes around plant roots*. D.Phil. thesis, University of Oxford.

MOORBY, H., NYE, P. H. & WHITE, R. E. (1985). The influence of nitrate nutrition on H$^+$ efflux by young rape plants (*Brassica napus* cv. emerald). *Plant and Soil*, **84**, 403–415.

NYE, P. H. & TINKER, P. B. (1977). *Solute Movement in the Soil–Root System. Studies in Ecology 4.* Blackwell Scientific Publications, Oxford.

PATE, J. S. (1983). Patterns of nitrogen metabolism in higher plants and their ecological significance. In: *Nitrogen as an Ecological Factor* (Ed. by J. A. Lee, S. McNeill & I. H. Rorison), pp. 225–255. Blackwell Scientific Publications, Oxford.

PETERKIN, J. H. (1981). *Plant growth and nitrogen nutrition in relation to temperature*. Ph.D. thesis, University of Sheffield.

RAVEN, J. A. (1985). Regulation of pH and generation of osmoregularity in vascular plants: a cost–benefit analysis in relation to efficiency of use of energy, nitrogen and water. *New Phytologist*, **101**, 25–78.

RAVEN, J. A. (1986). Biochemical disposal of excess H$^+$ in growing plants? *New Phytologist*, **104**, 175–206.

RAVEN, J. A. & SMITH, F. A. (1976). Nitrogen assimilation and transport in vascular land plants in relation to intra cellular pH regulation. *The New Phytologist*, **76**, 415–431.

READ, D. J. (1984). The structure and function of the vegetative mycelium of mycorrhizal roots. In: *The Ecology and Physiology of the Fungal Mycelium.* (Ed. by D. H. Jennings & A. D. M. Rayner), pp. 215–240. Cambridge University Press, Cambridge.

ROBINSON, D. & RORISON, I. H. (1983). A comparison of the responses of *Lolium perenne* L., *Holcus lanatus* L. and *Deschampsia flexuosa* (L.) Trin. to a localized supply of nitrogen. *New Phytologist*, **94**, 263–273.

RORISON, I. H. & GUPTA, P. L. (1974). The growth of seedlings in response to variable phosphorus supply. In: *Plant Analysis and Fertilizer Problems* (Ed. by J. Wehrmann), pp. 378–382. German Society of Plant Nutrition, Hanover.

RORISON, I. H. & ROBINSON, D. (1984). Calcium as an environmental variable. *Plant, Cell and Environment*, **7**, 381–390.

RORISON, I. H. & ROBINSON, D. (1986). Mineral nutrition. In: *Methods in Plant Ecology*, 2nd Edn (Ed. by P. D. Moore & S. B. Chapman), pp. 145–213. Blackwell Scientific Publications, Oxford.

RORISON, I. H., GUPTA, P. L. & SPENCER, R. E. (1983a). Nutrient balance and soil exhaustion in relation to plant growth and uptake by contrasted species. In: *Annual Report of the Unit of Comparative Plant Ecology (NERC), 1983*, pp. 26–27. University of Sheffield, Sheffield.

RORISON, I. H., PETERKIN, J. H. & CLARKSON, D. T. (1983b). Nitrogen source, temperature and plant growth. In: *Nitrogen as an Ecological Factor* (Ed. by J. A. Lee, S. McNeill & I. H. Rorison), pp. 189–209. Blackwell Scientific Publications, Oxford.

RORISON, I. H., SUTTON, F. & HUNT, R. (1986). Local climate, topography and plant growth in Lathkill Dale NNR. I. A twelve-year summary of solar radiation and temperature. *Plant, Cell and Environment*, **9**, 49–56.

SLOBODKIN, L. D. (1986). Minimalism in art and science. *American Naturalist*, **127**, 257–265.

SMITH, S. E., ST JOHN, B. J., SMITH, F. A. & BROMLEY, J. L. (1986). Effects of mycorrhizal infection on plant growth nitrogen and phosphorus nutrition in glasshouse-grown *Allium cepa* L. *New Phytologist*, **103**, 359–373.

THURSTON, J. M. (1969). The effect of liming and fertilizers on the botanical composition of permanent grassland, and on the yield of hay. In: *Ecological Aspects of the Mineral Nutrition of Plants* (Ed. by I. H. Rorison), pp. 3–10. Blackwell Scientific Publications, Oxford.

TILMAN, D. (1982). *Resource Competition and Community Structure*. Princeton University Press, Princeton, New York.

TILMAN, D. (1986). Nitrogen-limited growth in plants from different successional stages. *Ecology*, **67**, 555–563.

TROELSTRA, S. R., VAN DIJK, K. & BLACQUIERE, T. (1985). Effects of N source on proton excretion, ionic balance and growth of *Alnus glutinosa* (L.) Gaertner: comparison of N_2 fixation with single and mixed source of NO_3 and NH_4. *Plant and Soil*, **84**, 361–385.

VERESOGLOU, D. S. & FITTER, A. H. (1984). Spatial and temporal patterns of growth and nutrient uptake of five coexisting grasses. *Journal of Ecology*, **72**, 259–272.

New Phytol. (1987) **106** (Suppl.), 93–111

METAL TOLERANCE

By A. J. M. BAKER

Department of Botany, University of Sheffield, Sheffield S10 2TN, UK

Summary

This paper highlights major developments in the field of heavy metal tolerance in plants over the last 15 years. Advances in experimental and theoretical aspects are considered. The value of both intra- and interspecific studies in assessing the ecological significance of adaptive strategies is stressed.

Key words: Metal tolerance, metal toxicity, adaptation.

Introduction

The phenomenon of heavy metal tolerance in plants has attracted the interests of plant ecologists, plant physiologists and evolutionary biologists for more than 50 years. Early observational studies of plants growing on metal-contaminated soils have posed intriguing questions about the nature, scale and mechanisms of adaptation involved. It has, however, only been in the last 25 years that detailed experimental studies have provided some insight into the complex responses of plants to heavy metal toxins. Additionally, recent studies have focused on the time-scale of evolutionary response of plant populations in the face of the powerful selective forces of metal toxicities. As tolerance has proved an easily measurable character, it has been used extensively to investigate genecological phenomena and formulate evolutionary theory. 'Spin-off' into applied areas of ecology has been the development of metal-tolerant cultivars for use in the revegetation of metal-liferous wastes and the refinement of geobotanical and biogeochemical techniques of mineral exploration.

Several major reviews map out the progress of studies relating to heavy metal tolerance in plants. Ernst's monograph 'Schwermetallvegetation der Erde' (Ernst, 1974a) remains the standard source of reference on plants and plant communities of metal-contaminated soils on a global scale. Landmark reviews and papers on evolutionary aspects are Antonovics, Bradshaw & Turner (1971), Antonovics (1975) and Bradshaw (1976, 1984). Genetical aspects are also covered by Macnair (1981), methodology by Wilkins (1978) and mechanisms of tolerance by Turner (1969), Ernst (1974b, 1975, 1976), Wainwright & Woolhouse (1975), Farago (1981), Thurman (1981), Peterson (1983) and Woolhouse (1983).

It is not the purpose of this paper to re-review the extensive literature on metal tolerance but instead, by focusing on a number of recent studies, ventilate new developments and challenge some of the early conclusions which have largely been handed down as dogma. It is perhaps particularly appropriate to adopt this approach at this 25th Anniversary Symposium of the Unit of Comparative Plant Ecology where research frontiers are being highlighted.

0028-646X/87/05S093 + 19 $03.00/0

WHAT IS METAL TOLERANCE?

Strategies for survival on metal-contaminated soils

Heavy metals in the plant environment operate, according to Levitt (1980), as *stress* factors in that they cause physiological reaction change (*strain*) and in so doing can reduce vigour, or in the extreme, totally inhibit plant growth. '*Sensitivity*' describes the effects of a stress which result in injury or death of the plant. '*Resistance*' then refers to the reaction of a plant to heavy metal stress in such a way that it can survive and reproduce and in so doing contribute to the next generation. Resistance to heavy metals can be achieved by either of two strategies; *avoidance*, by which a plant is protected externally from the influences of the stress and *tolerance*, by which a plant survives the effects of internal stress. Tolerance is therefore conferred by the possession of specific physiological mechanisms which collectively enable it to function normally even in the presence of high concentrations of potentially toxic elements. A genetic basis to tolerance is therefore implied, in that these mechanisms are heritable attributes of tolerant mutants or genotypes. Türesson (1922) first drew attention to the superior fitness of adapted genotypes in that he believed that should a plastic (phenotypic) response be possible, so much may be demanded of a plant physiologically that overall vigour and hence fitness will be reduced drastically. Adapted genotypes are preadjusted without these constraints. However, the 'costs' of tolerance may be great and manifested in terms of the reduced fitness of tolerant plants when grown in normal (uncontaminated) situations (Cook, Lefèbvre & McNeilly, 1972; Cox & Hutchinson, 1981; Bradshaw, 1984). Ernst (1976) also suggests that the slower growth rates and lower biomass production of many tolerant plants by comparison with their non-tolerant counterparts are a corollary of the energy expenditure for operation of the mechanisms of tolerance involved. Such are the 'costs' of metal tolerance.

Metal-contaminated soils such as those developed on metalliferous mine spoils and naturally enriched soils in the vicinity of mineral veins may have been available for plant colonization for many years. Such areas can support relatively rich and diverse plant communities (Fig. 1), which may or may not be phytogeographically distinct from surrounding vegetation on uncontaminated soil. A satisfactory classification of taxa found on metal-contaminated soils is that of Lambinon & Auquier (1963) and is based on the degree of restriction to such sites. Metallophytes (absolute and local) are found only on metal-contaminated soils; pseudometallophytes occur on both contaminated and normal soils in the same region. Within this group two classes can be recognized; electives and indifferents in order of decreasing abundance and vigour on contaminated soil (Antonovics *et al.*, 1971). Accidental metallophytes usually include ruderals and annuals which appear only sporadically and show reduced vigour. In the terminology used above, metallophytes and pseudometallophytes achieve resistance by a true tolerance strategy resulting from population differentiation, whereas accidentals probably avoid the effects of metal stress.

Few detailed comparative studies are available to illustrate differences in tolerances of species under field conditions and the effects of gradients in metal toxicity on the structure and species composition of vegetation (see e.g. Simon, 1978; Clark & Clark, 1981; Thompson & Proctor, 1983). Tolerance and avoidance strategies are clearly shown by colonists of Tideslow Rake, Derbyshire (Fig. 1), a physically and chemically heterogeneous area of mine spoil. It supports an almost

Fig. 1. Tideslow Rake, Derbyshire, a site of former lead workings now derelict for over 200 years and colonized extensively by metal-tolerant races of a wide range of plant species.

continuous cover of vegetation but the distribution of most species is clearly non-random. In a detailed study (Shaw, 1984; Shaw, Rorison & Baker, 1984), distribution of species was investigated in relation to soil heterogeneity and nutrient status. Figure 2 summarizes the data for six of the chemical factors studied for five herbaceous species and for bare ground. The species fall into two main groups: those such as *Minuarita verna**, *Koeleria macrantha*, *Festuca ovina*, *Hieracium pilosella* and *Thymus praecox* ssp. *arcticus* which are characteristic of the shallow, coarser, nutrient-poor soils and those including *Dactylis glomerata* and

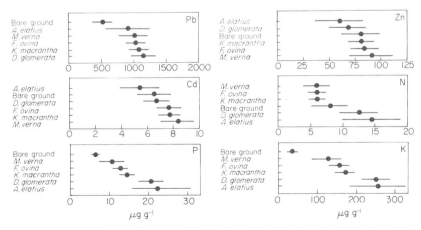

Fig. 2. Concentrations of Pb, Zn, Cd, N, P and K in soil samples taken from beneath *Arrhenatherum elatius* ($n = 14$). *Dactylis glomerata* ($n = 25$), *Minuartia verna* ($n = 18$), *Festuca ovina* ($n = 38$), *Koeleria macrantha* ($n = 39$) and bare ground at Tideslow Rake, Derbyshire (means with 95 % confidence limits). Extractions were made in 1 M ammonium acetate (pH 7·0), for Pb, Zn, Cd, P and K, and in water for NO_3-N. (From Shaw, Rorison & Baker, 1984.)

* Except where authorities are given, nomenclature follows Clapham, Tutin & Warburg (1981).

Arrhenatherum elatius found where the soil is deeper and richer in nutrients and organic matter. Further, *A. elatius* and *D. glomerata* occurred in sample quadrats with significantly lower species density. Bare spoil presents an inhospitable environment in terms of low nutrient status and high pH and an open, coarse texture, as well as the elevated concentrations of Pb, Zn and Cd expected on lead-mining spoil. The data are consistent with the 'hump-backed' model of Grime (1973) describing the impact of a gradient of increasing stress and/or disturbance upon potential species diversity in herbaceous vegetation. At Tideslow Rake, the indications are that, of those plant nutrients measured, it is N, P and K that are the major factors determining distribution of species, including that of *M. verna* which is generally considered to be a metallophyte. Tolerance of a wide range of metal concentrations and nutrient-poor soil allows growth of metal-tolerant races of certain species (mainly pseudometallophytes) on some parts of the Rake, while their establishment is prevented in other areas by competition from faster-growing species which appear to be confined to the more nutrient-rich and slightly less contaminated areas. Other avoiders include ruderal and winter annual species which are confined to areas of non-toxic spoil and limestone overburden, the latter most prominently on south-facing slopes of mounds of spoil.

The quantification and measurement of metal tolerance

The causality of the correlations between species distribution and soil chemistry suggested for Tideslow Rake can only be established by comparative experiments performed under standardized conditions. If metal tolerance is defined purely in terms of survival of individuals to the following generation when grown on metal-enriched soils, growth experiments are required to assess relative tolerance. Such long-term trials are frequently impractical and the experimental ecologist is forced to employ short-term or rapid screening tests. The virtues of this comparative approach have been discussed by Grime (1965). The effects of metal toxicity on individuals or species differing in their tolerance are normally so clear-cut that short-term survival tests using seedlings or vegetative propagules will prove an effective screening procedure. Beyond the seedling establishment phase, relative growth rate, or more simply yield, are suitable measures of performance and hence tolerance (Davis & Beckett, 1978). Bradshaw (1952), when attempting to grow tillers of a pasture population of *Agrostis capillaris* ($= A.$ *tenuis* Sibth.) in lead mine spoil, showed conclusively that a lack of tolerance was manifested by an inhibition of root growth. Wilkins (1957) subsequently developed the so-called 'root elongation method' which is based on quantification of the inhibitory effects of metal ions on root growth when supplied in simple metal salt solutions. The technique has been used widely and has been modified and refined by various workers [see Wilkins (1978) for a full appraisal of the method]. The advantages of the method are that it is simple (plants are allowed to root in metal-amended and control solutions), rapid (frequently only a few days growth in treatment solutions is required) and easy to perform. Indices of tolerance derived from ratios between data for treatment and control solutions can then be calculated and used to characterize individuals or populations. There is no doubt that the technique has been the one important tool enabling large-scale screening for tolerance to be achieved, so facilitating genecological investigations involving large numbers of individuals and their progeny.

The root elongation test has, however, been used indiscriminately for testing tolerance by various workers without due attention to its limitations. Although originally developed for grasses where the root system is fibrous, the test has been applied in studies on plants as diverse in root morphology as *Mimulus guttatus* (Allen & Sheppard, 1971; Macnair, 1977), *Armeria maritima* (Lefèbvre, 1968), *Plantago lanceolata* (Wu & Antonovics, 1975; Pollard, 1980) and *Leucanthemum vulgare* (Whitebrook, 1986). Both adult and seedling material have been tested using variants of the same technique but, at least for copper tolerance in *A. capillaris*, McNeilly & Bradshaw (1968) established a high degree of correlation between plants tested as seedlings and subsequently tested as adults, implying no change in sensitivity during development.

One basic assumption of the technique is that there exists an underlying relationship between index of tolerance as measured experimentally and some estimate of the activity of the metal concerned, as assessed by simple chemical analysis of soil in the rooting zone of the plants tested (total or extractable metal concentrations). Few workers have actually confirmed this fundamental relationship before using the method. However, good correlations have been established by Wilkins (1960), Karataglis (cited in Bradshaw, 1976) and Wigham, Martin & Coughtrey (1980) for studies with Pb, Cu and Cd, respectively, albeit for different species. A tacit assumption is perhaps therefore justified.

A further prerequisite for the test is that during the course of measurements the rate of root elongation in both treatment and control solutions remains constant. This is particularly critical in studies where a 'sequential' method (consecutive measurements on the same root in control followed by metal treatment solutions) is employed. 'Parallel' methods (simultaneous measurements of replicate individuals in control and treatment solutions) are less sensitive to changes in root elongation rate with time. Rates of root elongation do decline with time and so the period over which linear growth is maintained should be established before embarking upon tests, and all measurements restricted to this phase of growth. Humphreys & Nicholls (1984) have drawn attention to another problem where tolerance indices are based on ratios of root growth in treatment and control solutions. They suggest that the components of such indices may be under independent genetical control and hence unjustified conclusions can be drawn if the indices are not used with caution.

Another fundamental issue to be resolved is the metal concentration for screening. The concentration chosen has generally been one which brings about maximum separation between tolerant and non-tolerant genotypes. In view of varying responses to differing concentrations, some workers have preferred to employ a range of concentrations, as regression or probit analysis is then possible (see e.g. Craig, 1977; Wilkins, 1978; Nicholls & McNeilly, 1979). The last-mentioned authors were thus able to detect variation in both index of tolerance and sensitivity of rooting to increased [Cu] in seven populations of *A. capillaris*. They showed that these properties were phenotypically independent characters and suggested that rooting sensitivity therefore provided a further measure of copper tolerance. A similar approach, using treatments at only two concentrations, was adopted by Macnair (1977, 1981) to screen out copper-tolerant individuals of *M. guttatus* from large samples of rooted cuttings.

The background solution in which metal ions are supplied is another important variable. Traditionally, a single-salt solution of calcium nitrate (supplied at 0·5 or 1·0 g l^{-1}) has been used and metals added as either the nitrate or sulphate in order to permit the use of higher concentrations of metal ions without fear of precipitation (Wilkins, 1978). Attention has been drawn to the likely problems of nutrient deficiency effects in this basal solution, particularly for seedling material. Possible interactive effects are also masked. Accordingly, some workers have chosen to use a dilute nutrient solution as background or one which has been modified by the removal of phosphates and/or sulphates (Wilkins, 1978; Karataglis, 1980; Shaw, Rorison & Baker, 1982; Baker et al., 1986). Johnston & Proctor (1981) have further attempted to mirror field conditions by using a background solution which simulated the composition of soil water in a toxic serpentine soil. The pH of the test solution must also be adjusted appropriately and some means of aeration is desirable.

In a comparative study of tolerance to Pb, Zn and Cd of races of 21 spp. from contaminated and uncontaminated soil in Derbyshire (Shaw, 1984), tests were carried out in complete (0·1-strength Rorison Solution; Hewitt, 1966, table 30c) culture solution, complete culture −P and 0·5 g l⁻¹ calcium nitrate solution. Differing responses to the three metals supplied in the three different background solutions were apparent (Shaw et al., 1982). Figure 3 presents the data obtained for root extension of *F. ovina* and *Lotus corniculatus*; Table 1 records the dosage of lead required to produce 50 % inhibition of root extension. In both species, the

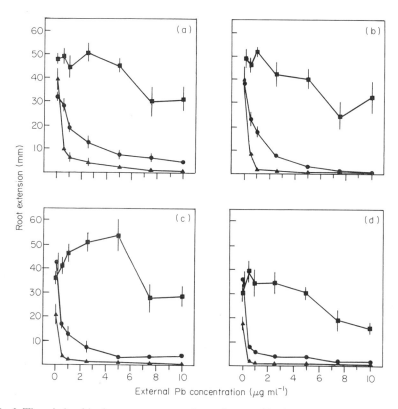

Fig. 3. The relationships between root extension and external lead concentration in the test solution for a lead-rake (a) and control (b) population of *Festuca ovina* and similar populations of *Lotus corniculatus*, (c) and (d) respectively. Lead was supplied in complete nutrient solution (■), in complete nutrient solution – phosphate (▲) and in 0·5 g l⁻¹ calcium nitrate solution (●). Mean root extensions of seedlings are shown with 95 % confidence limits attached. (From Shaw, Rorison & Baker, 1982.)

Table 1. *Concentrations of lead* (μg ml^{-1}) *in different background solutions required to bring about a 50 % inhibition of root extension in lead-rake (*Tideslow*) and non-rake populations of* Festuca ovina *and* Lotus corniculatus *(data of Shaw, 1984)*

	Festuca ovina		*Lotus corniculatus*	
Treatment	Rake population	Non-rake population	Rake population	Non-rake population
Complete nutrients	> 10	*c.* 7·5	> 10	*c.* 10
$-$P	0·4	0·25	0·25	0·25
Ca(NO$_3$)$_2$	1·5	0·75	0·38	0·30

ED$_{50}$ (Craig, 1977) was generally higher in the lead-rake population than in the non-rake population. All populations were most sensitive to lead when supplied in nutrient solution $-$P and least sensitive in the complete nutrient solution. The relative levels of tolerance are also important; for example, the higher tolerance (and, in fact, stimulation of root elongation) of *L. corniculatus* in the complete nutrient solution compared with *F. ovina*, and the greater tolerance of *F. ovina* in calcium nitrate. The latter is possibly a reflection of the differing response of the two species to the control solution. Extension of roots of *L. corniculatus* was inhibited by the lack of P from the complete nutrient solution but this was more than compensated for by the extra Ca and N in the calcium nitrate treatment. There was less separation between controls in *F. ovina*, with greatest root extension in complete nutrient and least in the calcium nitrate. Such experiments point to the need to specify the conditions of measurement very precisely in defining tolerance by the root elongation method [cf. also, the differing responses of species to changes in supply of nutrients (Rorison, 1987)].

Other variants of the root elongation test include total root length (rather than measurements on the longest root generated) and root biomass (see e.g. Baker, 1984) and total plant biomass or growth rate (Verkleij & Bast-Cramer, 1985; Ingroville & Smirnoff, 1986). The expected extra precision of these integrated measures of response is not always realized. Baker (1978) drew attention to cautious use of the root elongation method in tolerance studies. In the populations of *Silene maritima* he studied, the long-term characteristics of metal uptake (accumulation in root and root/shoot partition ratios for zinc) could be related to soil zinc status more simply than could indices of Zn tolerance derived from root elongation measurements on large seedling samples. Techniques of short-term measurement of metal tolerance unrelated to root elongation phenomena have also been devised. The so-called 'protoplasmic resistance' method of Repp (1963) has been used both in laboratory (Gries, 1966) and field studies (Ernst, 1972). In this, the ability of cells (generally in epidermal strips) to recover from plasmolysis after initial metal treatment is used to define critical metal concentrations. More recently, pollen tube growth in liquid media to which metals have been added has been used to measure metal tolerance (Searcy & Mulcahy, 1985). These authors found good agreement between indices of tolerance to Zn and Cu expressed in pollen sources of *Silene dioica*, *S. alba* and *M. guttatus* and those obtained on the clones of parent plants using the root elongation method. Homer, Cotton & Evans (1980) who describe a rapid tolerance assay based on chlorophyll fluorescence (whole leaf at room temperature) again found agreement with conventional measurements of tolerance by root elongation in clones of *Phalaris arundinacea*. The potential of this attractive and fast method has not yet been fully explored.

EVOLUTION OF METAL TOLERANCE

Processes involved in adaptation

Some taxa are found on metal-contaminated soils, others are not. This basic observation poses intriguing questions about the evolutionary processes involved in adaptation to metalliferous environments. Bradshaw (1984) concludes that it is the restriction in supply of genetic variability which directly limits the evolutionary processes and the extent of adaptation. Most species, he believes, are in a state of *genostasis*, where a lack of appropriate genetic variability precludes further evolutionary change. This has been demonstrated elegantly by Gartside & McNeilly (1974) and Ingram (cited in Bradshaw, 1984) by screening for copper-tolerant individuals of a range of grass species from large seedling samples derived from non-tolerant populations. There has been no case reported of a species that evolves tolerance but which does not possess variability for tolerance in its normal populations. However, for some species which possess this variability, tolerant populations have not yet been demonstrated. The grasses *D. glomerata* and *Lolium perenne* are rarely found on metal-contaminated soils but some sites, such as Tideslow Rake (Fig. 1), do support both species. As has been suggested earlier, these species are restricted in their micro-distribution at such chemically hetero-geneous sites to weakly contaminated, more fertile soils. Their absence from the majority of mine sites probably then reflects their strict nutritional requirements and their rare occurrences at contaminated sites are instances of biological inertia, although it remains to be ascertained if these species do in fact possess tolerance to Zn, Pb and Cd in their normal populations, as has been shown to be the case for Cu (Bradshaw, 1984).

Early reports of heavy metal tolerance in vascular plants suggested that the phenomenon was a feature of genera and species in the families, Gramineae (=Poaceae), Caryophyllaceae and Labiatae (= Lamiaceae). However, it is now clear that tolerance has arisen independently in the full spectrum of families; no obvious phylogenetic relationships are apparent. It is perhaps surprising to find that one of the families best represented is the Leguminosae (= Fabaceae), as selection for metal tolerance here presumably involves independent events in the plant and its rhizobial symbiont (see Shaw, Rorison & Baker, 1985; Wu & Kruckeberg, 1985). The prevalence of reports of metal tolerance in members of the Poaceae probably merely reflects the disproportionately large size of this family.

Endemism in metallophyte floras

That distinctive metal-tolerant floras, such as those of calamine soils in Western Europe (Ernst, 1974a), ultramafic (Brooks, 1987) and cupriferous soils in south-central Africa (Wild & Bradshaw, 1977; Brooks & Malaisse, 1985) have evolved poses further questions about the processes of adaptation and the time-scale over which they have occurred. There is no doubt that the evolution of metal tolerance can be a rapid process (see below) and this is clearly important in newly contaminated environments but, in the examples cited above, metal-contaminated soils have frequently been available for plant colonization for thousands of years and they now support distinctive metallophyte floras rich in endemic taxa. For example, the extensive mineralization of copper in Upper Shaba, Zaïre, comprises about 100 copper–cobalt ore deposits totalling some 20 km² disseminated in a metallogenic province nearly 1000 times greater in area. A number of distinct metalliferous habitats exist such as 'copper clearings' and isolated and ancient (natural) copper hillocks which bear specific plant assemblages, notably of small annual herbs and caespitose grasses. About 220 taxa are present on these outcrops of which 42 are endemic, some on a single hillock (Malaisse, 1983). The flora thus comprises two separate components, endemic taxa and species found outside the metalliferous outcrops, notably in the high plateaux steppe–savannahs.

There is still considerable debate about the origin of this endemism (see e.g. Antonovics *et al.*, 1971). Some believe that the so-called mine taxa are 'palaeo-endemics', relict species of formerly wide distribution now confined to particular areas (as apparently is the case for *M. verna* and *T. alpestre* in Europe), while others suggest they are 'neo-endemics', species that have originated recently in given areas in response to peculiar environmental conditions. The weight of evidence in support of either theory is not conclusive. A major problem discussed by Antonovics *et. al.* (1971) is that such taxa cannot frequently be distinguished morphologically from their relatives from which they are not isolated reproductively. They suggest that mine taxa could be the result of parallel evolution on different outcrops from relatives in the surrounding vegetation and that evolution of both metal tolerance and morphological differences has proceeded via 'mine ecotypes'. On this basis, mine taxa would best be regarded as 'neo-endemics'. Recent experimental studies on the *Silene burchelli* Otth. complex in Zaïre (Baker *et al.*, 1983; Malaisse, Colonval-Elenkov & Brooks, 1983) gives support to this theory. The complex is represented by *S. burchelli* var. *angustifolia* Sond., which is widespread on the high plateaux, an ecotype of this species on a cupriferous outcrop at Luita, and *S. cobalticola* Duvign. et Plancke on highly copper–cobalt mineralized deposits at Mindigi. It is an ecophyletic series providing a gradient of physiological characteristics in terms of copper and cobalt tolerance (Fig. 4), and morphological and anatomical features. In interpreting the present-day, often disjunct, distribution of mine taxa, the involvement of human agencies of dispersal cannot be ignored ('transport endemism'; Antomovics *et al.*, 1971). Thus, isolated and recent mine spoils may support populations of metal-tolerant taxa similar to more ancient areas, a fact which further gives some circumstantial support to the neo-endemic hypothesis.

In the present context, 'serpentine' vegetation developed over ultramafic soils is particularly interesting both in terms of phylogenetic relationships and endemism. Ultramafic soils are typically infertile (low N and P status), usually have unfavourable Mg/Ca ratios and may be enriched with Ni, Cr and sometimes Co. On these soils, both in temperate regions and the tropics, a number of endemic taxa and adapted races of more widespread species are found. The degree of endemism may be very high, particularly in some tropical island floras such as that of New Caledonia. On this island, 400 km long × 40 km wide, ultramafic soils cover about one-third of the island surface, occurring principally over the southern massif (55000 km²) but also on a series of smaller isolated outcrops. In marked contrast to the flora of the copper–cobalt mineralization of south-central Africa, the serpentine flora of approximately 1500 species comprises largely woody perennials, epiphytic orchids and Cyperaceae growing in xerophytic scrub at low altitude (maquis) or in rain forest at mid or higher altitudes. Endemism in the

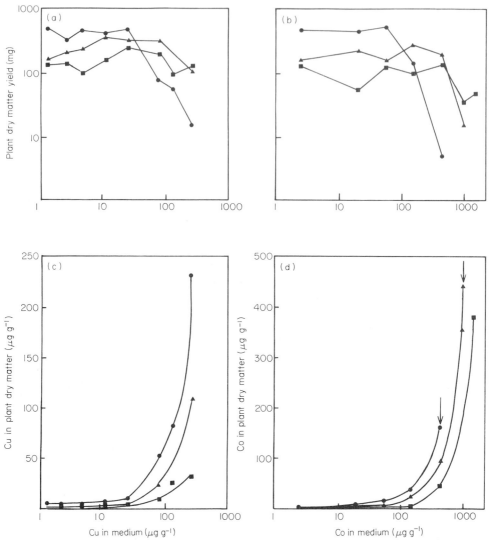

Fig. 4. The effects of external concentrations of copper and cobalt in the growth medium on the yield and metal uptake of *Silene burchelli* var. *angustifolia* (●), a copper/cobalt tolerant ecotype of *S. burchelli* (▲) and *S. cobalticola* (■), grown for 8 weeks from seedlings under glasshouse conditions. The vertical arrows indicate the limit of tolerance or the metal concentration in the substrate above which no seedlings survived. (Modified from Baker *et al.*, 1983.)

island flora is 76 % and over 79 % of the maquis species are restricted to ultramafic soils; 99 % of these are endemics. Similar figures for the rain forest formations are 31 and 98 %, respectively (Morat, Veillon & MacKee, 1984; Morat *et al.*, 1986). Particularly in the latter situation, there is no clear indication of Ni toxicity prevailing in spite of high concentrations (up to 6000 μg g^{-1} Ni) in the soil. However, analysis of the foliage of many of these serpentine taxa reveals extremely high concentrations of Ni in the dry matter, often exceeding 1000 μg g^{-1}, or 0·1 %. This 'hyperaccumulation' (Brooks *et al.*, 1977) of Ni seems characteristic of serpentine endemics not only in New Caledonia but elsewhere in the world. The most extreme example from the New Caledonian flora is the tree *Sebertia acuminata* Pierre ex Baill. which is large (up to 10 m) and vigorous on ultramafic soils and can accumulate up to 25·74% Ni in its sap on a dry weight basis (Jaffré *et al.*, 1976). The large genus *Psychotria* (Rubiaceae) is also well represented in New Caledonia, where about 60 species are recorded. Only one, however, is completely restricted to ultramafic soils, *P. douarrei* (G. Beauvisage) Däniker, and this also emerges as a hyperaccumulator of Ni (Baker, Brooks & Kersten, 1985). In one area of rain forest developed on ultramafic soil, *P. douarrei* grows within metres of another species, *P. baillonii* Schlechter, which is more typical of non-nickeliferous soils developed over shales and schists. If the leaves of this putative Ni-tolerant ecotype of *P. baillonii* are analyzed, only low ($< 100\ \mu g\ g^{-1}$) Ni accumulations are found. Thus, two sympatric and closely related species have evolved completely different mechanisms of Ni tolerance (accumulation and exclusion *sensu* Baker, 1981, see below). As with the example cited from Zaïre (*S. burchelli* complex), an ecophyletic series may be suggested, proceeding from *P. baillonii*, to an Ni-tolerant ecotype of this species, through to the serpentine endemic, *P. douarrei*. Adaptation could thus have proceeded by a neo-endemic process, although palaeo-endemism seems more plausible in other endemics in such ancient and undisturbed ecosystems. Wild & Bradshaw (1977) arrived at similar conclusions for endemism in the flora of copper, ultramafic and other anomalous soils in Zimbabwe. Here, most endemic species appear to be palaeo-endemics as they are best regarded as relict outliers of genera that occur elsewhere. However, many other species occur on and off anomalous areas without obvious differentiation but differentiation must clearly have occurred or is still occurring. Palaeo-endemics and neo-endemics thus coexist on these metalliferous soils. Wild & Bradshaw (1977) argue that the reason why most tolerant populations are probably neo-endemic in origin might be because of climatic changes eradicating earlier colonists, so allowing the immigration of others. Circumstantial evidence for this theory comes from the fact that the endemic species today are only found on anomalous soils of large area where the developing endemic could always find suitable habitats during periods of climatic change by, for example, migrating to higher or lower altitudes.

Hyperaccumulation of nickel is a relatively rare response of serpentine plants to ultramafic soils. So far it has only been demonstrated in a restricted range of species between the latitudes of 48° N and 32° S. However, it is clearly a tolerance phenomenon which has evolved independently in species from a range of families on a global scale. Within the Magnoliopsida (Cronquist, 1981), some 104 of the 144 hyperaccumulating species discovered to date are in the subclass Dilleniidae (Brooks, 1987). It is significant that 28 of these species are found in the closely related families Flacourtiaceae and Violaceae of the single order Violales. The other main group of species is within the family Brassicaceae (= Cruciferae) (at least 76 taxa). Here, to a greater extent than in the first group, the frequency of hyperaccumulators is exaggerated by the restriction of this property to very few genera, notably *Alyssum* (50 species) and *Thlaspi* (about 20 species). Thus, the present evidence suggests that hyperaccumulation of nickel has probably evolved in parallel in two distinct major phylogenetic and biogeographical groups, (1) a few genera of the Brassicaceae in the temperate zone of the northern hemisphere and (2) a more diverse group of tropical genera and species, largely from the families Violaceae and Flacourtiaceae. The chemical evidence suggests that at least two different mechanisms of nickel accumulation and detoxification are involved in these two groups. Furthermore, within the tropical areas hyperaccumulation appears to be a feature of families generally regarded as primitive and therefore may be a primitive character.

The rapid evolution of tolerance

Evolution of tolerance can occur within one generation in plant populations as a result of the powerful selective forces of metal toxicity, provided appropriate genetic variability is available on which selection can act (Wu & Bradshaw, 1972; Bradshaw, 1984). Recently contaminated sites such as road verges and areas affected by fall-out from metal smelters have provided some convincing examples of rapid evolution of tolerance (for a review see Baker et al., 1986). The number of tolerant individuals so selected may be small and it may take many generations to build up tolerance in the population. The selective forces and the environments in which they operate may be very different on mineralized soils and those recently contaminated by aerial deposition of metal particulates (Table 2). In the latter case,

Table 2. *Some features of mineralized and aerially contaminated soils in relation to selective forces for metal tolerance*

Mine wastes and mineralized soils	Aerially polluted soils
May have been present 200 to 2000 years or more	Pollution of recent (< 50 years) origin
Metal concentrations often very high	Metal concentrations low (background) initially
Metal concentrations unchanging with time	Continuous deposition leads to cumulative rise in metal concentrations
Metal concentrations may be highly heterogeneous spatially	Little spatial variation in metal concentrations
Uniform distribution of metals through soil profile	Accumulation of metals in surface (organic) horizon
Deficient in nutrients and organic matter; often poor physical structure	'Normal' soil
Selective forces for metal tolerance strong and stable	Selective forces weak but increasing with time

the system is out of equilibrium and the selective forces are likely to increase progressively with continued deposition. It is therefore not surprising to find that tolerance in plant populations from such sites may be different from conventional tolerance of mine plants. Thus, J. A. C. Verklij (pers. comm., 1985), in a study of Zn tolerance in *A. capillaris* populations in the vicinity of a major Zn smelter in the Netherlands, referred to 'hypertolerance' in this material, as it was shown to exhibit the greatest level of Zn tolerance in any population screened to date. Such a phenomenon can probably be explained by the very high concentrations of Zn accumulating on the surface of the soil as a result of continuous deposition of particulates. This will serve to 'filter out' only the most tolerant genotypes at seed germination. Subsequently, on establishment, these plants will root well beneath the most contaminated soil horizon and soil analysis within this rhizosphere zone will be a poor indicator of the conditions in which they were originally selected.

Baker et al. (1986) have drawn attention to the possible importance of environmentally induced metal tolerances in the adaptation of plant populations to toxic metalliferous soils. A strictly genetic basis for tolerance has always been presumed, if not demonstrated unequivocally (Macnair, 1981) but, in the study by Baker *et al.* (1986), the Cd tolerance of a range of populations of the grasses *Holcus lanatus*, *A. capillaris*, *Deschampsia cespitosa* and *Festuca rubra* declined overall by 13 % when clones were transplanted from their original soils into uncontaminated soil. In detailed tests with *H. lanatus*, there was an indication that clones from mine sites were more stable in their Cd tolerance than those from an aerially polluted site when tested in this way. Tolerance could also be partially induced in non-tolerant clones by growth in Cd-enriched soil. Such tolerance may be important in the initial survival of individuals in newly contaminated sites. However, the extent to which inducible tolerance is a peculiarity of responses to Cd has yet to be resolved.

Metal tolerance without the evolution of tolerant races
    The possibility that some plants may possess a *constitutional* tolerance to heavy metals was originally dismissed by Antonovics *et al.* (1971). However, there are now convincing reports which suggest that species can differ widely in their thresholds of tolerance (Wu & Antonovics, 1976). Comparative studies by McNaughton *et al.* (1974) and Taylor & Crowder (1984) on populations of *Typha latifolia* from sites affected by fall-out from Zn and Cu/Ni smelters, respectively, and control populations from uncontaminated soils provided no evidence for population differentiation even though the soils were heavily contaminated with metals. A constitutional tolerance is therefore suggested. A similar conclusion was reached by Reeves & Baker (1984) when comparing the characteristics of metal uptake by serpentine and non-serpentine populations of *Thlaspi goesingense* Hálácsy when grown on a range of Zn-, Ni- and Co-enriched soils. The limestone (non-serpentine) plants showed similar uptake and accumulation of these metals to the serpentine plants, suggesting the existence of a non-specific metal detoxi-fication system in the species. More recently, Fiedler (1985) concluded from a study of heavy metal accumulation in the genus *Calochortus* (Liliaceae) that the ability to tolerate excessively high levels of Ni, Co and Cu may be a constitutional property of the whole genus.

THE SPECIFICITY OF METAL TOLERANCE

In many of the major reviews on metal tolerance published to date (e.g. Turner, 1969; Antonovics *et al.*, 1971; Antonovics, 1975; Wainwright & Woolhouse, 1975), the specificity and independence of individual tolerances has been stressed. Thus, plants growing in soils multiply contaminated with heavy metals can be demonstrated to possess tolerances to all metals present at potentially toxic concentrations. This phenomenon of *multiple tolerance* is well documented (Gregory & Bradshaw, 1965; Coughtrey & Martin, 1978; Cox & Hutchinson, 1980; Wong, 1982). Reports of '*co-tolerance*', whereby tolerance to one metal confers some degree of tolerance to another not present or enriched in the soil, are fewer (Allen & Sheppard, 1971; Cox & Hutchinson, 1979, 1981; Symeonidis, McNeilly & Bradshaw, 1985; Verkleij & Bast-Cramer, 1985). However, when earlier work (such as that of Gregory & Bradshaw, 1965) is reviewed in the light of these reports, it is apparent that the possibility of low-level tolerance to metals other than the primary contaminant has largely been overlooked. The recent work by Symeonidis *et al.* (1985) with cultivars of *A. capillaris* has also confirmed such non-specific low-level tolerances. Verkleij & Bast-Cramer (1985) revealed the complexity of co-tolerance in metal-tolerant populations of *Silene cucubalus* Wib. [= *S. vulgaris* (Moench) Garcke]. Populations from soils enriched only with Cu showed a marked co-tolerance for Zn but the opposite did not apply; none of the Zn-tolerant populations were co-tolerant for Cu. It is clear that much work has still to be done using both genetical and physiological studies to resolve this conundrum.

THE MULTIPLICITY OF MECHANISMS AND SYNDROME OF TOLERANCE

There is now assembled a substantial body of literature relating to physiological and biochemical aspects of heavy metal tolerance in plants (for major reviews, see the Introduction). Whilst the emphasis of such studies has been largely on comparative experiments on tolerant and non-tolerant clones of grasses such as *A. capillaris*, *D. cespitosa* and *Anthoxanthum odoratum*, detailed work with dicotyledonous species such as *S. vulgaris*, and interspecific screening experiments in metallophyte floras has also provided an insight into tolerance mechanisms. What has emerged is a complex picture where few generalizations are possible. It is apparent, however, that mechanisms are largely *internal* in that metals are rarely not absorbed by plants from metalliferous substrates and so resistance is achieved by true tolerance (*sensu* Levitt, 1980). It also emerges that this tolerance is manifested by a suite of physiological and biochemical adaptations developed to varying degrees for different metals in different species and populations. Baker (1981) suggested two basic strategies of tolerance; metal exclusion, whereby metal uptake and transport is restricted, and metal accumulation, where there is no such restriction and metals are accumulated in a detoxified form. Detoxification may result from cell-wall binding, active pumping of ions into vacuoles, complexing by organic acids and possibly by specific metal-binding proteins. Other more subtle properties such as enzymic adaptations and effects on membrane permeability can also be detected.

Metal exclusion or at least restricted uptake may result from the presence of mycorrhizal fungi in or on the root system of plants growing in metalliferous soils. The importance of mycorrhizal infection in metal tolerance has largely been overlooked but recent studies in ericaceous plants (Bradley, Burt & Read, 1982) and seedlings of the woody species, *Betula* (Brown & Wilkins, 1985), have demonstrated the crucial role of the fungal symbiont in tolerance in both ericoid and ectomycorrhizal systems.

There is no doubt that we still have a poor understanding of the mechanisms of metal tolerance involved at the subcellular level. The application of more sophisticated analytical technology, such as X-ray microprobe analysis, has opened up a new frontier (Mullins, Hardwick & Thurman, 1985). Similarly, the use of cell suspensions and isolated protoplasts generated from callus cultures of tolerant and non-tolerant plants allows a novel approach in mechanistic studies (Poulter *et al.*, 1985). There is increasing interest in the possible importance of metallothioneins and other low molecular weight, specific metal-binding proteins in metal tolerance in plants as suggested from animal studies (see e.g. Rauser & Curvetto, 1980; Rauser, 1984; Lolkema *et al.*, 1984; Schultz & Hutchinson, 1985; Robinson & Thurman, 1985, 1986; Robinson & Jackson, 1986). Further, Grill, Winnacker & Zenk (1985) have reported the induction of a set of novel heavy metal-complexing peptides (phytochelatins) in cell suspension cultures in a range of unrelated species upon heavy metal treatment. Much work is clearly required to establish the role of such inducible compounds.

A further property of metal-tolerant plants which has received little attention but which clearly relates to mechanism is the stimulatory effects of low-level metal treatment. Such effects are manifested by an apparent 'need' for a metal both in short-term (root elongation) and long-term (growth) studies. Grime & Hodgson (1969) attempted to explain stimulatory responses to Fe and Al in the rooting behaviour of calcifuge plants as a corollary of a rather non-specific metal-complexing system involved in tolerance to these polyvalent cations. Physiological requirements for Fe were therefore elevated above normal nutritional levels. The same may apply to metals such as Zn and Cu but it is difficult to invoke the same argument for non-essential metals such as Pb and Cd.

CONCLUSION

Antonovics (1975) entitled his stimulating review of metal tolerance in plants 'Perfecting an evolutionary paradigm'. In this he was reflecting upon recent work since the publication of his earlier review (Antonovics *et al.*, 1971). Eleven years later it is questionable to what extent this paradigm has been perfected further. We now have a fuller understanding of the evolutionary processes involved but the studies of mechanism referred to above do not give a clear picture of the nature of adaptation to the powerful selective forces of metal toxicity. The concepts of constitutional and inducible tolerance, although based on limited experimental evidence at present, clearly challenge earlier ideas. Similarly, the incidences of metal co-tolerance and low-level background tolerance throw some doubts on the specificity of heavy metal tolerance. There is no doubt, however, that metal tolerance in plants will continue to intrigue all those plant scientists attempting to understand the nature and scale of plant adaptation to the environment and will remain an evolutionary paradigm *par excellence*.

ACKNOWLEDGEMENTS

I wish to record my thanks to Professor A. D. Bradshaw and Drs R. R. Brooks, R. D. Reeves and J. Proctor for their helpful comments and suggestions during the preparation of this paper.

REFERENCES

ALLEN, W. R. & SHEPPARD, P. M. (1971). Copper tolerance in some Californian populations of the monkey flower, *Mimulus guttatus*. *Proceedings of the Royal Society of London, Series B*, **177**, 177–196.

ANTONOVICS, J. (1975). Metal tolerance in plants: perfecting an evolutionary paradigm. *Proceedings of the International Conference on Heavy Metals in the Environment, Toronto, Canada*, **2**, 169–186.

ANTONOVICS, J., BRADSHAW, A. D. & TURNER, R. G. (1971). Heavy metal tolerance in plants. *Advances in Ecological Research*, **7**, 1–85.

BAKER, A. J. M. (1978). Ecophysiological aspects of zinc tolerance in *Silene maritima* With. *New Phytologist*, **80**, 635–642.

BAKER, A. J. M. (1981). Accumulators and excluders — strategies in the response of plants to heavy metals. *Journal of Plant Nutrition*, **3**, 643–654.

BAKER, A. J. M. (1984). Environmentally-induced cadmium tolerance in the grass *Holcus lanatus* L. *Chemosphere*, **13**, 585–598.

BAKER, A. J. M., BROOKS, R. R., PEASE, A. J. & MALAISSE, F. (1983). Studies on copper and cobalt tolerance in three closely related taxa within the genus *Silene* L. (Caryophyllaceae) from Zaïre. *Plant and Soil*, **73**, 377–385.

BAKER, A. J. M., BROOKS, R. R. & KERSTEN, W. J. (1985). Accumulation of nickel by *Psychotria* species from the Pacific Basin. *Taxon*, **34**, 89–95.

BAKER, A. J. M., GRANT, C. J., MARTIN, M. H., SHAW, S. C. & WHITEBROOK, J. (1986). Induction and loss of cadmium tolerance in *Holcus lanatus* L. and other grasses. *New Phytologist*, **102**, 575–587.

BRADLEY, R., BURT, A. J. & READ, D. J. (1982). The biology of mycorrhiza in the Ericaceae. VIII. The role of mycorrhizal infection in heavy metal resistance. *New Phytologist*, **91**, 197–209.

BRADSHAW, A. D. (1952). Populations of *Agrostis tenuis* resistant to lead and zinc poisoning. *Nature* **169**, 1098.

BRADSHAW, A. D. (1976). Pollution and evolution. In: *Effects of Air Pollutants on Plants* (Ed. by T. A. Mansfield), pp. 135–159. Cambridge University Press, Cambridge.

BRADSHAW, A. D. (1984). Adaptation of plants to soils containing toxic metals – a test for conceit. In: *Origins and Development of Adaptation, CIBA Foundation Symposium 102* (Ed. by D. Evered & G. M. Collins), pp. 4–19. Pitman, London.

BROOKS, R. R. (1987). *Serpentine and its Vegetation*. Dioscorides Press, Portland, Oregon.

BROOKS, R. R. & MALAISSE, F. (1985). *The Heavy Metal-tolerant Flora of South-Central Africa*. Balkema, Rotterdam.

BROOKS, R. R., LEE, J., REEVES, R. D. & JAFFRÉ, T. (1977). Detection of nickeliferous rocks by analysis of herbarium specimens of indicator plants. *Journal of Geochemical Exploration*, **7**, 49–57.

BROWN, M. T. & WILKINS, D. A. (1985). Zinc tolerance of mycorrhizal *Betula*. *New Phytologist*, **99**, 101–106.

CLAPHAM, A. R., TUTIN, T. G. & WARBURG, E. F. (1981). *Excursion Flora of the British Isles*, 3rd Edn. Cambridge University Press, Cambridge.

CLARK, R. K. & CLARK, S. C. (1981). Floristic diversity in relation to soil characteristics in a lead mining complex in the Pennines, England. *New Phytologist*, **87**, 799–815

COOK, S., LEFÈBVRE, C. & McNEILLY, T. (1972). Competition between metal tolerant and normal plant populations on normal soil. *Evolution*, **26**, 366–372.

COUGHTREY, P. J. & MARTIN, M. H. (1978). Tolerance of *Holcus lanatus* to lead, zinc and cadmium in factorial combination. *New Phytologist*, **81**, 147–154.

COX, R. M. & HUTCHINSON, T. C. (1979). Metal co-tolerances in the grass *Deschampsia cespitosa*. *Nature*, **279**, 231–233.

COX, R. M. & HUTCHINSON, T. C. (1980). Multiple metal tolerances in the grass *Deschampsia cespitosa* (L.) Beauv. from the Sudbury smelting area. *New Phytologist*, **84**, 631–647.

COX, R. M. & HUTCHINSON, T. C. (1981). Multiple and co-tolerance to metals in the grass *Deschampsia cespitosa*: adaptation, preadaptation and 'cost'. *Journal of Plant Nutrition*, **3**, 731–741.

CRAIG, G. C. (1977). A method of measuring heavy metal tolerance in grasses. *Transactions of the Rhodesian Scientific Association*, **58**, 9–16.

CRONQUIST, A. (1981). *An Integrated System of Classification of Flowering Plants*. Columbia University Press, New York.

DAVIS, R. D. & BECKETT, P. H. T. (1978). Critical levels of twenty potentially toxic elements in young spring barley. *Plant and Soil*, **49**, 395–408.

ERNST, W. (1972). Ecophysiological studies on heavy metal plants in South Central Africa. *Kirkia*, **8**, 125–145.

ERNST, W. (1974a). *Schwermetallvegetation der Erde*. Fischer-Verlag, Stuttgart.

ERNST, W. (1974b). Mechanismen der Schwermetallresistenz. *Verhandlungen der Gesellschaft für Ökologie, Erlangen 1974*, pp. 189–197.

ERNST, W. H. O. (1975). Physiology of heavy metal resistance in plants. *Proceedings of the International Conference on Heavy Metals in the Environment, Toronto, Canada*, **2**, 121–136.

ERNST, W. (1976). Physiological and biochemical aspects of metal tolerance. In: *Effects of Air Pollutants on Plants* (Ed. by T. A. Mansfield), pp. 115–133. Cambridge University Press, Cambridge.

FARAGO, M. E. (1981). Metal tolerant plants. *Co-ordination Chemistry Reviews*, **36**, 155–182.

FIEDLER, P. L. (1985). Heavy metal accumulation and the nature of edaphic endemism in the genus *Calochortus* (Liliaceae). *American Journal of Botany*, **72**, 1712–1718.

GARTSIDE, D. W. & McNEILLY, T. (1974). The potential for evolution of heavy metal tolerance in plants. II. Copper tolerance in normal populations of different plant species. *Heredity*, **32**, 335–348.

GREGORY, R. P. G. & BRADSHAW, A. D. (1965). Heavy metal tolerance in populations of *Agrostis tenuis* Sibth. and other grasses. *New Phytologist*, **64**, 131–143.

GRIES, B. (1966). Zellphysiologische Untersuchungen über die Zinkresistenz bei Galmeiökotypen und Normalformen von *Silene cucubalus* Wib. *Flora*, **B 156**, 271–290.

GRILL, E., WINNACKER, E.-L. & ZENK, M. H. (1985). Phytochelatins: the principal heavy-metal complexing peptides of higher plants. *Science*, **230**, 674–676.

GRIME, J. P. (1965). Comparative experiments as a key to the ecology of flowering plants. *Ecology*, **46**, 513–515.

GRIME, J. P. (1973). Control of species density in herbaceous vegetation. *Journal of Environmental Management*, **1**, 151–167.

GRIME, J. P. & HODGSON, J. G. (1969). An investigation of the ecological significance of lime-chlorosis by means of large-scale comparative experiments. In: *Ecological Aspects of the Mineral Nutrition of Plants* (Ed. by I. H. Rorison), pp. 67–99. Blackwell Scientific Publications, Oxford.

HEWITT, E. J. (1966). *Sand and Water Culture Methods Used in the Study of Plant Nutrition*, 2nd Edn. Technical Communication No. 22, Commonwealth Agricultural Bureaux, Farnham Royal.

HOMER, J. R., COTTON, R. & EVANS, E. H. (1980). Whole leaf fluorescence as a technique for measurement of tolerance of plants to heavy metals. *Oecologia*, **45**, 88–89.

HUMPHREYS, M. O. & NICHOLLS, M. K. (1984). Relationships between tolerance to heavy metals in *Agrostis capillaris* L. (*A. tenuis* Sibth.). *New Phytologist*, **95**, 177–190.

INGROUILLE, M. J. & SMIRNOFF, N. (1986). *Thlaspi caerulescens* J. & C. Presl. (*T. alpestre* L.) in Britain. *New Phytologist*, **102**, 219–233.

JAFFRÉ, T., BROOKS, R. R., LEE, J. & REEVES, R. D. (1976). *Sebertia acuminata*: a hyperaccumulator of nickel from New Caledonia. *Science*, **193**, 579–580.

JOHNSTON, W. R. & PROCTOR, J. (1981). Growth of serpentine and non-serpentine races of *Festuca rubra* in solutions simulating the chemical conditions in a toxic serpentine soil. *Journal of Ecology*, **69**, 855–869.

KARATAGLIS, S. S. (1980). Behaviour of *Agrostis tenuis* populations against copper ions in combination with the absence of some macro-nutrient elements. *Bericht der Deutschen Botanischen Gesellschaft*, **93**, 417–424.

LAMBINON, J. & AUQUIER, P. (1963). La flore et la végétation des terrains calaminaires de la Wallonie septentrionale et de la Rhénanie aixoise. Types chorologiques at groupes écologiques. *Nature mosana*, **16**, 113–131.

LEFÈBVRE, C. (1968). Note sur un indice de tolérance au zinc chez des populations d'*Armeria maritima* (Mill.) Willd. *Bulletin de la Société Royale de Botanique de Belgique*, **102**, 5–11.

LEVITT, J. (1980). *Responses of Plants to Environmental Stresses*, 2nd Edn, vol. 2. Academic Press, New York.

LOLKEMA, P. C., DONKER, M. H., SCHOUTEN, A. J. & ERNST, W. H. O. (1984). The possible role of metallothioneins in copper tolerance in *Silene cucubalus*. *Planta*, **162**, 174–179.

MACNAIR, M. R. (1977). Major genes for copper tolerance in *Mimulus guttatus*. *Nature*, **268**, 428–430.

MACNAIR, M. R. (1981). Tolerance of higher plants to toxic materials. In: *Genetic Consequences of Man Made Change* (Ed. by J. A. Bishop & L. M. Cooke), pp. 177–207. Academic Press, London & New York.

MALAISSE, F. (1983). Phytogeography of the copper and cobalt flora of Upper Shaba, Zaïre, with emphasis on its endemism, origin and evolution mechanisms. *Bothalia*, **14**, 173–180.

MALAISSE, F., COLONVAL-ELENKOV, E. & BROOKS, R. R. (1983). The impact of copper and cobalt ore bodies upon the evolution of some plant species from Upper Shaba, Zaïre. *Plant Systematics and Evolution*, **142**, 207–221.

MCNAUGHTON, S. J., FOLSOM, T. C., LEE, T., PARK, F., PRICE, C., ROEDER, D., SCHMITZ, J. & STOCKWELL, C. (1974). Heavy metal tolerance in *Typha latifolia* without the evolution of tolerant races. *Ecology*, **55**, 1163–1165.

MCNEILLY, T. & BRADSHAW, A. D. (1968). Evolutionary processes in populations of copper tolerant *Agrostis tenuis* Sibth. *Evolution*, **22**, 108–118.

MORAT, PH., VEILLON, J.-M. & MACKEE, S. (1984). Floristic relationships of New Caledonian rain forest phanerogams. In: *Biogeography of the Tropical Pacific* (Ed. by F. J. Radovsky, P. H. Raven & S. H. Sohmer), pp. 71–128. *Association of Systematics Collections and Bernice P. Bishop Museum Special Publications* No. 72, Honolulu.

MORAT, PH., JAFFRÉ, T., VEILLON, J.-M. & MACKEE, H. S. (1986). Affinités floristiques et considérations sur l'origine des maquis miniers de la Nouvelle-Calédonie. *Adansonia*, **2**, 133–182.

MULLINS, M., HARDWICK, K. & THURMAN, D. A. (1985). Heavy metal location by analytical electron microscopy in conventionally fixed and freeze-substituted roots of metal tolerant and non tolerant ecotypes. *Proceedings of the International Conference on Heavy Metals in the Environment, Athens, Greece*, vol. 2, pp. 43–46. CPC Consultants, Edinburgh.

NICHOLLS, M. K. & MCNEILLY, T. (1979). Sensitivity of rooting and tolerance to copper in *Agrostis tenuis* Sibth. *New Phytologist*, **83**, 653–664.

PETERSON, P. J. (1983). Adaptation to toxic metals. In: *Metals and Micronutrients: Uptake and Utilization by Plants* (Ed. by D. A. Robb & W. S. Pierpoint), pp. 51–69. Academic Press, London.

POLLARD, A. J. (1980). Diversity of metal tolerance in *Plantago lanceolata* L. from the southeastern United States. *New Phytologist*, **86**, 109–117.

POULTER, A., COLLIN, H. A., THURMAN, D. A. & HARDWICK, K. (1985). The role of the cell wall in the mechanism of lead and zinc tolerance in *Anthoxanthum odoratum* L. *Plant Science*, **42**, 61–66.

RAUSER, W. E. (1984). Copper-binding protein and copper tolerance in *Agrostis gigantea*. *Plant Science Letters*, **33**, 239–247.

RAUSER, W. E. & CURVETTO, N. R. (1980). Metallothionein occurs in roots of *Agrostis* tolerant to excess copper. *Nature*, **287**, 563–564.

REEVES, R. D. & BAKER, A. J. M. (1984). Studies on metal uptake by plants from serpentine and non-serpentine populations of *Thlaspi goesingense* Hálácsy (Cruciferae). *New Phytologist*, **98**, 191–204.

REPP, G. (1983). Die Kupferresistenz des Protoplasmas höherer Pflanzen auf Kupfererzböden. *Protoplasma*, **57**, 643–659.

ROBINSON, N. J. & JACKSON, P. J. (1986). 'Metallothionein-like' metal complexes in angiosperms; their structure and function. *Physiologia Plantarum*, **67**, 499–506.

ROBINSON, N. J. & THURMAN, D. A. (1985). Copper-binding protein in *Mimulus guttatus*. *Proceedings of the International Conference on Heavy Metals in the Environment*, Athens, Greece, vol. 2, pp. 47–50. CPC Consultants, Edinburgh.

ROBINSON, N. J. & THURMAN, D. A. (1986). Involvement of a metallothionein-like copper complex in the mechanism of copper tolerance in *Mimulus guttatus*. *Proceedings of the Royal Society of London*, Series B, **227**, 493–501.

RORISON, I. H. (1987). Mineral nutrition in time and space. In: *Frontiers of Comparative Plant Ecology* (Ed. by I. H. Rorison, J. P. Grime, R. Hunt, G. A. F. Hendry & D. H. Lewis), *New Phytologist*, **106**, (Suppl.), 79–92. Academic Press, New York & London.

SCHULTZ, C. L. & HUTCHINSON, T. C. (1985). Copper tolerance in the grass *Deschampsia cespitosa*: is metallothionein involved? *Proceedings of the International Conference on Heavy Metals in the Environment*, Athens, Greece, vol. 2, pp. 51–54. CPC Consultants, Edinburgh.

SEARCY, K. B. & MULCAHY, D. L. (1985). The parallel expression of metal tolerance in pollen and sporophytes of *Silene dioica* (L.) Clairv., *S. alba* (Mill.) Krause and *Mimulus guttatus* DC. *Theoretical and Applied Genetics*, **69**, 597–602.

SHAW, S. C. (1984). *Ecophysiological studies on heavy metal tolerance in plants colonizing Tideslow Rake, Derbyshire*. Ph.D. thesis, University of Sheffield, Sheffield, UK.

SHAW, S. C., RORISON, I. H. & BAKER, A. J. M. (1982). Physiological mechanisms of heavy metal tolerance in plants. *Annual Report, Unit of Comparative Plant Ecology* (NERC), *University of Sheffield*, pp. 13–15.

SHAW, S. C., RORISON, I. H. & BAKER, A. J. M. (1984). Physiological mechanisms of heavy metal tolerance in plants. *Annual Report, Unit of Comparative Plant Ecology* (NERC), *University of Sheffield*, p. 25.

SHAW, S. C., RORISON, I. H. & BAKER, A. J. M. (1985). Heavy metal tolerance among legumes. *Annual Report, Unit of Comparative Plant Ecology* (NERC), *University of Sheffield*, pp. 18–19.

SIMON, E. (1978). Heavy metals in soils, vegetation development and heavy metal tolerance in plant populations from metalliferous areas. *New Phytologist*, **81**, 175–188.

SYMEONIDIS, L., McNEILLY, T. & BRADSHAW, A. D. (1985). Differential tolerance of three cultivars of *Agrostis capillaris* L. to cadmium, copper, lead, nickel and zinc. *New Phytologist*, **101**, 309–315.

TAYLOR, G. J. & CROWDER, A. A. (1984). Copper and nickel tolerance in *Typha latifolia* clones from contaminated and uncontaminated environments. *Canadian Journal of Botany*, **62**, 1304–1308.

THOMPSON, J. & PROCTOR, J. (1983). Vegetation and soil factors on a heavy metal mine spoil heap. *New Phytologist*, **94**, 297–308.

THURMAN, D. A. (1981). Mechanism of metal tolerance in higher plants. In: *Effect of Heavy Metal Pollution on Plants* (Ed. by N. W. Lepp), vol. 2, pp. 239–249. Applied Science Publishers, Barking.

TÜRESSON, G. (1922). The genotypical response of the plant species to the habitat. *Hereditas*, **3**, 211–350.

TURNER, R. G. (1969). Heavy metal tolerance in plants. In: *Ecological Aspects of the Mineral Nutrition of Plants* (Ed. by I. H. Rorison), pp. 399–410. Blackwell Scientific Publications, Oxford.

VERKLEIJ, J. A. C. & BAST-CRAMER, W. B. (1985). Cotolerance and multiple heavy-metal tolerance in *Silene cucubalus* from different heavy-metal sites. *Proceedings of the International Conference on Heavy Metals in the Environment*, Athens, Greece, vol. 2, pp. 174–176. CPC Consultants, Edinburgh.

WAINWRIGHT, S. J. & WOOLHOUSE, H. W. (1975). Physiological mechanisms of heavy metal tolerance in plants. In: *The Ecology of Resource Degradation and Renewal* (Ed. by M. J. Chadwick & G. T. Goodman), pp. 231–257. Blackwell Scientific Publications, Oxford.

WHITEBROOK, J. (1986). *Heavy metal tolerance and tetraploidy in* Leucanthemum vulgare *L. (Compositae)*. Ph.D. thesis, University of Bristol.

WIGHAM, H., MARTIN, M. H. & COUGHTREY, P. J. (1980). Cadmium tolerance of *Holcus lanatus* L. collected from soils with a range of cadmium concentrations. *Chemosphere*, **9**, 123–125.

WILD, H. & BRADSHAW, A. D. (1977). The evolutionary effects of metalliferous and other anomalous soils in South Central Africa. *Evolution*, **31**, 282–293.

WILKINS, D. A. (1957). A technique for the measurement of lead tolerance in plants. *Nature*, **180**, 37–38.

WILKINS, D. A. (1960). The measurement and genetical analysis of lead tolerance in *Festuca ovina*. *Report of the Scottish Plant Breeding Station, 1960*, pp. 85–98.

WILKINS, D. A. (1978). The measurement of tolerance to edaphic factors by means of root growth. *New Phytologist*, **80**, 623–633.

WONG, M. H. (1982). Metal cotolerance to copper, lead, and zinc in *Festuca rubra*. *Environmental Research*, **29**, 42–47.

WOOLHOUSE, H. W. (1983). Toxicity and tolerance in the responses of plants to metals. In: *Encyclopedia of Plant Physiology*, New Series, vol. 12C (Ed. by O. L. Lange, P. S. Nobel, C. B. Osmond & H. Ziegler), pp. 245–300. Springer-Verlag, Berlin.

WU, L. & ANTONOVICS, J. (1975). Experimental ecological genetics in *Plantago*. I. Induction of roots and shoots on leaves for large-scale vegetative propagation and metal tolerance testing in *P. lanceolata*. *New Phytologist*, **75**, 277–282.

Wu, L. & Antonovics, J. (1976). Experimental ecological genetics in *Plantago*. II. Lead tolerance in *Plantago lanceolata* and *Cynodon dactylon* from a roadside. *Ecology*, **57**, 205–208.

Wu, L. & Bradshaw, A. D. (1972). Aerial pollution and the rapid evolution of copper tolerance. *Nature*, **238**, 167–169.

Wu, L. & Kruckeberg, A. L. (1985). Copper tolerance in two legume species from a copper mine habitat. *New Phytologist*, **99**, 565–570.

New Phytol. (1987) **106** (Suppl.), 113–130

CLIMATIC TOLERANCE AND THE DISTRIBUTION OF PLANTS

By J. GRACE

Department of Forestry and Natural Resources, University of Edinburgh, Edinburgh EH9 3JU, UK

SUMMARY

Boundaries to the distribution of life forms and species often coincide with isometric lines of climatological variables. Prominent examples of this phenomenon are the restriction of megaphanerophytes to areas where the mean temperature of the warmest month does not fall below about 10 °C, and the restriction of species within the British Isles to areas of specific summer or winter warmth. In the interpretation of these patterns, it is important to realise that the climate at plant surfaces may differ appreciably from that of the atmosphere as a whole, to an extent which depends on structural attributes of the vegetation. In particular, chamaephytes experience a much higher surface temperature than phanerophytes when comparisons are made in bright sunlight and this contributes to their success in mountains. Temperatures of flowers may similarly be elevated. In the British Isles, temperatures in the summer months may be decisive in determining the success of many species, as physiological processes in C_3-plants display near-linear relationships with temperatures in the range which prevails for most of the time. The life-cycle of native plants responds to environmental cues in such a way as to synchronize the development of the plant with the succession of seasons. When species are transplanted to different phytogeographical zones, as in forestry and horticulture, they frequently fail because of asynchrony.

Key words: Plant distribution, tree line, life forms, temperature climatological variables.

CLIMATIC VARIABLES

The climate of the British Isles is discussed extensively by Chandler & Gregory (1976) and additional maps are given by White & Smith (1982). Most climatological variables display large gradients across space and time. Many exert a strong influence on the growth of plants, either directly, e.g. temperature, or indirectly, e.g. the combination of variables which determine the supply of water for vegetation.

The maps in Figure 1 summarize the British climate. Solar radiation varies from a high mean value of 11 MJ m^{-2} d^{-1} in the south to about two-thirds of this value in central Scotland [Fig. 1(a)]. This variation is small in relation to the variation between months and quite trivial relative to the variation between open sites and the woodland floor.

Isotherms run from north to south in the winter and, in January, the east is two degrees colder than the west [Fig. 1(b)]. In July, the isotherms run broadly from east to west, some southern places being four degrees warmer than the north coast of Scotland [Fig. 1(c)]. Around the coast, especially the west coast, the seasonal amplitude is much less than that inland and the climate is said to be oceanic, in contrast to the more continental regime of the south-eastern and central parts of England.

The variation in temperature within a small area caused by topography can be large in relation to the geographical ranges. In a Derbyshire dale, the summer mean temperature was 3 °C higher on a south-facing slope than on a north-facing slope

0028-646X/87/05S113+18 $03.00/0

Fig. 1. Isometric lines showing the distribution of climatic variables over the British Isles. (a) Mean solar radiation (MJ m^{-2} d^{-1}); (b) mean air temperature in January (°C), adjusted to sea level; (c) mean air temperature in July (°C), adjusted to sea level; (d) mean wind speed (m s^{-1}); (e) mean water vapour pressure at 1550 h in January (10^2 Pa); (f) mean water vapour pressure at 1500 h in July (10^2 Pa); (g) mean annual 'actual' evapotranspiration (10^2 mm); (h) mean annual rainfall (10^2 mm). Sources: Chandler & Gregory (1976), Gloyne (1964).

(Rorison, Sutton & Hunt, 1986), this variation being equivalent to a latitudinal shift of hundreds of kilometres.

Wind speeds measured at coastal sites in Britain are very high and damage to property and crops, especially tree crops, is common [Fig. 1(d); Gloyne, 1964]. The influence of wind on vegetation is catastrophic in extreme gales, but at other times is less obvious, acting mainly on the coupling between surface and air temperatures. These influences are generally referred to in biological literature as *exposure*, although the phenomenon is ill-defined and the use of the term is to be deprecated (Grace, 1977).

Plant growth responds markedly to the availability of water, as is evident from agricultural yields observed in those years when water is in short supply (Carter, 1978). The availability of water to the plant is not easy to assess, as it is influenced by rainfall, topography, soil, atmospheric humidity and wind speed. Rainfall varies markedly from high mean values of 1000 to 2000 mm year^{-1} on the west side of the country to 600 to 800 mm year^{-1} on the east. The use of mean values may not be entirely appropriate, as drought may exert its effect on survival of species in only a few years in a century. Maps showing the probability of low rainfall are given by Chandler & Gregory (1976). On this basis, the driest areas are in Cambridge-shire, Northamptonshire and Bedfordshire (counties known for market gardening). In this region, the probability of receiving less than 500 mm of rain in any year is about 0·3, whilst on the western half of the country it is nil. The atmospheric humidity, expressed as water vapour pressure, is a variable of considerable interest as it is used in most calculations of water balance and the stomata of many species respond directly to its variations [Fig. 1(e), (f)]. The driving gradient for evaporation is the difference between surface and atmospheric water vapour pressure, and this normally increases strongly with temperature, simply because the saturation vapour pressure is a strong function of temperature. It also increases with radiation, as surface temperatures are a function of absorbed radiation. An approximate map of actual evapotranspiration prepared from hydrological sources shows that rates in the south-eastern corner of Britain are perhaps twice those of the north [Fig. 1(g)]. Another very useful map which takes into account not only losses in water by evapotranspiration but also gains owing to rain, is given by Green (1964).

Altitude causes considerable variation which is superimposed on the patterns just described. Air temperatures decline by about 7 ± 1 °C km^{-1}, and wind speeds increase with altitude (Taylor, 1976). Rates of evaporation and transpiration usually decline with altitude, especially in British conditions because of the increasing cloud cover and consequent decline in solar radiation. Attempts to quantify effects of altitude on crop performance have often been made. In the grass crop in southern Scotland, floral development was delayed by 0·04 d m^{-1} and yield fell by about 0·1% m^{-1} (Hunter & Grant, 1971). In the uplands of Wales, only part of the lower productivity could be attributed to shorter and poorer summers at high altitude. When a mild autumn was followed by a severe winter, the density of tillers was reduced substantially in the springtime (Munro & Davies, 1973).

CLIMATE AND THE PHYSIOGNOMY OF VEGETATION

On a geographical scale, the importance of climatic variables in moulding the vegetation is quite apparent. One of the earliest attempts to describe relationships between vegetation and climate was that of Raunkiaer (1909, 1934) who proposed a functional classification of plants on the basis of *life forms*. He pointed out that tropical forests contained a wide spectrum of life forms but, in extreme climates, the spectrum became reduced. Tundra and alpine regions lack trees and the dominant life forms are dwarf shrubs (chamaephytes). Above 1000 m in the Scottish Grampians, 27% of the species are chamaephytes compared to a mere 9% in the flora of the world as a whole (Tansley, 1939). The biological spectrum of cold, windy places can be broadened by creating shelter, as Whitehead (1954, 1959) showed by building low stone walls on a mountain plateau in Italy. At the sheltered site, new species were recruited over the years, leading to a taller type of vegetation. In the Swiss Alps, much of the pattern in the mosaic of vegetation types of the Dischma Valley could be accounted for by topographic shelter from the wind (Nägeli, 1971).

Trees are restricted to those areas of the world where the mean summer temperature exceeds 10 °C. This can be demonstrated by simply plotting climatological data of high altitude tree-line stations from all over the world (Fig. 2). The sites differ markedly in winter conditions but display nearly identical summer temperatures. Other authors have come to the same conclusion when discussing the northern extent of trees (Mikola, 1962). It is quite surprising to find such uniformity of behaviour, as different taxa are involved in different countries. It is also intriguing to note that dwarf shrubs have life-cycles not very different from those of trees yet they always progress to higher altitudes and latitudes.

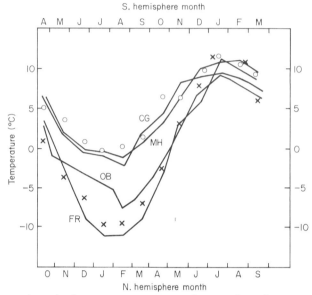

Fig. 2. The annual march of mean temperature near the natural tree line at several geographical locations. OB, Obergurgl, Austria (2070 m); FR, Front Range, Colorado (3350 m); ○, Craigieburn Mountains, New Zealand (1340 m); CG, Cairngorm Mountains, Scotland (760 m); ×, Haugastol, Hardangervid, Norway (988 m); MH, Moor House, England (558 m). For sources, see Grace (1977), from which this figure is reproduced by permission of Academic Press.

SPECIES DISTRIBUTION AND CLIMATE

There have been several attempts to relate the distribution of species to isometric lines of climatological variables. Salisbury (1926) showed that *Rubia peregrina** (madder) occurs to the south of the 40 °F mean January isotherm in northern Europe. Another early attempt was that of Iversen (1944) who made careful observations on *Ilex aquifolium* during a period of very cold winters, 1939 to 1942. He was able to show extensive damage by frost in those years, and pointed out that the species was confined to those areas where the mean temperature of the coldest month was in excess of -0.5 °C. In both these examples, the observations strongly suggest that the distribution is limited by the inability of the plant to tolerate low temperatures. In the British Isles, as we have seen, winter isotherms run from north to south and the highest temperatures are in the west. There are several species, known to be sensitive to frost, which occur only in the extreme west: examples are *Erica erigena, Daboecia cantabrica, Pinguicula grandiflora* and *Juncus acutus*. There are also very 'tender' exotics, planted in long-established botanic gardens on the west of Scotland: many Chilean and South African plants grow at Inverewe in Rosshire. From this extreme westerly distribution, characterized by the most 'tender' natives and exotics, a sequence of species may be compiled using provisional distribution maps from Fitter (1978). These species display distribution limits which follow winter isotherms (Fig. 3). It seems very likely that they display increasing tolerance of low winter temperatures, though experimental evidence is so far lacking.

---

* Nomenclature for species follows Clapham, Tutin & Warburg (1981).

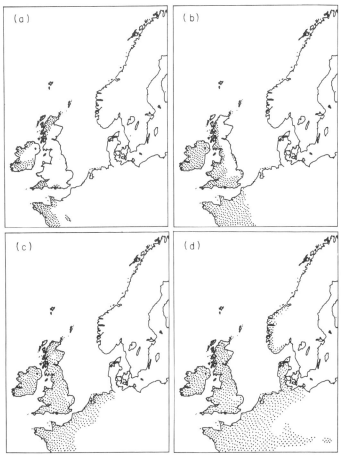

Fig. 3. Four species displaying a westerly distribution in northern Europe. (a) *Pinguicula lusitanica*;
(b) *Hypericum androsaemum*; (c) *Hyacinthoides non-scripta*; (d) *Ilex aquifolium*.

In contrast, there are other species with a pronounced northern or north-eastern distribution, including *Linnaea borealis* and *Trientalis europaea*. It is not easy to account for northern patterns. Perring & Walters (1962) suggest that these species may have a winter chilling requirement for germination which is not met at southern latitudes. Crawford & Palin (1981), who present data on the respiration rates of overwintering organs of two northern species compared with related southern types, conclude that the northern species display values that are damagingly high if they are maintained at temperatures characteristic of southern latitudes.

Many species appear to be restricted by low summer temperatures. The distribution of *Cirsium acaule* (stemless thistle) near to its northern limit is mainly on south-facing slopes (Pigott, 1974). Pigott showed that distribution is limited by failure to set seed and that reproductive success can be modified experimentally by imposing a marginally more favourable microclimate. In *Tilia cordata* (small-leaved lime), individuals at the northern limit of distribution do not set seed unless the summer is an especially warm one (Pigott & Huntley, 1978, 1980, 1981). Just as the case of tolerance to winter cold, it is possible to draw up a series according

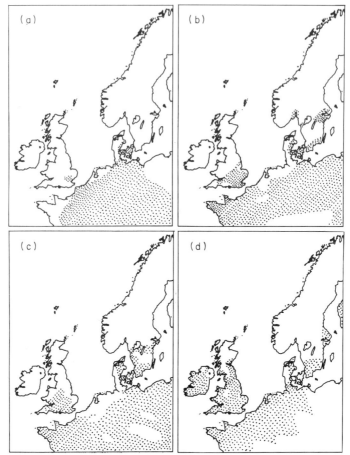

Fig. 4. Four species displaying a southerly distribution in northern Europe. (a) *Primula elatior*;
(b) *Lactuca serriola*; (c) *Cirsium acaule*; (d) *Leontodon hispidus*.

to tolerance of low temperatures in the summer (Fig. 4). Low temperature in this
case, is, however, likely to be acting quite differently, i.e. by preventing the species
from completing its life-cycle rather than by killing foliage and other organs.

There are inherent dangers in drawing conclusions from pattern of distribution
and, at best, they should be used only to form hypotheses about how species may
be limited by climate, or by extreme weather (Dahl, 1951, 1963; Willis, 1986).
One problem which has seldom been addressed is the role of competition or
association between species at the limits of their distribution (Woodward & Pigott,
1975; Woodward, 1975; Pigott, 1984). Another is the intercorrelation between
variables and the need to understand the complex interplay between individual
variables and the pattern of plant development. This requires a time-consuming
and expensive experimental programme, preferably coupled with a parallel
programme of field observations using reciprocal transplant experiments. Finally,
we should not forget that the meteorological instruments furnishing these basic
data are exposed in a standard Stevenson screen on level ground, whereas
conditions at the plant surface are different and depend on the partition of absorbed
solar energy between convection and evaporation, and on the rate at which surface

layers of air are mixed with the atmosphere as a whole. Many of the factors that determine surface conditions are characteristics of the vegetation itself.

EPICLIMATES

At plant surfaces, the temperature and other conditions may differ considerably from those measured in the air above the vegetation. Monteith (1981) coined the term 'epiclimate' to describe conditions at the surface. It is common in cold climates for temperatures of leaf surfaces to exceed those of the air, whilst in hot climates leaves are often cooler than air (Linacre, 1964, 1967). The exact temperature attained depends on the weather variables (net radiation, wind speed and water vapour pressure of the air) and it also depends on those structural factors of the vegetation which determine the aerodynamic resistance, and on the stomatal resistance. Many authors have presented calculations of leaf temperature which apply strictly to the aerodynamics of individual isolated leaves although, in reality, the aerodynamic features of the shoots, inflorescences and vegetation as a whole are often more significant (C. Wilson, J. Grace & S. Allen, in preparation). One of the most important findings has been that the aerodynamic term is a strong function of the height of the vegetation (Monteith, 1973; Grace, 1981, 1983). As a consequence, epiclimates in forests are never very different from the general climate whilst those of dwarf vegetation may be totally different from it (Fig. 5; Cernusca, 1976; Körner & Cochrane, 1983), being considerably warmer in bright sunshine and appreciably cooler on clear nights. This effect is so pronounced that near-lethal high temperatures have been observed in alpine conditions on days with bright sunshine and low wind speeds (Larcher, 1980). Overall, however, the

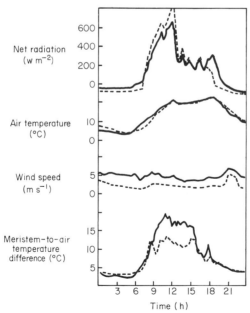

Fig. 5. The influence of plant stature on the temperatures of apical meristems near the tree line in the Cairngorm Mountains; (–––) 2 to 3 m tall stand of *Pinus sylvestris* at the tree line at 650 m; (——) less than 10 cm tall dwarf shrub community of *Arctostaphylos uva-ursa* at 700 m. Highest apical temperatures are recorded within the dwarf shrub community (C. Wilson, J. Grace & S. Allen, unpublished observations).

elevated temperatures of sunny days are likely to be advantageous in a cold climate, enabling species to extend to higher altitudes and latitudes.

Temperatures of flowers may be especially important to the completion of the life-cycle. In some species, the petals may focus the solar rays on to the gynoecium, or the inflorescence may display heliotropic movements, both of which lead to favourable temperatures for seed development (Kjellberg, Karlsson & Kerstensson, 1982; Gall, 1984). It may be significant that peduncles of alpine and arctic plants are often short and that inflorescences are borne close to the canopy in cushion plants, where temperatures are higher.

In desert climates, some species display features that promote transpirational cooling, with high rates of transpiration (Maximov, 1929). Such species may be 10 °C cooler than air (Lange, 1959; Smith, 1978; Althawadi & Grace, 1986) and thus survive at air temperatures which would otherwise be lethal. In the conditions of northern Europe, however, the very high transpiration rates that are required to sustain cooling are unlikely to occur as the leaf-to-air vapour pressure difference is only a fraction of that encountered in the desert. More usually, the leaves, apices and other organs are a few degrees warmer than air when the sun is shining (Landsberg, Butler & Thorpe, 1974; Russell & Grace, 1979). Even small differences in mean temperatures (~ 1 °C) are likely to be highly significant in an otherwise cold climate.

ABSOLUTE LIMITS TO SURVIVAL

The lowest temperatures which any plant material can tolerate depend on the extent of previous exposure to low temperatures. It is well known that the twigs of many tree species and the tissues of some herbaceous plants can survive after immersion in liquid nitrogen at -196 °C, but only if they are first subjected to a step-wise treatment of progressively lower temperatures. Sakai (1960) showed that once the step-wise cooling had reached -30 °C in mulberry twigs, such material survived immersion for short periods in liquid nitrogen. He also demonstrated that twigs of *Populus* and *Salix* spp. could be rooted and grew normally during the following summer after a 24 h exposure to -196 °C in the winter. Such observations have been repeated by others, working with boreal species of the genera *Abies*, *Picea*, *Pinus*, *Betula* and *Ribes* and also with herbaceous plants and dwarf shrubs (Larcher & Bauer, 1981).

The resistance to low temperatures in northern species in their natural environment has been assessed by collecting material for low temperature testing at various times of year. Tolerance varies on an annual cycle, depending not only on the development of a *hardening* response in relation to the prevailing seasonal temperatures (Pisek & Schiessl, 1947), but also on the position on the plant within the profile of a stand of vegetation (Larcher & Mair, 1969).

On a broader scale, the cold tolerance of winter buds is related to the climate of the region (Fig. 6; Sakai & Eiga, 1985). In tropical species, death occurs under test conditions at temperatures which are above freezing. On the other hand, in Alaska and eastern Siberia, the native vegetation can tolerate temperatures as low as -70 °C.

There are also genotypic variations in tolerance within species. Seedlings of the wild tomato derived from seed collected at various altitudes in Peru and Ecuador, were exposed to 0 °C for 7 d at an early stage of growth (Patterson, Paul & Smillie, 1978). Populations from sea level were killed, whereas 80% of those collected at

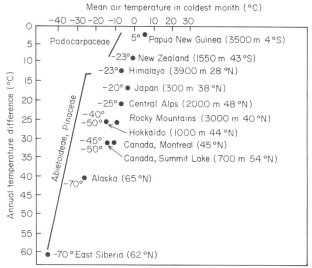

Fig. 6. Hardiness of winter buds of conifers. Numerals in the figure indicate the lowest temperature which the material tolerated. The axes refer to the temperature regime of the collection site: the mean air temperature of the coldest month and the difference between the mean air temperature of the coldest month and warmest month. Reproduced from Sakai & Eiga (1985) with permission from Norwegian University Press.

3 km survived. Similar trends were observed by Sakai & Eiga (1985) in tests on clones of *Abies sachalinensis* taken from various altitudes in Japan, using a test temperature of $-43\,°C$ and in latitudinal races of *Pinus sylvestris* in Sweden (Ericsson & Andersson, 1985).

In seeking to understand the mechanisms of low temperature tolerance, it is important to distinguish between three groups of plants. (1) Chill-sensitive, which are killed at temperatures above freezing, (2) freeze-sensitive, which are not killed until ice forms in the tissues, (3) freeze-tolerant, which can withstand ice in their tissues. In the first group, even short periods of chilling (2 h at 5 °C) cause visible breakages in thylakoid, plastid and mitochondrial membranes. Less direct, and now controversial, evidence for changes in membrane characteristics in these plants comes from the use of Arrhenius plots of selected biological features. These plots often show an abrupt change of slope at the temperatures known to cause injury (Steponkus, 1981, 1984). Recent studies have used techniques of electron spin resonance and fluorescence to investigate changes in state of the thylakoid membranes at low temperatures (Raison *et al.*, 1980) but still there is no consensus on the chain of events which kills chill-sensitive species.

The formation of ice in plant tissue occurs below 0 °C. Supercooling is commonly considerable (-4 to -14 °C; Lindow *et al.*, 1978), and thus many plants avoid freezing. The extent of avoidance may depend on the presence of ice-nucleating materials, including certain bacteria (Lindow *et al.*, 1978).

In freeze-sensitive plants, ice forms intracellularly and death is presumed to occur by gross disruption of cellular contents. In freeze-tolerant material, ice which forms in intercellular spaces and within xylem rather than intracellularly displays a lower chemical potential than liquid water and so water is withdrawn from cells to the extracellular locations of ice. Thus, the cells dehydrate and collapse as the volume of ice grows.

In an ecological context, it seems that most species display adequate cold
hardiness when growing in their natural habitat, except in years with unusually
low temperatures or unusually rapid declines in temperature. Natural selection will
operate during these periods and especially at the edge of the distribution range.
There are taxonomic variations in the limit to which this process of evolution can
occur. Larcher & Bauer (1981) note that low winter temperatures may be
responsible for the northerly and altitudinal limits to the distribution of broad-
leaved evergreen woody plants, for the northern limit of temperate deciduous
forests in North America and Eurasia and the altitudinal limits for survival of
some species on tropical mountains.

Frost damage is much more evident in exotics which originate in locations with
a different annual cycle of temperature from that of the site where the plants are
growing. In forestry, this is an especially important problem because of the
longevity of the crop and the consequently high probability of encountering an
extreme winter or extremely low temperatures in the spring (Cannell & Smith,
1984, 1986).

In Britain, the lowest temperatures on record occur in the months of January
and February and most are in Scotland, for example, Aberdeenshire and Ber-
wickshire ($-29\,°C$). It seems likely that the lowest values will tend to occur on the
eastern side of the country with a pattern similar to that of mean temperatures in
January (Fig. 1). None of the extremely low values quoted by Manley (1952) occur
on the west.

Physiological Performance within the Survival Limits

At high photon flux density, rates of net photosynthesis by particular species
display a temperature optimum which is related to their place of origin. C_4-species,
which include many tropical grasses, have an optimal temperature in the range 35
to $47\,°C$, whereas cool temperature species are usually C_3 and display maximal
rates between 15 and $25\,°C$ (Fig. 7).

In the case of *Spartina anglica*, a C_4-grass which is widely distributed in coastal
regions of Britain, the optimum is about $30\,°C$ (Long & Woolhouse, 1978). When
comparison is made between the photosynthetic performance of this grass and

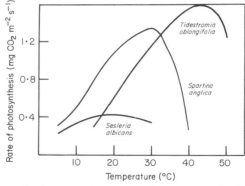

Fig. 7. Rates of photosynthesis measured at a range of temperatures for *Sesleria albicans* (Teesdale
provenance), a C_3-grass of north temperate distribution; *Spartina anglica*, a north temperature
C_4-grass distributed in coastal regions, and *Tidestromia oblongifolia*, a C_4-plant of the hot desert
in North America. Redrawn from Lloyd & Woolhouse (1976), Long & Woolhouse (1978) and
Björkman (1981).

more typical north temperature grasses with C_3 photosynthesis (Lloyd & Woolhouse, 1976), it is evident that the rate in the C_4-species over the range 5 to 10 °C is nearly as high as that of many C_3-grasses (Fig. 7). C_4-species are thus not necessarily limited in distribution by the responses to temperature that are evident in the tropical members which have received most attention (see also Caldwell, Osmond & Nott, 1977).

Many species show population differentiation with respect to temperature. Races from high altitudes have lower temperature optima than those from low altitude, both when rates are measured in material which has been transplanted to a phytotron (Slatyer, 1977) or where measurements have been made *in situ* (Benecke & Havranek, 1980). There are often seasonal changes in the temperature optimum, broadly suggesting acclimation to the pattern of temperature prevailing at that time of year (Mooney, Björkman & Collatz, 1978).

At extreme temperatures, the photosynthetic system is damaged. At the upper limit, irreversible inactivation of chloroplastic membranes occurs (Björkman, 1981), although species from different localities also differ in the heat stability of their soluble photosynthetic enzymes. At the low limit, chilling also operates on thylakoid membranes. After *Zea mays* was chilled for 6 h at 5 °C and then returned to 20 °C, the quantum yield of oxygen evolution was reduced by 45 % (Baker, East & Long, 1983). Even in hardy species, low temperatures reduce the capacity for photosynthesis. Neilson, Ludlow & Jarvis (1972) made comparisons between rates of photosynthesis of spruce shoots which were unhardened and those which were collected from the forest in February. Measurements were made at 20 °C before and after a chilling treatment. Shoots collected in February, described as 'partially hardened', showed no impairment of photosynthesis if chilled to −4 °C for 1 h but photosynthesis was completely prevented by a 1 h chilling cycle at −5 °C. Unhardened material was unaffected by chilling at 0 °C but impaired by 20 % after exposure to −2 °C.

The carbon balance of a species may depend critically on the temperature regime not only because temperature influences photosynthesis but also because of the effect of temperature on respiration (Mooney & Billings, 1965; Grace & Marks, 1978; Crawford & Palin, 1981). Grace & Marks (1978) discuss the temperature limits of the distribution of *Calluna vulgaris* and *Rubus chamaemorus* using computer models of carbon balance based upon physiological measurements. In *Rubus*, the carbon balance became negative if the temperature was increased from that corresponding to the study area in the English Pennines, where the species is close to its southern limit. In contrast, the carbon balance of *Calluna* became negative if the temperature was decreased because temperature has such a large influence on photosynthesis. The same sensitivity to temperature was demonstrated using observations on annual growth rings.

The expansion of leaves is determined by temperature acting at the meristems. This has been demonstrated by experiments in which the temperature of leaf, root and meristem have been varied independently (Watts, 1971). In dwarf vegetation, as we have seen, meristems might have temperatures which differ considerably from those of leaves because of strong temperature gradients near the ground. This may be especially important in grasses, where the vegetative meristems are at the soil surface. The onset of growth in the spring may thus be greatly influenced by topography, as a result of different insolation and surface temperatures of north- and south-facing slopes.

In C_3-species, the response of leaf expansion to changes in temperature is steep

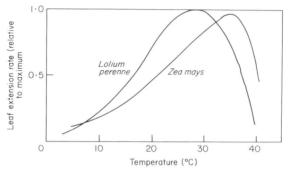

Fig. 8. Effect of meristem temperature on the rate of leaf expansion in a C_3-grass (*Lolium perenne*) and a tropical C_4-grass (*Zea mays*). Redrawn from Peacock (1975) and Watts (1971), respectively.

over the range of temperatures which is commonly experienced in north temperature countries (Fig. 8) and small changes in temperature usually exert a large influence on growth (Arnold, 1974). The relationship in the range 5 to 20 °C is approximately linear and this probably accounts for the widespread success in using 'degree days', 'accumulated temperature' or 'heat sum', in agricultural and forest meteorology, as a predictor of growth (Sucoff, 1971; Landsberg, 1974; Milford & Riley, 1980; Russell *et al.*, 1982).

Leaf extension, which is normally not directly dependent on the rate of photosynthesis, will continue for days after a plant is transferred to darkness, utilizing assimilates which are stored in various parts of the plant. Temperature acts on both the rate of photosynthesis and the component processes of growth such as cell division and expansion. Its effect on phloem transport seems remarkably small (Watson, 1975), although loading and unloading are very temperature-sensitive. The responses of the various processes to temperature are presumably similar and co-ordinated, or there would otherwise be extreme seasonal variations in the pool sizes of sugar and other constituents. In fact, it seems that photosynthesis is often less reduced by low temperatures than is cell division. Assimilates normally accumulate as storage products in plant tissues with the onset of winter (see Hendry, 1987).

There are many other physiological parameters which may be important in determining survival. Of particular interest is the observation by Pigott & Huntley (1981) that the growth of pollen tubes in *T. cordata* requires an especially high threshold temperature (Fig. 9). It would be of great interest to see if this is generally the case in C_3-species which fail to set seed at their northern limits of distribution.

Although techniques for measuring plant water potential are well developed, there have been few attempts to evaluate the role of water stress in determining survival of plants in the dry parts of their range. One recent development here is the ultrasonic technique of detecting cavitation in the xylem conduits. The cavitation process may not always be reversible and so one episode of water stress may predispose plants to stress when a second episode arises (Pẽna & Grace, 1986).

CLIMATIC CONTROL OF THE LIFE-CYCLE

Climate impinges on the plant at each point in the life-cycle. Some insight into the complexity of this on the performance of plants is obtained from a brief

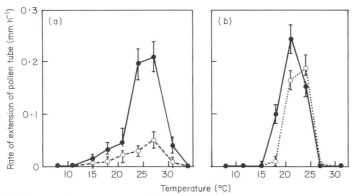

Fig. 9. Influence of temperature on the rate of extension of pollen tubes of *Tilia cordata* (means and 95% confidence limits) from (a) Shrawley in 0·29 and 0·58 mol dm⁻³ sucrose (●, ○, respectively), (b) Cringlebarrow in 0·58 mol dm⁻³ (●) and Rusland in 0·29 mol dm⁻³ (○). From Pigott & Huntley (1981).

consideration of the growth cycle of a hypothetical conifer. In winter, the resting bud contains the leaves which were formed as primordia in the previous summer (Cannell, Thompson & Lines, 1976). Leaves formed in previous years display positive rates of net photosynthesis during the more favourable spells of winter weather but no growth occurs and so stored photosynthate accumulates. Bradbury & Malcolm (1978) found that seedlings of *Picea sitchensis* in a nursery at south Scotland actually doubled their weight in winter whilst remaining dormant but this may have been an extreme result consequent on the mild winter of that year. During winter, the plant displays considerable tolerance of low temperatures, usually much more than is required for survival (Larcher & Bauer, 1981). There is a tendency for the leaf water potential to decline as winter progresses, especially at high altitude sites where the ground is frozen, and especially in sunny weather. At the tree line, critically low potentials may occur, leading to death of a large proportion of the leaves (Tranquillini, 1982). In addition, branches may be lost because of breakage by wind, snow or under the weight of rime ice. Bud break does not occur until specific thermal conditions have been met. Usually, a chilling requirement is required during the winter, after which a certain 'temperature sum' is required to elicit bud break.

Once bud break has occurred in the spring, extension growth of stems and new needles depends strongly on temperature (Hiley & Cuncliffe, 1922; Little, 1980) and possibly humidity (Grace, Malcolm & Bradbury, 1975) but the number of needles has already been determined by the previous year's conditions. During summer, photosynthesis is a function of absorbed radiation, present temperature and previous temperature (Ludlow & Jarvis, 1971; Neilson *et al.*, 1972). The foliage is no longer frost-hardy and during early summer may be killed by late frosts (Cannell & Smith, 1986). Extension growth stops when the day length falls below a certain threshold (Malcolm & Pymar, 1975). The cuticle continues to develop as the summer progresses (Lange & Schulze, 1966).

Analyses of annual increments in cores taken from stems usually show strong relationships with temperatures in the second half of the summer, with a somewhat weaker relationship with water availability (Hustich, 1947; Schweingruber, Braker & Schar, 1979; Hughes *et al.*, 1984). Near the limits of distribution, cone production is erratic and seed viability is low (Miller & Cummins, 1982).

This survey of the growth cycle has shown that climatic variables operate on conifers in numerous ways. Similar patterns, involving the regulation of the plant's development by signals of photoperiod and temperature, have been demonstrated in a wide range of agricultural species (Aitken, 1974). In wild species, we may suppose no less interplay between the genotype and the environment, each species responding in a characteristic manner to stimuli, climatic variables and degrees of stress (Haslam, 1975; Bannister, 1978, 1981; Woodward & Jones, 1984; Woodward, 1986).

GENERAL DISCUSSION

Distribution patterns are probably determined by failure at one part of the life-cycle caused by extremes of climate—either extreme values of specific variables such as occur only once or twice in a century, or unusual combinations which collectively uncouple the usual relationships between different parts of the life-cycle. For this reason, it is unusual to be able to observe failure 'in action'. There are, however, several ways of approaching the problem. In woody perennials, the annual growth rings represent a rich store of information and statistical relationships with climatic variables may often be apparent. The elucidation of these relationships is the subject matter of dendrochronology, a discipline which has been mainly applied in the reconstruction of past climates rather than in the understanding of climatic tolerances of plants. In mat flowering perennials, long-term measurements of colony diameter may provide similar but less extensive information (Hunt, Hope-Simpson & Snape, 1985). Alternatively, in other groups of plants, long-term measurements of vitality stretching over a decade or so may also be revealing (e.g. Willis, 1986), although they suffer inevitably from a lack of a 'control'.

Reciprocal transplant experiments are an alternative approach, although the results may sometimes be too complex to interpret easily and different controlling factors may apply in different years.

Ultimately, the physiological reasons for failure may take years of research to identify. It is not adequate simply to demonstrate that two closely related species differ in one single physiological or biochemical attribute, as survival is likely to depend on a whole suite of attributes acting together. Unless the physiologist is guided by field observations, or unless comparisons are being made of closely related species of different distribution (Grime, 1965), the task is impossible for there are hundreds of physiological processes and biochemical reactions to be considered, all of which are likely to be sensitive to climatic variables but only a few of which are likely to determine survival.

One substantial difficulty is the evaluation of the ecological significance of a lack of vigour measured as a reduced rate of photosynthesis or reduced growth rate. The problem here is that small reductions in growth may be crucial in determining the competitive balance. But how small is small? The biotic complexity which exceeds the complexity of the physical environment is much harder to study. Failure in the wild may be caused by pathogens which invade tissues that have been exposed to climatic stresses and consequently become predisposed to attack.

ACKNOWLEDGEMENT

I would like to dedicate this article to the memory of a colleague, Maurice Caborn, who died during the period when this article was being prepared. His interest in climatic and microclimatic effects on vegetation did much to stimulate my own.

REFERENCES

AITKEN, Y. (1974). *Flowering Time, Climate and Genotype.* Melbourne University Press, Victoria, Australia.
ALTHAWADI, A. M. & GRACE, J. (1986). Water use by the desert curcurbit *Citrullus colocynthus* (L). Schrad. *Oecologia*, **70**, 475–480.
ARNOLD, S. M. (1974). The relationship between temperature and seedling growth of two species which occur in Upper Teesdale. *New Phytologist*, **73**, 333–340.
BAKER, N. R., EAST, T. M. & LONG, S. P. (1983). Chilling damage to photosynthesis in young *Zea mays*. II. Photochemical function of thylakoids *in vivo*. *Journal of Experimental Botany*, **34**, 189–197.
BANNISTER, P. (1978). Flowering and shoot extensions in heath plants of different geographical origin. *Journal of Ecology*, **66**, 117–131.
BANNISTER, P. (1981). Phenology and carbohydrate content of heathers of southern origin grown in Scotland. *Transactions of the Botanical Society of Edinburgh*, **43**, 271–279.
BENECKE, U. & HAVRANEK, W. M. (1980). Gas-exchange of trees at altitudes up to timberline, Craigieburn range, New Zealand. In: *Mountain Environments and Subalpine Tree Growth* (Ed. by U. Benecke & M. R. Davis), pp. 195–212. Forest Research Institute of New Zealand, Rotorua.
BJÖRKMAN, O. (1981). The response of photosynthesis to temperature. In: *Plants and their Atmospheric Environment* (Ed. by J. Grace, E. D. Ford & P. G. Jarvis), pp. 273–301. Blackwell Scientific Publications, Oxford.
BRADBURY, I. K. & MALCOLM, D. C. (1978). Dry matter accumulation by *Picea sitchensis* saplings during winter. *Canadian Journal of Forest Research*, **8**, 207–213.
CALDWELL, M. M., OSMOND, C. B. & NOTT, D. L. (1977). C_4 pathway photosynthesis at low temperatures in cold-tolerant *Atriplex* species. *Plant Physiology*, **60**, 157–164.
CANNELL, M. G. R. & SMITH, R. I. (1984). Spring frost damage on young *Picea sitchensis*. II. Predicted dates of budburst and probability of frost damage. *Forestry*, **57**, 177–197.
CANNELL, M. G. R. & SMITH, R. I. (1986). Climatic warming, spring budburst and frost damage on trees. *Journal of Applied Ecology*, **23**, 177–191.
CANNELL, M. G. R., THOMPSON, S. & LINES, R. (1976). An analysis of inherent differences in shoot growth within some north temperate conifers. In: *Tree Physiology and Yield Improvement* (Ed. by M. G. R. Cannell & F. T. Last), pp. 173–205. Academic Press, London.
CARTER, E. S. (1978). The effect of drought on British agriculture. *Proceedings of the Royal Society of London*, **A363**, 43–54.
CERNUSCA, A. (1976). Bestandesstruktur, Bioklima und Energiehaushalt von alpinen Zwergstrauchbeständen. *Oecologia Plantarum*, **11**, 71–102.
CHANDLER, T. J. & GREGORY, S. (1976). *The Climate of the British Isles.* Longman, London.
CLAPHAM, A. R., TUTIN, T. G. & WARBURG, E. F. (1981). *Excursion Flora of the British Isles*, 3rd Edn. Cambridge University Press, Cambridge.
CRAWFORD, R. M. M. & PALIN, M. A. (1981). Root respiration and temperature limits to the north–south distribution of four perennial maritime plants. *Flora*, **171**, 338–354.
DAHL, E. (1951). On the relation between summer temperature and the distribution of alpine vascular plants in the lowlands of Fennoscandia. *Oikos*, **3**, 22–52.
DAHL, E. (1963). On the heat exchange of a wet vegetation surface and the ecology of *Koeniga islandica*. *Oikos*, **14**, 190–211.
ERICSSON, T. & ANDERSSON, B. (1985). Scots pine breeding for North-Sweden by early tests. In: *Plant Production in the North* (Ed. by A. Kaurin, O. Junttila & J. Nilson), pp. 322–328. Norwegian University Press, Oslo.
FITTER, A. (1978). *An Atlas of the Wild Flowers of Britain and Northern Europe.* Collins, London.
GALL, A. (1984). *A study of the temperatures and energy balances of arctic and alpine flowers.* B.Sc. thesis, University of Edinburgh.
GLOYNE, R. W. (1964). Some characteristics of the natural wind and their modification by natural and artificial obstructions. *Scientific Horticulture*, **17**, 7–19.
GRACE, J. (1977). *Plant Response to Wind.* Academic Press, London.
GRACE, J. (1981). Some effects of wind on plants. In: *Plants and Their Atmospheric Environment* (Ed. by J. Grace, E. D. Ford & P. G. Jarvis), pp. 125–130. Blackwell Scientific Publications, Oxford.
GRACE, J. (1983). *Plant–Atmospheric Relationships.* Chapman & Hall, London.
GRACE, J. & MARKS, T. C. (1978). Physiological aspects of bog production at Moor House. In: *Production Ecology of British Moors and Montane Grasslands* (Ed. by O. W. Heal & D. F. Perkins), pp. 38–51. Springer–Verlag, Berlin.
GRACE, J., MALCOLM, D. C. & BRADBURY, I. (1975). The effect of wind and humidity on leaf diffusive resistance in Sitka spruce seedlings. *Journal of Applied Ecology*, **12**, 931–940.
GREEN, F. H. W. (1964). A map of annual average potential water deficit in the British Isles. *Journal of Applied Ecology*, **1**, 151–158.
GRIME, J. P. (1965). Comparative experiments as a key to the ecology of flowering plants. *Ecology*, **46**, 513–515.

HASLAM, S. M. (1975). The performance of *Phragmites communis* Trin. in relation to temperature. *Annals of Botany*, **39**, 881–888.

HENDRY, G. A. F. (1987). The ecological significance of fructan in a contemporary flora. In: *Frontiers of Comparative Plant Ecology* (Ed. by I. H. Rorison, J. P. Grime, R. Hunt, G. A. F. Hendry & D. H. Lewis), *New Phytologist*, **106** (Suppl.), 201–216. Academic Press, New York & London.

HILEY, W. E. & CUNCLIFFE, N. (1922). *An Investigation into the Height Growth of Trees and Meteorological Conditions*. Oxford Forestry Memoirs 1, Clarendon Press, Oxford.

HUGHES, M. K., SCHWEINGRUBER, F. H., CARTWRIGHT, D. & KELLY, P. M. (1984). July–August temperature at Edinburgh between 1721 and 1975 from tree-ring density and width data. *Nature*, **308**, 341–344.

HUNT, R., HOPE-SIMPSON, J. F. & SNAPE, J. B. (1985). Growth of the dune wintergreen (*Pyrola rotundifolia* ssp. *maritima*) at Braunton Burrows in relation to weather factors. *International Journal of Biometeorology*, **29**, 323–334.

HUNTER, R. F. & GRANT, S. A. (1971). The effect of altitude on grass growth in east Scotland. *Journal of Applied Ecology*, **8**, 1–20.

HUSTICH, I. (1947). Climate fluctuation and vegetation growth in northern Finland during 1890–1939. *Nature*, **160**, 478–479.

IVERSEN, J. (1944). *Viscum, Hedera* and *Ilex* as climate indicators. *Geologiska föreningens i Stockholm förhandlinger*, **66**, 463–483.

KJELLBERG, B., KARLSSON, S. & KERSTENSSON, I. (1982). The effects of heliotropic movements of flowers of *Dryas octopetala* on gynoecium temperature and seed development. *Oecologia*, **54**, 10–13.

KÖRNER, CH. & COCHRANE, P. (1983). Influence of plant physiognomy on leaf temperature on clear midsummer days in the Snowy Mountains, south-eastern Australia. *Acta Oecologia/Oecologia Plantarum*, **4**, 117–124.

LANDSBERG, J. J. (1974). Apple fruit bud development and growth; analysis and an empirical model. *Annals of Botany*, **38**, 1013–1023.

LANDSBERG, J. J., BUTLER, D. R. & THORPE, M. R. (1974). Apple bud and blossom temperature. *Journal of Horticultural Science*, **49**, 227–239.

LANGE, O. L. (1959). Untersuchungen über Wärmehaushalt und Hitzeresistenz maurentanischer Wüsten- und Savannenpflanzen. *Flora*, **147**, 595–651.

LANGE, O. L. & SCHULZE, E.-D. (1966). Untersuchungen über die Dickenentwicklung der kutikularen Zellwandschichten bei der Fichtennadel. *Forstwissenschaftliches Centralblatt*, **85**, 27–38.

LARCHER, W. (1980). *Physiological Plant Ecology*, 2nd Edn. Springer-Verlag, Berlin.

LARCHER, W. & BAUER, H. (1981). Ecological significance of resistance to low temperature. In: *Physiological Plant Ecology 1, Encyclopedia of Plant Physiology, Vol. 12a* (Ed. by O. L. Lange, P. S. Nobel, C. B. Osmond & H. Ziegler), pp. 403–437. Springer-Verlag, London.

LARCHER, W. & MAIR, B. (1969). Die Temperaturresistenz als ökophysiologisches Konstitutionsmerkmal. 1. *Quercus ilex* und andere Eichenarten des Mittelmeergebietes. *Oecologia Plantarum*, **4**, 347–376.

LINACRE, E. T. (1964). A note on a feature of leaf and air temperatures. *Agricultural Meteorology*, **1**, 66–72.

LINACRE, E. T. (1967). Further notes on a feature of leaf and air temperatures. *Archiv für Meteorologie, Geophysik und Bioklimatologie*, **15**, 422–436.

LINDOW, S. E., ARNEY, D. C., UPPER, C. D. & BARCHET, W. R. (1978). The role of bacterial ice nuclei in frost injury to sensitive plants. In: *Plant Cold Hardiness and Freezing Stress* (Ed. by P. H. Li & A. Sakai), pp. 249–263. Academic Press, New York.

LITTLE, C. H. A. (Ed.) (1980). *Control of Shoot Growth in Trees*. Maritimes Forest Research Centre, New Brunswick, Canada.

LLOYD, N. D. H. & WOOLHOUSE, H. W. (1976). The effect of temperature on photosynthesis and transpiration in populations of *Sesleria caerulea*. *New Phytologist*, **77**, 553–559.

LONG, S. P. & WOOLHOUSE, H. W. (1978). The response of net photosynthesis to light and temperature in *Spartina townsendii* (*sensu lato*), a C_4 species from a cool temperate climate. *Journal of Experimental Botany*, **29**, 803–814.

LUDLOW, M. M. & JARVIS, P. G. (1971). Photosynthesis in Sitka spruce [*Picea sitchensis* (Bong.) Carr.]. I. General characteristics. *Journal of Applied Ecology*, **8**, 925–953.

MALCOLM, D. C. & PYMAR, C. F. (1975). The influence of temperature on the cessation of height growth of Sitka spruce [*Picea sitchensis* (Bong.) Carr] provenances. *Silvae Genetica*, **24**, 129–132.

MANLEY, G. (1952). *Climate and the British Scene*. Collins, London.

MAXIMOV, N. A. (1929). *The Plant in Relation to Water*. Allen & Unwin, London.

MIKOLA, P. (1962). Temperature and tree growth near the northern timber line. In: *Tree Growth* (Ed. by T. T. Kozlowski), pp. 265–274. Ronald Press, New York.

MILFORD, G. F. R. & RILEY, J. (1980). The effects of temperature on leaf growth of sugar beet varieties. *Annals of Applied Biology*, **94**, 431–443.

MILLER, G. R. & CUMMINS, R. T. (1982). Regeneration of Scots pine *Pinus silvestris* at a natural tree line in the Cairngorm Mountains, Scotland. *Holarctic Ecology*, **5**, 27–34.

MONTEITH, J. L. (1973). *Principles of Environmental Physics*. Arnold, London.

MONTEITH, J. L. (1981). Coupling of plants to the atmosphere. In: *Plants and their Atmospheric Environment* (Ed. by J. Grace, E. D. Ford & P. G. Jarvis), pp. 1–29. Blackwell Scientific Publications, Oxford.

MOONEY, H. A. & BILLINGS, W. D. (1965). Effects of altitude on carbohydrate content of mountain plants. *Ecology*, **46**, 750–751.

MOONEY, H. A., BJÖRKMAN, O. & COLLATZ, G. J. (1978). Photosynthetic acclimation to temperature in the desert shrub, *Larrea divaricata* I. Carbon exchange characteristics of intact leaves. *Plant Physiology*, **61**, 406–410.

MUNRO, J. M. M. & DAVIES, D. A. (1973). Potential pasture production in the uplands of Wales. 2. Climatic limitations on production. *Journal of the British Grassland Society*, **28**, 161–169.

NÄGELI, W. (1971). *Der Wind als Standortsfaktor bei Anforstungen in der subalpinen Stufe (Stillbergalp im Dischmatal Kanton Graubünden)*. Mitteilungen 47, Schweizerische Anstalt für das Förstliches Versuchswesen, 147 pp.

NEILSON, R. E., LUDLOW, M. M. & JARVIS, P. G. (1972) Photosynthesis in Sitka spruce [*Picea sitchensis* (Bong.) Carr.]. II. Response to temperature. *Journal of Applied Ecology*, **9**, 721–745.

PATTERSON, B. D., PAULL, R. & SMILLIE, R. M. (1978). Chilling resistance in *Lycopersicon hirsutum* Humb. & Bonpl., a wild tomato with a wide altitudinal distribution. *Australian Journal of Plant Physiology*, **5**, 609–617.

PEACOCK, J. M. (1975). Temperature and leaf growth in *Lolium perenne*. II. The site of temperature perception. *Journal of Applied Ecology*, **12**, 115–124.

PEÑA, J. & GRACE, J. (1986). Water relations and ultrasound emissions of *Pinus sylvestris* L. before, during and after a period of water stress. *New Phytologist*, **103**, 515–524.

PERRING, F. H. & WALTERS, S. M. (1962). *Atlas of the British Flora*. Nelson, London.

PIGOTT, C. D. (1974). The response of plants to climate and climatic change. In: *The Flora of a Changing Britain* (Ed. by F. Perring), pp. 32–44. Classey, London.

PIGOTT, C. D. (1984). The flora and vegetation of Britain: ecology and conservation. *New Phytologist*, **98**, 119–128.

PIGOTT, C. D. & HUNTLEY, J. P. (1978). Factors controlling the distribution of *Tilia cordata* at the northern limits of its geographical range. I. Distribution in north-west England. *New Phytologist*, **81**, 429–441.

PIGOTT, C. D. & HUNTLEY, J. P. (1980). Factors controlling the distribution of *Tilia cordata* at the northern limits of its geographical range. II. History in north-west England. *New Phytologist*, **84**, 145–164.

PIGOTT, C. D. & HUNTLEY, J. P. (1981). Factors controlling the distribution of *Tilia cordata* at the northern limits of its geographical range. III. Nature and causes of seed sterility. *New Phytologist*, **87**, 817–839.

PISEK, A. & SCHIESSL, R. (1947). Die Temperaturbeeinflussbarkeit der Frosthärte von Nadelhölzern und Zwergsträuchern an der alpinen Waldgrenze. *Bericht des Naturwissenschaftlich-medizinischen Vereins in Innsbruck*, **47**, 33–52.

RAISON, J. K., BERRY, J. A., ARMOND, P. A. & Pike, C. S. (1980). Membrane properties in relation to the adaptation of plants to temperature stress. In: *Adaptation of Plants to Water and High Temperature Stress* (Ed. by N. C. Turner & P. J. Kramer), pp. 261–273. Wiley, New York.

RAUNKIAER, C. (1909). Formationsundersogelse og Formationsstatisk. *Botanisk Tidsskrift*, **30**, 20–132.

RAUNKIAER, C. (1934). *The Life Forms of Plants and Statistical Plant Geography*. (Translation.) Oxford University Press.

RORISON, I. H., SUTTON, F. & HUNT, R. (1986). Local climate topography and plant growth in Lathkill Dale NNR. I. A twelve-year summary of solar radiation and temperature. *Plant, Cell and Environment*, **9**, 49–56.

RUSSELL, G. & GRACE, J. (1979). The effect of shelter on the yield of grasses in southern Scotland. *Journal of Applied Ecology*, **16**, 319–330.

RUSSELL, G., ELKS, R. P., BROWN, J., MILBOURN, G. M. & HAYTER, A. M. (1982). The development and yield of autumn- and spring-sown barley in south east Scotland. *Annals of Applied Biology*, **100**, 167–178.

SAKAI, A. (1960). Survival of the twig of woody plants at −196 °C. *Nature*, **4710**, 393–394.

SAKAI, A. & EIGA, S. (1985). Physiological and ecological aspects of cold adaptation of boreal conifers. In: *Plant Production in the North* (Ed. by A. Kaurin, O. Juntila & J. Nilson), pp. 157–170. Norwegian University Press, Oslo.

SALISBURY, E. J. (1926). The geographic distribution of plants in relation to climatic factors. *Geographical Journal*, **57**, 312–335.

SCHWEINGRUBER, F. H., BRAKER, O. U. & SCHAR, E. (1979). Dendroclimatic studies on conifers from central Europe and Great Britain. *Boreas*, **8**, 427–452.

SLATYER, R. O. (1977). Altitudinal variation in the photosynthetic characteristics of snow gum, *Eucalyptus pauciflora* Sieb. ex Sprent. III. Temperature response. *Australian Journal of Plant Physiology*, **4**, 301–312.

SMITH, W. K. (1978). Temperatures of desert plants: another perspective on the adaptability of leaf size. *Science*, **201**, 614–616.

STEPONKUS, P. L. (1981). Response to extreme temperatures. Cellular and subcellular bases. In: *Physio-*

logical Plant Ecology 1, Encyclopedia of Plant Physiology, Vol. 12a (Ed. by O. L. Lange, P. S. Nobel, C. B. Osmond & H. Ziegler), pp. 372–402. Springer-Verlag, Berlin.

STEPONKUS, P. L. (1984). Role of the plasma membrane in freezing injury and cold acclimation. *Annual Review of Plant Physiology*, **35**, 543–584.

SUCOFF, E. (1971). Timing and rate of bud formation in *Pinus resinosa. Canadian Journal of Botany*, **49**, 1821–1832.

TANSLEY, A. G. (1939). *The British Islands and their Vegetation.* Cambridge University Press, Cambridge.

TAYLOR, J. A. (1976). Upland climates. In: *The Climate of the British Isles* (Ed. by J. J. Chandler & S. Gregory), pp. 264–286. Longman, London.

TRANQUILLINI, W. (1982). Frost-drought and its ecological significance. In: *Physiological Plant Ecology II, Encyclopedia of Plant Physiology, Vol. 12b* (Ed. by O. L. Lange, P. S. Nobel, C. B. Osmond & H. Zeigler), pp. 379–400. Springer-Verlag, Berlin.

WATSON, B. T. (1975). The influence of low temperatures on the rate of translocation in the phloem of *Salix viminalis* L. *Annals of Botany*, **39**, 889–900.

WATTS, W. R. (1971). Role of temperature in the regulation of leaf extension in *Zea mays. Nature*, **229**, 46–47.

WHITE, E. J. & SMITH, R. I. (1982). *Climatological Maps of Great Britain.* Institute of Terrestrial Ecology, Cambridge.

WHITEHEAD, F. H. (1954). A study of the relation between growth form and exposure on Monte Maiella, Italy. *Journal of Ecology*, **42**, 180–186.

WHITEHEAD, F. H. (1959). Vegetational changes in response to alterations of surface roughness on M. Maiella, Italy. *Journal of Ecology*, **47**, 603–606.

WILLIS, A. J. (1986). Plant diversity and change in a species-rich dune system. *Transactions of the Botanical Society of Edinburgh*, **44**, 291–308.

WOODWARD, F. I. (1975). The climate control of the altitudinal distribution of *Sedium rosea* (L.) Scop. and *S. telephium* L. II. The analysis of plant growth in controlled environments. *New Phytologist*, **74**, 335–348.

WOODWARD, F. I. & JONES, N. (1984). Growth studies of selected plant species with well-defined European distributions. 1. Field observations and computer simulations on plant life cycles at two altitudes. *Journal of Ecology*, **72**, 1019–1030.

WOODWARD, F. I. & PIGOTT, C. D. (1975). The climatic control of altitudinal distributions of *Sedum rosea* (L.) Scop. and *S. telephium* L. 1. Field observations. *New Phytologist*, **74**, 323–334.

WOODWARD, F. I. (1986). Ecophysiological studies on the shrub *Vaccinium myrtillus* L. taken from a wide altitudinal range. *Oecologia*, **70**, 580–586.

New Phytol. (1987) **106** (Suppl.), 131–160

COMPARATIVE STUDIES OF LEAF FORM: ASSESSING THE RELATIVE ROLES OF SELECTIVE PRESSURES AND PHYLOGENETIC CONSTRAINTS

By THOMAS J. GIVNISH

Department of Botany, University of Wisconsin, Madison, WI 53706, USA

SUMMARY

Leaves or their functional analogues provide outstanding opportunities for comparative studies. Here, I use leaves to illustrate the crucial role of ecological, biogeographic and phylogenetic comparisons in generating and testing hypotheses regarding the adaptive significance of morphological variation, the relative importance of selective pressures *vs* phylogenetic constraints, and the rise of adaptations within lineages. The complementary roles of comparative studies and optimality models are stressed throughout.

The first section of this paper reviews 23 ecological patterns in leaf form, physiology and arrangement which have been uncovered by comparative studies. Three general sets of energetic trade-offs, involving the economics of gas exchange, support, and biotic interactions, appear likely to influence the evolution of leaves and underlie these trends. The first of these trade-offs is illustrated with an analysis of the adaptive significance of leaf size, in both terrestrial and aquatic plants. The resulting predictions are compared with the actual trends observed, and the relative strengths and limitations of the approach are discussed.

The second section addresses the role of selective pressures and phylogenetic constraints in determining features of leaf form and phenology in forest herbs. Ecological comparisons of 74 species from a site in the Virginia Piedmont show that members of each temporal photosynthetic guild display evolutionary convergence in several aspects of leaf form and arrangement. These convergences can each be understood in terms of models that assume that selection favours plants whose form and physiology tend to maximize whole-plant growth. Phylogenetic comparisons indicate that congeners of guild members share the same leaf phenology as the guild members themselves. This remarkable finding suggests that phenology is evolutionarily rather non-labile within genera but that, within guilds, species in several different genera and families converge strongly in other leaf traits. In this case, phylogenetic constraints appear to be important mainly in determining which lineages evolve particular phenologies and leaf adaptations, not whether they arise. The section concludes with a critical discussion of the capacity of two classes of evolutionary models, based primarily on functional considerations or phylogenetic constraints, to produce truly deductive predictions.

The third section briefly reviews an analysis of adaptive radiation in leaf shape among violets of eastern North America. Each ecological group of species displays the leaf shape expected on functional grounds and separate lineages (as judged by means of phylogeny independent of leaf shape) show parallel trends. Ecological and phylogenetic comparisons, in combination with optimality models, provide insights into the sequence of habitats invaded and leaf forms evolved.

I conclude with comments on the advantages and limitations of comparative studies, and speculate on avenues for future research on leaf form. An integrated approach, involving comparisons at ecological, phylogenetic and biogeographic levels, and complemented by optimality analyses and detailed populational studies involving biomechanics, physiological ecology and ecological genetics, is strongly advocated.

Key words: Leaf form, evolutionary convergence, phylogenetic constraints, adaptive radiation, optimality analysis.

INTRODUCTION

Studies comparing the ecology of individual plant species usually have one of four goals: (1) to generate hypotheses regarding the adaptive significance of variations in plant form, physiology and behaviour, the factors limiting species distributions and the basis for trends in community structure; (2) to test such hypotheses developed on other grounds; (3) to evaluate the relative importance of selection *vs* phylogenetic constraints in determining species and community patterns, and (4) to analyze adaptive radiation and the rise of adaptations within lineages.

In this paper, I will illustrate each of these approaches, focusing on leaves. Leaves (or their functional analogues) provide outstanding opportunities for comparative studies. As the primary site of photosynthesis, they are of fundamental functional importance to green plants. Furthermore, their extraordinary variation in form, physiology and phenology (both within and between species) has implications for not only carbon exchange but also water loss, allocation to above- *vs* below-ground tissues, branching pattern, whole-plant growth and interactions with competitors, predators and mutualists. Most importantly, from the point of view of comparative studies, the plants that bear leaves provide the ample numbers of comparisons needed to distinguish the effects of selection, phylogeny and chance. The angiosperms alone comprise over 250000 species in more than 10000 genera and 300 families, and occupy a wide range of ecological conditions on several different continents. Such diversity permits far greater power in testing theories of adaptation among plants than do most animal groups.

The first section of this paper focuses on the generation and testing of hypotheses regarding some general constraints on leaf form, with principal emphasis on leaf size. The second addresses the role of selective pressures *vs* phylogenetic constraints in determining features of leaf form and phenology in forest herbs. The third briefly analyzes an example of adaptive radiation in leaf form among violets of eastern North America. I conclude with comments on the advantages and limitations of comparative studies and their role in concert with other approaches to the study of leaf form and speculate on avenues for future research.

GENERATING AND TESTING HYPOTHESES

Introduction

A major theme that has emerged in the 15 years since the publication of Horn's (1971) *Adaptive Geometry of Trees* is the complementary role played by optimality analyses and comparative studies. Natural selection should generally favour plants whose form and physiology tend to maximize their net rate of growth, because such plants often have the most resources with which to reproduce and compete for additional space (Horn, 1971; Orians & Solbrig, 1977; Givnish, 1979, 1986a). Optimality models based on this principle can produce detailed, often quantitative predictions of how plant form, physiology and behaviour should vary with environmental conditions.

Such deductions can be tested quite powerfully by comparative studies of the actual trends observed across species: convergence, both within a habitat and in trends across habitats, is often the most persuasive evidence that an observed pattern is the result of selection, rather than drift, founder effects, pleiotropy, epistasis or the historical availability of sufficient and appropriate genetic variability. Because they make precise predictions related to specific environmental factors, optimality models are a better means of generating hypotheses than comparative studies themselves, which alone can only demonstrate broad correlations between plant behaviour and ecological factors. However, the uncovering of such correlations plays a critical role in generation of hypotheses by suggesting the importance of particular energetic trade-offs, which can be incorporated in more precise optimality analyses.

In this section, I first detail some of the broad ecological patterns in leaf form, physiology and arrangement that have been uncovered by comparative studies. Second, I outline three general sets of energetic trade-offs that are likely to influence the evolution of leaves and should be considered in any cost–benefit analysis. Finally, I illustrate one of these trade-offs with an analysis of the adaptive significance of leaf size, in both terrestrial and aquatic plants.

Patterns

Comparative studies have revealed the existence of well-marked ecological trends in some 23 aspects of leaf form, physiology and pattern of arrangement among terrestrial species of diverse lineage. These trends include the following patterns.

(1) Effective leaf size (i.e. the width of a leaf or its lobes or leaflets) tends to increase along gradients of increasing rainfall, humidity and/or soil fertility, and to decrease with increasing irradiance (Schimper, 1898; Bews, 1925; Raunkiaer, 1934; Beard, 1944, 1955; Shields, 1950; Cain *et al.*, 1956; Anderson, 1961; Ashton, 1964; Beadle, 1966; Webb, 1968; Brünig, 1970, 1983; Sarmiento, 1972; Givnish, 1979, 1984; Cowling & Campbell, 1980; Dolph & Dilcher, 1980a,b; Peace & MacDonald, 1981; Chiariello, 1984). Effective leaf size also tends to decrease with elevation on mountains in regions receiving high rainfall at low elevation and to increase and then decrease with elevation in more arid regions (Brown, 1919; Grubb *et al.*, 1963; Grubb, 1977; Buckley, Corbett & Grubb, 1980; Sugden, 1982; Tanner & Kapos, 1982; Givnish, 1984). Finally, even when growing under similar conditions, juvenile trees often bear broader leaves than do mature individuals of the same species (Büsgen & Münch, 1929; Richards, 1952).

(2) Leaf thickness tends to increase with decreasing rainfall, humidity and/or soil fertility, and to increase with increasing irradiance and/or leaf lifetime (Schimper, 1898; Shields, 1950; Beadle, 1966; Björkman *et al.*, 1972; Grubb, 1977; Sobrado & Medina, 1980; Medina, 1984). Leaf thickness tends to increase with elevation on mountains receiving high rainfall (Grubb, 1977; Buckley *et al.*, 1980; Tanner & Kapos, 1982).

(3) Leaf absorptance in the visible spectrum (400 to 700 nm) tends to decrease – that is, leaves tend to be more highly reflective or glaucous – in sites that are sunnier, more arid or less fertile (Ehleringer & Mooney, 1978; Sobrado & Medina, 1980; Ehleringer, 1981; Ehleringer *et al.*, 1981; Givnish, 1984; Ehleringer & Werk, 1986; Nobel, 1986).

(4) Leaf inclination from the horizontal tends to be greater in sunnier, more arid, or less fertile sites (Büsgen & Münch, 1929; Shields, 1950; Ehleringer & Forseth, 1980; Sobrado & Medina, 1980; Medina, 1984; Ehleringer & Werk, 1986).

(5) Amphistomatous leaves (i.e. those with stomata on both surfaces) are frequent in sunny and/or dry sites, whereas hypostomatous leaves (i.e. those with stomata only on the lower surface) predominate elsewhere (Wood, 1934; Parkhurst, 1978).

(6) Stomatal conductance generally increases with increasing humidity, soil moisture supply and mesophyll photosynthetic capacity (Cowan, 1977; Cowan & Farquhar, 1977; Schulze & Hall, 1982; Field & Mooney, 1986), with the latter being conditioned by irradiance, leaf nitrogen content, leaf water potential and leaf

age. Stomatal conductance usually increases with irradiance until heat load and/or leaf water potential become limiting (Cowan, 1982, 1986).

(7) Mesophyll photosynthetic capacity (i.e. maximum leaf photosynthetic rate at a given concentration of carbon dioxide in the mesophyll) tends to increase with increasing supplies of light, water and/or nutrients (Mooney & Gulmon, 1979; Björkman, 1981; Mooney et al., 1981; Field & Mooney, 1986). Photosynthetic capacity per unit leaf mass tends to be greater in deciduous leaves than in evergreen leaves among plants growing in the same area (Chabot & Hicks, 1982).

(8, 9) Leaves tend to have a higher protein chlorophyll ratio and a lower chlorophyll a/chlorophyll b ratio in more sunlit environments (Björkman, 1968, 1981).

(10) Plants with C_4 or CAM photosynthesis, though less dominant than C_3-plants in most terrestrial habitats, become relatively more common in drier and/or hotter areas or seasons of growth, and on saline soils (Osmond, 1978; Pearcy & Ehleringer, 1984).

(11) Evergreen leaves are common in habitats with nutrient-poor soils and/or little seasonal variation in the favourability of conditions for photosynthesis (i.e. in aseasonal or winter-rainfall climates) (Givnish, 1984). Plants with deciduous leaves predominate elsewhere, principally in deserts and semi-deserts, seasonal tropical forests, upper storeys of rain forests and temperate forests of eastern Asia, eastern North America and northern Europe (Beard, 1944, 1955; Richards, 1952; Mooney & Dunn, 1970; Walter, 1973; Chabot & Hicks, 1982).

(12) Leaves with non-entire margins (i.e. toothed or lobed leaves) are most common in dicots of the north temperate zone and forest understoreys everywhere (Bailey & Sinnott, 1916; Wolfe, 1978; Givnish, 1979).

(13) Lobed leaves are common only in north temperate trees and in tropical trees of early succession (Givnish, 1978a).

(14) Leaves with long, acuminate drip tips are common in wet rain forests and cloud forests, particularly among understorey species (Richards, 1952; Grubb, 1977; Dean & Smith, 1979; Williamson, 1981).

(15) Leaves with cordate (i.e. heart-shaped) bases are common among vines, forest herbs and aquatic herbs (Givnish & Vermeij, 1976).

(16) Trees with compound leaves are most common in arid and semi-arid habitats that favour the deciduous habit, at low elevations and in gap-phase succession (Givnish, 1978b, 1979, 1984; Stowe & Brown, 1981).

(17) Leaves tend to be borne in a spiral phyllotaxis on erect twigs in sunny environments and in a distichous phyllotaxis on horizontal twigs in the shade (Leigh, 1972).

(18 to 21) Reddish leaf undersides, lens-shaped epidermal cells and blue irridescence are often associated with the extreme shade of rain forest understoreys (Haberlandt, 1912; Lee & Lowry, 1975; Lee, 1977, 1986; Lee, Lowry & Stone, 1979; Bone, Lee & Norman, 1984). Such conditions are also associated with some rare instances (notably among the Begoniaceae and Gesneriaceae) of stomata arranged in clusters rather than singly (Skog, 1976; Givnish, 1984; S. Hoover, pers. comm.).

(22, 23) Asymmetric leaf bases and anisophylly (i.e. unequal leaves at each node in species with opposite leaves) are also common in rain forest understoreys and other shady habitats (Givnish, 1984).

Although these trends seem quite general, most have been established only qualitatively. Rarely have investigators quantified variation in both a leaf trait and correlated environmental parameters to check whether the same relationship between leaf form and environment holds in different regions. Notable exceptions include the demonstration by Ehleringer *et al.* (1981) of identical relationships between leaf absorptance and rainfall among *Encelia* (Asteraceae) native to North and South America (Fig. 1); and the analysis by Givnish (1984) showing a convergent relationship between average leaf width and rainfall among tree species in lowland tropical America and Australia (Fig. 2).

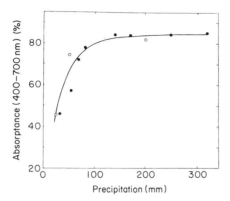

Fig. 1. Leaf absorptance in *Encelia* species native to arid and semi-arid regions of North and South America, as a function of annual rainfall (after Ehleringer *et al.*, 1978). *Encelia* species: ●, California; ○, Chile.

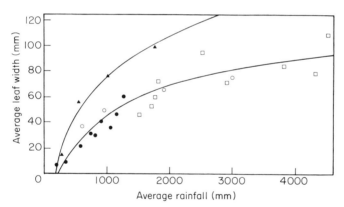

Fig. 2. Average leaf width at low elevations as a function of annual rainfall in tropical regions of Australia, Central America and South America (after Givnish, 1984). Curves represent the relationships $y = 32 \cdot 7 \ln (x/244 \cdot 5)$ for sites in southern South America, Central America and Australia ($r^2 = 0 \cdot 82$, $P < 0 \cdot 001$ for 20 df), and $y = 45 \cdot 0 \ln (x/188 \cdot 6)$ for sites in northern South America with a less marked dry season ($r^2 = 0 \cdot 98$, $P > 0 \cdot 01$ for 2 df). The less seasonal sites in northern South America support broader leaves at a given rainfall. ▲, Northern South America; ●, southern South America; ○, Australia; □, Central America.

Furthermore, although each of the trends mentioned appears to apply to species in many different families and genera, only rarely (e.g. Bailey & Sinnott, 1916) has an attempt been made to analyze explicitly the importance of phylogeny. Some leaf traits – such as the anatomy and fine venation of marginal teeth (Hickey & Wolfe, 1975) – do appear to be nothing more than phylogenetic idiosyncrasies, characterizing particular families or orders, and showing little if any pattern in the ecological distribution of species bearing them.

Finally, although trends in the average expression of a given trait (e.g. leaf size) across habitats may be well-marked, there is often considerable divergence in the expression of that trait within a habitat (see, for example, the data of Brown, 1919; Bews, 1925; Grubb et al., 1963; Howard, 1969; Sarmiento, 1972). Such within-habitat variation could elicit one of two views:

(1) that it reflects adaptation to variations in ecological conditions within a habitat and so provides excellent material for further tests of any hypothetical advantage of variation in the trait, or

(2) that such variation belies the alleged importance of variation in the trait and/or the importance of the trade-offs alleged to underlie patterns in its expression across habitats.

Unfortunately, in very few cases have there been attempts to analyze whether within-habitat variation in a trait can be dissected into regular patterns that reflect the trends seen across habitats and that potentially can be understood in terms of the same principles thought to underlie those trends. Almost all of these attempts address the variation seen in the leaves of species found in different vertical layers within a forest or in leaves found at different positions within the crown of a given individual (e.g. for trends in leaf size, thickness, lobation and complexity, see Brown, 1919; Büsgen & Münch, 1929; Cain et al., 1956; Loveless & Asprey, 1957; Talbert & Holch, 1957; Asprey & Loveless, 1958; Chiariello, 1984). Only three attempts have been made to try to account for the considerable variation in leaf form and/or arrangement among species coexisting within a layer; these focus on the incidence of mono- vs multilayered crowns (Horn, 1971), simple vs compound leaves (Givnish, 1978a, 1984) and various anti-herbivore defences (Coley, 1983) among species that differ in successional status. The significance of within-layer variation in many other aspects of leaf form, physiology and arrangement remains a major ecological issue.

Underlying trade-offs

In considering the net contribution of traits of a leaf or canopy to whole-plant carbon gain, three basic kinds of energetic trade-offs are likely to arise and influence the evolution of that trait. These involve the economics of gas exchange, the economics of support and the economics of biotic interactions.

(1) The economics of gas exchange (Givnish, 1986b) arise from the ineluctable association of carbon gain with water loss: any passive structure that permits the passage of large, slow-moving CO_2 molecules will also allow the diffusion of smaller, faster molecules of water vapour. Thus, the photosynthetic benefit of any trait that increases the rate at which CO_2 can diffuse into a leaf must be weighed against the energetic costs associated with increased water loss. Such transpirational costs might include a reduction in mesophyll photosynthetic capacity caused by decreased leaf water potential, an increased allocation of energy to unproductive roots and xylem, and/or a shortened period of photosynthetic activity (Givnish & Vermeij, 1976; Orians & Solbrig, 1977; Givnish, 1979, 1984, 1986c). A complementary trade-off results from the inevitable conflict between leaf photosynthetic capacity and the energetic costs of constructing and maintaining tissue capable of high photosynthetic rates (Mooney & Gulmon, 1979; Gulmon & Chu, 1981). Highly productive leaves require large inputs of nitrogen, phosphorus and other mineral nutrients to create the pools of enzymes and pigments needed to sustain high rates of CO_2 uptake (Field & Mooney, 1986).

Generally speaking, variations in a given trait that increase the photosynthetic capacity of leaf or canopy result in photosynthetic benefits that rise, then plateau, as the costs of transpiration and/or nutrient capture associated with that trait increase, because ultimately other factors or traits limit photosynthesis. As a result, selection should favour an intermediate level of expression of the trait which maximizes the difference between its photosynthetic benefits and related costs of water and/or nutrient uptake. Furthermore, as the energetic costs of absorbing a given amount of water or nutrients increase as sites become drier or less fertile, the optimum expression of the trait should shift to outcomes that result in lower photosynthetic capacity and lower associated costs under fixed conditions (e.g. lower stomatal conductance, absorptance, or leaf nitrogen content). Trade-offs in the economics of gas exchange have been implicated in the evolution of a variety of traits that influence both photosynthesis and transpiration, including effective leaf size (Givnish & Vermeij, 1976; Givnish, 1984), stomatal conductance (Cowan, 1977, 1986; Cowan & Farquhar, 1977; Givnish, 1986c), leaf absorptance (Ehleringer & Mooney, 1978; Ehleringer & Werk, 1986), leaf orientation (Ehleringer & Forseth, 1980; Ehleringer & Werk, 1986; Nobel, 1986), leaf nitrogen content and mesophyll photosynthetic capacity (Mooney & Gulmon, 1979; Field, 1981; Gulmon & Chu, 1981), leaf protein/chlorophyll ratio (Björkman *et al.*, 1972; Björkman, 1981; Cowan, 1986), chlorophyll *a*/chlorophyll *b* ratio (Björkman *et al.*, 1972; Björkman, 1981) and internal leaf architecture (Parkhurst, 1986).

(2) The economics of support arise because, among the leaf and crown forms that have equivalent effects on photosynthesis and transpiration, many differ in the efficiency with which the leaves can be mechanically supported (Givnish, 1986d). Such differences imply trade-offs between photosynthetic benefits and mechanical costs. Such trade-offs, in turn, have been implicated in the evolution of several aspects of leaf shape and arrangement, including parallel *vs* pinnate venation (Givnish, 1979), leaf apical and basal angles (Givnish, 1979, 1984), tree branching angles (Borchert & Tomlinson, 1984), leaf lobation (Givnish, 1978b), leaf dentition (Givnish, 1978a, 1979), compound *vs* simple leaves (Givnish, 1979, 1984) and asymmetric leaf bases, anisophylly, and opposite *vs* alternate leaf arrangement (Givnish, 1984). Both the economics of support and of gas exchange appear to be involved in the evolution of spiral *vs* distichous phyllotaxes (Givnish, 1984).

(3) The economics of biotic interactions arise because many traits that enhance a plant's potential rate of growth – such as high leaf nitrogen content, heavy allocation to foliage, low allocation to defensive compounds or mutualists, or an erect growth habit – may also increase its potential attractiveness to herbivores, implying a trade-off between photosynthetic benefits and biotic costs (Janzen, 1974; Chew & Rodman, 1979; Mooney & Gulmon, 1982; Givnish, 1986e). Such trade-offs, together with those associated with gas exchange and support, may underlie visual mimicry or divergence in leaf form (Gilbert, 1975; Barlow & Wiens, 1977; Rausher, 1978, 1980; Givnish, 1984; Ehleringer *et al.*, 1986), allocation to defensive compounds (Janzen, 1974; McKey *et al.*, 1978; Chew & Rodman, 1979; Mooney & Gulmon, 1982) and leaf flushing (McKey, 1979; Coley, 1983), and influence such traits as leaf toughness and pubescence (Coley, 1983) and leaf nitrogen content (Lincoln *et al.*, 1982). Analysis of the costs and benefits associated with biotic interactions is complicated greatly by the complexity of potential ecological effects of interactions among herbivores, the predators of herbivores and plant competitors and mutualists (see Givnish, 1986e). For example, having small amounts of defensive compounds, high palatability and low

stature may decrease a plant's potential rate of energy capture but may yield a relative advantage if it subsidizes grazers that damage its competitors far more (e.g. see Steneck, 1982). In addition, Price *et al.* (1980) have shown that the benefit of allocations to anti-herbivore defences may depend as much on the presence of the herbivores' predators or parasitoids as on the reduction of plant palatability or digestability. For example, in soybean (*Glycine max*), the so-called 'resistant' morphs suffer far greater damage from Mexican bean beetles than more easily digested forms when grown in the absence of the pentastomid bug that attacks the beetle. Under these conditions, beetle larvae simply have to eat more leaf tissue of the 'resistant' morph in order to mature, although they take much longer to do so than on other morphs. In the presence of its parasitoid, however, the slowing of the beetle's larval development by the resistant morph allows the parasitoid to reduce the beetle population and consequent damage to the plant.

Leaf size: tests and generation of hypotheses

Qualitative as well as quantitative cost–benefit analyses have been advanced for several leaf and canopy traits (e.g. see Horn, 1971; Givnish & Vermeij, 1976; Cowan & Farquhar, 1977; Orians & Solbrig, 1977; Ehleringer & Mooney, 1978; Honda & Fisher, 1978; Parkhurst, 1978; Mooney & Gulmon, 1979, 1982; King, 1981; Chabot & Hicks, 1982; Givnish, 1982, 1986c,f; Cowan, 1986). One of the best-studied of these is leaf size in terrestrial plants. As indicated previously, the characteristic width of leaves, lobes or leaflets tends to increase in moister environments (e.g. see Fig. 1). Paradoxically, leaf width tends to decline in less fertile sites, even when an abundant supply of moisture is available. Narrow leaves are found in rain forests on sterile podsols (Ashton, 1964; Webb, 1968; Brünig, 1970; Whitmore, 1984), in montane rain forests and cloud forests with highly leached soils (Grubb, 1977) and in bogs (Small, 1972; Walter, 1973).

Givnish & Vermeij (1976) and Givnish (1979, 1984, 1986c) provided an explanation for these general patterns and the paradoxical convergence of leaves adapted to dry conditions and nutrient poverty, based on an analysis of the costs and benefits associated with variations in leaf width. To isolate the effects of leaf width *per se*, comparisons were made of idealized plants that have the same total leaf mass, leaf thickness, stomatal conductance and photosynthetic biochemistry but differ in the size of the individual leaves into which the canopy is subdivided. Variations in leaf size then influence the whole-plant rate of growth mainly through its effects on the conductance of the leaf boundary layer, which in turn affects leaf heat exchange, uptake of carbon dioxide and loss of water vapour.

The thickness of the boundary layer, the film of stagnant air surrounding the leaf, increases with the width of a leaf, or its lobes or leaflets, and decreases with increasing wind speed (Gates, 1965, 1980). Thicker boundary layers tend to impede heat exchange between a leaf and the surrounding air and to slow the diffusion of carbon dioxide and water vapour in and out of the leaf. The partial blocking of heat transfer influences leaf temperature, however, and has additional indirect effects on heat exchange, photosynthesis and transpiration by influencing the magnitude of leaf-atmosphere differences in heat, CO_2 and water vapour.

The net result of these effects in leaves of moderate stomatal conductance in a sunny environment is that broader leaves tend to achieve higher temperatures and rates of transpiration, with the extent of the increases dependent on both environmental factors (e.g. wind speed, relative humidity) and leaf factors (e.g.

stomatal conductance, leaf orientation) (Gates & Papian, 1971; Geller & Smith, 1980; Chiariello, 1984).

Increased temperature may increase photosynthesis, although the energetic benefits obtained thereby should plateau and then decline at greater leaf sizes and temperatures, as boundary layer conductance becomes limiting relative to stomatal or mesophyll conductance, and/or as the thermal photosynthetic optimum is exceeded (Givnish & Vermeij, 1976). Increases in transpiration, on the other hand, incurs energetic costs that result either from the consequent decline, at a given allocation ratio to leaf *vs* root tissue, in leaf water potential, and hence, mesophyll photosynthetic capacity; and/or an increased allocation to unproductive roots (Givnish, 1984, 1986c). The optimal leaf width maximizes the difference between these benefits and costs. Optimal leaf width should decrease in less humid or drier habitats, as the rate of transpiration or the root cost of supplying a given rate of transpiration increases. Similarly, smaller leaves should also be favoured in less fertile sites, because scarce soil nutrients are more costly to absorb and thus limit the level of photosynthetic enzymes in leaves and the extent to which increases in temperature (or other limiting factors) can enhance photosynthesis.

This theory explains many of the most obvious ecological trends in leaf size that emerge from comparative studies, at least in qualitative terms, but so might some other theory. An important step in the evaluation of any optimality model is to consider alternative models and see if their predictions differ from each other or from the patterns actually observed, as was done for leaf size by Givnish & Vermeij (1976). The next obvious step – in this case as well as other applications of optimality analysis – is to make the theory quantitative and to check whether there is a quantitative agreement with the predictions of the model. Indeed, a crude but quantitative model for optimal leaf size and stomatal conductance successfully predicts the mean values of these traits actually observed by Taylor (1972) under 'typical' summer conditions at the Michigan Biological Station (Givnish, 1986c). An even more powerful test would be to refine this analysis and repeat it for a gradient of sites to see whether observed and predicted leaf widths continue to concur. Finally, the most powerful test of an optimality model for any trait would be to conduct a study of differences in resource allocation, carbon balance and competitive ability associated with differences among individuals in a population in the expression of that trait. Such detailed physiological studies, however, must always be complemented by broad comparative studies because (1) the immense effort they require means that few such intensive studies will ever be conducted and (2) intensive studies on single populations are necessarily subject to being influenced as strongly by aspects of genetic variation, linkage, epistasis and pleiotropy peculiar to that population as by the selective pressures likely to be associated with variation in the trait in question across populations generally.

What factors determine optimal leaf width in underwater plants? In such plants, transpirational costs and thermal effects are likely to be negligible. However, boundary layer thickness and conductance are likely to be of fundamental direct importance, because carbon dioxide diffuses some 10^4 times more slowly in water than in air. Thus, selection should favour effectively narrow leaves in submersed species. Indeed, submersed freshwater angiosperms and pteridophytes have achieved effectively narrow lamina in a large number of ways, including narrow awl- and ribbon-shaped leaves, pinnately and palmately divided leaves and (in *Aponogeton*) fenestrate leaves with numerous 'windows' perforating an otherwise broad lamina (Sculthorpe, 1967). Furthermore, amphibious species that produce both aerial (either floating or emergent) leaves and submersed leaves generally produce much narrower lamina below water (Sculthorpe, 1967).

Traditional explanations for effectively narrow lamina in submersed leaves are, however, not based on the direct impact of boundary layer resistance on CO_2 uptake but rather on two alternative hypotheses. The 'drag' hypothesis suggests that narrow underwater leaves reduce drag in an incompressible medium and the likelihood of being torn or swept away. The 'surface area' hypothesis is that narrowly divided leaves have much higher surface area/volume ratios than do broad leaves and thereby counteract some of the diffusive limitations on photosynthesis underwater.

One way of testing the 'boundary layer' hypothesis against these alternative explanations for narrow leaves underwater is to test its corollaries. These include the following.

(1) Leaves should be narrower in still *vs* flowing water (precisely opposite to the prediction that emerges from the drag hypothesis).

(2) Leaves should be narrower in more fertile waters, reflecting the greater potential limitation that boundary layer conductance might place on photosynthesis when mesophyll photosynthetic capacity is high (the drag hypothesis would predict no change in leaf width with fertility *per se*).

(3) Leaves should be narrower in warmer waters, again reflecting the greater potential limitation by boundary layer conductance when internal photosynthetic capacity is high and external concentrations of dissolved CO_2 are low (the drag hypothesis would predict an increase in leaf width, because warm water is less viscous and exerts less drag). The 'surface area' hypothesis would be negated by any finding that narrow leaves do not have appreciably higher surface/volume ratios than broad leaves of comparable thickness.

All three of these predictions are supported by a superb set of comparative data compiled by Madsen (1986). He compared the proportions of submersed species with entire *vs* dissected leaves in over 150 local floras. Entire leaves, such as those of *Potamogeton*, have characteristic widths of the order of several millimetres, whereas dissected leaves, such as those of *Myriophyllum* or *Cabomba*, have characteristic widths of the order of tenths of millimetres. Madsen (1986) found that lakes had a significantly higher incidence of submersed species with non-entire, extremely narrow leaves than did streams (Fig. 3). Furthermore, the incidence of extremely narrow, dissected leaves increases from a low proportion in nutrient-poor, oligotrophic lakes, to a higher one in richer mesotrophic lakes, to

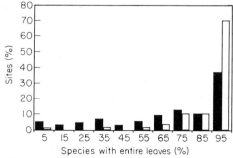

Fig. 3. Distribution of proportions of submersed species with entire *vs* dissected leaves in lake and stream floras (after Madsen, 1986). ■, All lakes; □, all streams.

the highest in mineral-rich eutrophic lakes (Fig. 4). Finally, Madsen (1986) also found that species active in warmer water – either as a result of latitudinal or seasonal differences between species – had narrower leaves, as expected. These results seem clearly to support the 'boundary layer' theory and negate the 'drag' hypothesis, at least as a means of explaining differences in leaf width between

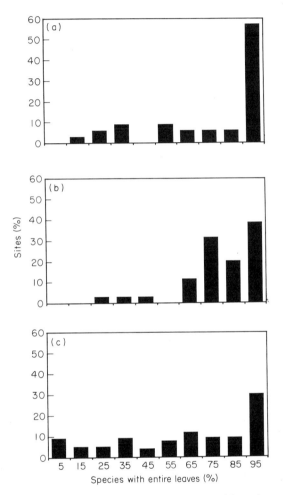

Fig. 4. Distribution of proportions of submersed species with entire *vs* dissected leaves in floras of oligotrophic, mesotrophic and eutrophic lakes (after Madsen, 1986). (a) Oligotrophic; (b) mesotrophic; (c) eutrophic.

species. However, considerations of drag and laminar mechanical strength are undoubtedly involved in the determination of the minimum leaf width attained by submersed species. The unalloyed 'surface area' theory is also unlikely to be correct: even if leaves or leaf divisions are only as wide as they are thick, their surface/volume ratio can at most be twice that of an infinitely extended sheet of the same thickness.

It is interesting to note that seaweeds – particularly kelps (Phaeophyta) but also several green and red macroalgae – often have broad fronds and inhabit cold water with relatively high concentrations of CO_2 and bicarbonate. Under these conditions, and given that low temperatures are also likely to reduce photosynthetic capacity, large fronds may be less limiting to photosynthesis than under other conditions. Many macroalgae with broad fronds, however, have various protuberances, ruffles and ridges on their lamina when found in relatively still water but are relatively smooth in areas exposed to wave surge or tidal currents (Koehl, 1982, 1986). Koehl (1986) reports that protuberances increase turbulence and thus decrease boundary layer resistance and increase photosynthesis, albeit at the expense of greater mechanical drag. Thus, where currents are weak, ruffles should be favoured because the photosynthetic benefits of decreased boundary layer resistance are likely to outweigh any costs associated with increased drag (e.g. damage, need for greater reinforcement). Where currents are strong, smooth fronds should be favoured because the photosynthetic benefits of increased turbulence should be small and the costs of drag much higher. Thus, the presence of protuberances on fronds appears to be set by a balance of photosynthetic benefits and mechanical costs. The same seems likely to be true for the determination of optimal width of frond or lamina (see above), given the photosynthetic advantages of narrower lamina and the greater probability of breakage and/or the greater proportional allocation to unproductive structural material needed to prevent breakage in narrower lamina.

SELECTION *vs* PHYLOGENETIC CONSTRAINTS IN THE EVOLUTION OF LEAF FORM AND PHENOLOGY IN FOREST HERBS

Introduction

Herbaceous plants growing in temperate deciduous forests show a remarkable diversity in leaf form, foliage height and plant architecture. These forest herbs also show a striking range of seasonal patterns of leaf deployment and belong to a large number of plant families and genera. These features afford outstanding opportunities for studies of the significance of within-layer variation in leaf form and of the relative importance of selective pressures and phylogenetic constraints in setting patterns of species and communities.

To address these and related issues, John Terborgh, Donald Waller and I initiated a comparative study of leaf adaptations, leaf phenology, species distribution, temporal community structure and species richness in herbs distributed along a topographic gradient from oak woods, through mesic forest to flood plain forest in the Virginia Piedmont. Here, I recount aspects of this study (to be published in full detail elsewhere) that exemplify important approaches to within-layer divergence in plant form and the relative roles of selective pressures and phylogenetic constraints in determining species and community patterns. Botanical nomenclature follows Fernald (1950).

Study area

The transect studied runs perpendicular to the Potomac River near McLean, Virginia, following a ridge down the north slope of a dissected palisade, underlain mainly by mica schist. The mature forest is some 30 to 40 m in height, has not been substantially disturbed by man since the Civil War and has been described as part of the most diverse stand now found in the Piedmont (Radford, 1976). The transect drops 55 m in roughly 400 m from the uplands to the Potomac flood plain and entails gradients in soils, available moisture, canopy composition and irradiance which are, in many ways, a microcosm of the typical

Piedmont ridge-slope-valley bottom landscape (Givnish, Terborgh & Waller, in preparation).

Soils are thin, acid and clayey in the oak woods on ridges; deeper, less acid and loamy in the mesic forests on the slopes and deepest, silty-sandy, and circum-neutral in the floodplain. Paralleling these trends in soil depth and texture, summer moisture availability and the concentration of many nutrients increase down the gradient. Irradiance on the forest floor varies seasonally, reaching 40 to 60% of full sunlight in winter and 1 to 5% of full sunlight after the canopy closes in the upland woods. However, irradiance in summer reaches 10 to 15% full sunlight under the relatively open flood plain canopy. The deciduous canopy closes in mid-May and re-opens in late October. Winter daytime temperatures drop below freezing only occasionally and little of the roughly 1000 mm of annual precipitation falls as snow.

Definition of temporal guilds

At each of nine stations along this gradient, we established ten 1 m² marked plots and tracked coverage by each of some 74 herbaceous species over 13 sampling intervals during a yearly cycle. These data allowed us to classify the herbs into temporal photosynthetic guilds, based on seasonal pattern of leaf deployment. The principal guilds, which include three or more species, are:

winter annuals – such as *Veronica hederaefolia* – that germinate in late fall and persist almost to canopy closure;

spring ephemerals – such as *Dentaria laciniata* – that leaf out in early spring and senesce before canopy closure;

early summer species – such as *Arisaemia triphyllum* – that emerge in spring and persist through canopy closure for varying lengths of time but reach peak coverage before midsummer;

late summer species – such as *Laportea canadensis* – that emerge in mid to late spring and reach peak coverage after midsummer;

wintergreen species – such as *Tipularia discolor* – that emerge in early to mid autumn and persist through winter, senescing before canopy closure;

evergreen species – such as *Hepatica acutiloba* – that maintain active leaves throughout the year; and

dimorphic species – such as *Dentaria heterophylla* – that hold two different sets of leaves or leaf positions at two different seasons.

Convergence within temporal guilds

Do members of the same guild share characteristic adaptations to conditions during their season of photosynthetic activity? Several leaf traits show convergence within each guild. These include effective leaf size and thickness; chlorophyll/N ratio and leaf N content; stomatal distribution and conductance; photosynthetic light response; foliage height; leaf arrangement; leaf maculation and the number of seasonal leaf flushes. A few of these patterns are discussed in detail below.

Leaves tend to be relatively small and thick in spring ephemerals, which are exposed to high irradiances and low soil temperatures in early spring. Early summer species have broader and thinner leaves and experience both high and low irradiances. Late summer species have the broadest and thinnest leaves and experience mainly low irradiances (Fig. 5). Winter annuals, wintergreen, and evergreen species are exposed to high irradiances and display a range of leaf sizes and thickness within that found in spring ephemerals. The trend toward thinner

T. J. GIVNISH

and broader leaves in passing from spring ephemerals through early summer species
to late summer species parallels the general trend for shade-adapted leaves to be
thinner and broader than leaves adapted to greater irradiances; Givnish & Vermeij
(1976) outline the energetic trade-offs favouring these trends.

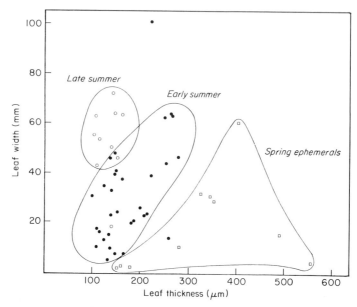

Fig. 5. Effective leaf width and leaf thickness for all non-graminoid members of the spring
ephemeral (□), early summer (●) and late summer (○) guilds, McLean, Virginia. Each point
represents the mean of 25 measurements on fully expanded leaves for each trait.

Figure 6 shows the fraction of coverage within a guild by species having various
leaf arrangements. Winter annuals are sprawling mats with small leaves; spring
ephemerals have basal leaves or leaves arranged on short, umbrella-like structures
(i.e. peltate or deeply cordate leaves, leaf whorls, or ternately compound leaves
on erect petioles or stems); early summer species have leaves arranged on tall
umbrella-like structures, or long arching stems; late summer species scatter their
leaves along an erect stem; and evergreen species mainly have basal leaves. The
warm-season leaves of dimorphic species are arranged in ways similar to those of
spring or early summer species, whereas cool-season leaves are arranged like those
of winter annuals or evergreen species.

Plants in the same guild often also show similarities in leaf height. Evergreen
species are generally 2 to 10 cm tall; spring ephemerals, 5 to 15 cm tall; early summer
species, 10 to 65 cm tall; and late summer species, 40 to 160 cm tall. Average plant
height increases down the gradient from about 15 cm in the oak woods to over a
metre in the flood plain.

Even leaf shape shows some interesting patterns across different guilds. For
example, among late summer species there is a striking convergence on thin,
ovate-lanceolate leaves with a dentate margin, involving species of *Aster*, *Eupatorium* and *Solidago* in the Asteraceae; *Urtica*, *Laportea* and *Pilea* in the Urticaceae;
Impatiens in the Balsaminaceae; and *Circaea* in the Onagraceae (Fig. 7). Even more
surprising than the convergence on a dentate margin, which seems explicable given

Fig. 6. Fraction of total coverage within each major guild by species having various growth forms, indicated by the relative height of bars under the figures illustrating each leaf arrangement. Growth forms include, from left to right: sprawling mats, basal leaves, short umbrellas, tall umbrellas, arching stems, erect stems, vines and graminoids. Growth forms of the cool-season (winter to early spring) and warm-season (late spring to summer) phases of dimorphic species are tabulated separately.

Fig. 7. Leaf silhouettes of late summer species with foliage scattered along erect stems. From left to right: *Laportea canadensis, Aster divaricatus, Urtica dioica, Circaea quadrisulcata, Pilea pumila, Eupatorium rugosum, Impatiens pallida* and *Solidago caesia*.

the analysis by Givnish (1978b) and the fact that late summer species have the thinnest leaves of any guild, is the remarkable similarity their leaves bear to those of many trees and shrubs. Possible reasons for this puzzling similarity, and divergence from 'bizarre' leaf shapes seen in many other forest herbs, are discussed below.

Significance of trends in leaf height and growth form

Features of leaves. Because leaves of comparable gas exchange properties can be held at different heights, be arranged on cordate, peltate, ternately compound, arching or erect surfaces and display different shapes, we can understand the significance of such growth forms only by considering other factors, such as the mechanical costs and competitive benefits associated with each growth form.

Leaf height. For example, what is the significance of leaf height and why does it increase along the gradient? Competition for light is an important selective pressure on leaf height in understorey herbs that die back to the ground each year and two sets of trade-offs are involved (Givnish, 1982). First, the proportion of a herb's resources annually allocated to leaves rather than stems should decrease with leaf height, reflecting the disproportionate increase in support tissue needed to ensure mechanical stability. This tends to reduce the growth rate of a plant and, acting alone, would favour plants that arrange their leaves flat on the ground. However, competition for light favours increased leaf height which prevents overtopping.

If a plant is much above the leaf height of its competitors, it should photosynthesize at some maximum rate that depends, in a complicated way, on environmental conditions. If it is much below the height of its competitors, it will photosynthesize at some minimum rate, depending on these same conditions *and* (in a very simple way) on the density of competing foliage. For, if competing foliage is sparse, it is unlikely that a plant will be next to and thus under a competitor. The expected rate of photosynthesis per unit leaf biomass, averaged over many individuals of a given height, will thus depend only weakly on relative leaf height. However, if the density of herbaceous foliage is great, a plant will probably be next to and thus under a competitor, so that the expected rate of photosynthesis will increase sharply with relative leaf height (Givnish, 1982).

In this evolutionary game, plants should grow taller until the expected photosynthetic gain each makes by being slightly taller than an opponent is just balanced by the structural cost of a decreased proportion of energy allocated to leaves. So, where ecological conditions harshly limit total herbaceous cover, there is little photosynthetic advantage to an increment in leaf height and the structural cost of leaf height favours short plants. In areas where more resources favour dense herbaceous cover, an increment in height confers a larger advantage and favours taller plants at evolutionary equilibrium. Indeed, using data from the Virginia transect, Givnish (1982) demonstrated a tight relationship between leaf height and the average herbaceous cover in which species within each of six height classes (0 to 5, 5 to 10, 10 to 20, 20 to 40, 40 to 80 and 80 to 160 cm) find themselves during their period of photosynthetic activity.

To determine whether that pattern corresponds to the quantitative predictions of the leaf height model, we need to know how the proportion of leaf *vs* stem tissue varies with absolute leaf height and how the expected rate of photosynthesis varies with relative leaf height. Such a quantitative test (Givnish, 1982, 1986f) showed a satisfactory agreement between the observed and predicted relationships of leaf height to herbaceous cover. An important finding on which this analysis was based is that late summer species require less support tissue to support a given weight of leaves at a given height than do early summer species. This appears to result, at least in part, from the fact that leaves of late summer species are scattered over a broad vertical interval rather than in an umbrella-like arrangement and thus have narrow crowns and short, relatively inexpensive horizontal lever arms, though at the expense of greater self-shading. Perhaps not coincidentally, late summer species tend to reach peak coverage in habitats which, like the flood plain, are better lit in summer than are those occupied by most early summer species (Givnish, 1982).

Leaf temperature could also be an important selective force on leaf height in winter-active herbs. Winter photosynthesis in deciduous forest understoreys is likely to be limited not by light but by temperature and winter-active species should place their leaves in the warmest microclimate in the winter forest, the forest floor. Indeed, evergreens, wintergreens, winter annuals and in the winter phases of dimorphic species have their leaves at or near ground level, both on the Virginia transect and throughout northeastern North America (Givnish, 1986f).

If an evergreen has its leaves close to the ground which maximizes winter photosynthesis, it cannot compete successfully in dense summer herb communities. This could be one of several factors, in addition to advantages in nutrient conservation and differences in growth rate, that restrict evergreen species to dry, acid woods that support small amounts of herb cover (Givnish, 1982).

Growth forms. Similar biomechanical and thermoregulatory considerations can be used to analyze the significance of the growth forms characteristic of each major guild. Winter annuals arrange their leaves in sprawling mats, which (1) increases leaf temperature at a season when air temperature is low; (2) reduces the fractional allocation of energy to support *vs* photosynthetic tissue, thus increasing the rate at which the cost of leaf construction can be amortized during their short season of growth; and (3) allows an indeterminate expansion of the canopy in connection with the annual habit.

Spring ephemerals have basal leaves or leaves arranged on short umbrellas. These also increase leaf temperature and reduce fractional allocation to support tissue, permitting rapid amortization of the costs of construction during their brief period of activity. In addition, a determinate pattern of growth allows rapid expansion of the leaf canopy to harvest efficiently a short, predictable seasonal 'window' of light.

Late summer species are less able to capitalize on the spring 'window' of light than any other guild and depend heavily on capture of light after the canopy closes. Coverage by such herbs is largely confined to habitats that are relatively well lit in summer, such as the flood plain (Givnish, 1982). Persistently high irradiance, often in combination with much water and nutrients, favours dense herbaceous cover which, in turn, favours tall plants with narrow, mechanically efficient crowns and indeterminate growth which allows continued growth in height in the face of many resources. Indeterminate growth also confers an ability to expand the crown opportunistically in response to what are unpredictable, year-to-year shifts in midsummer availability of light.

One problem faced by plants with leaves scattered along a vertical, indeterminate axis is that their lower leaves become more deeply shaded as new leaves are added above them. To maintain an efficient multilayered crown, abscission of the lower portions of the plant's crown as they become unprofitable is necessary. Such abscission, in turn, favours a clear distinction between ephemeral leaves and a permanent support skeleton, so that a minimum amount of structural material is shed (Givnish, 1982, 1986f). It is perhaps for this reason that the leaf shapes of late summer herbs and certain woody plants are so similar. The diversity of peltate, cordate and other leaf shapes in forest herbs of determinate growth may result from lack of selection against 'webbing' of the lamina to the supporting stems or veins when that lamina is ordinarily not shed without its support.

Cool temperatures, high irradiances, and low amounts of herb cover favour basal leaves in evergreen species and winter phases of dimorphic species. The warm-season phases of dimorphic species often have leaf forms and arrangements convergent on those of other summer-active herbs. The potential conflict in selective pressures on leaf height caused by winter air temperatures and summer herb cover may restrict evergreens to the sterile oak woods while favouring dimorphic species with tall summer shoots in habitats that support more herbaceous cover in summer (Givnish, 1982).

Early summer species arrange their foliage in a variety of umbrella-like structures, including peltate and deeply cordate leaves, leaf whorls and ternately compound leaves. An umbrella-like leaf arrangement has the advantage of (1) minimizing self-shading and maximizing interception of light under densely shaded summer conditions; and (2) increasing the efficiency of mechanical support by favouring short, roughly equal, radial lever arms whose cost increases as a high power of their length. One means of testing this hypothesis is to examine the allometry of support tissue in taxa that diverge from an umbrella-like leaf arrangement. Work along these lines is currently underway on several herbaceous genera. For example, in *Podophyllum peltatum* Givnish (1986f) was able to quantify the additional vein mass associated with leaf asymmetry and use it in a quantitative model for optimal branching angle in sexual shoots, whose predictions coincide with the mean observed branching angle.

Selective pressures vs *phylogenetic constraints*

Three important conclusions can be drawn from this study. Firstly, essentially every leaf trait affecting gas exchange or mechanical efficiency shows convergence within each major guild, involving species in several different genera and families. This alone strongly suggests that these traits are adaptations to growth conditions prevailing at different seasons, and the conclusion is strengthened because the trends agree with predictions of cost–benefit models often derived independently of the Virginia study. Secondly, this study is one of the first to analyze and account for much of the divergence in leaf form and arrangement in species found within the same layer at a given site. It is therefore important in rebutting the oft-held view that within-habitat, within-layer variation in leaf form is essentially random and non-adaptive. Thirdly, this study permits us to distinguish the roles played by selective pressures and phylogenetic constraints in determining species patterns, with an interesting result. Remarkably, the congeners, or closest relatives, of each member of a given guild almost always have the same phenology as the guild members themselves and occur in habitats that appear to favour that phenology (Givnish, Terborgh & Waller, in preparation). Congeners of winter annuals, for example, are either desert winter annuals or ruderal weeds native to areas of Mediterranean climate and winter rainfall. Congeners of spring ephemerals are themselves forest ephemerals, alpine herbs or perennial spring ephemerals of chaparral areas. Late summer species are primarily related to meadow and prairie species, with long periods of photosynthetic activity and growth culminating toward the end of a long warm season; the late summer genera are also the only group on the Virginia transect with strong tropical affinities. Early summer species have other summer-active, shade-adapted forest herbs as their closest relatives, are often taxonomically isolated and seem to be the group most highly specialized to life in forest understoreys. Congeners of evergreen herbs are other evergreen herbs, often native to coniferous boreal forests. Finally, dimorphic species are interesting in that they belong to genera that show considerable phenological versatility, with different species displaying different phenologies. Dimorphic genera also show strong northern affinities. Both of these features may be related to an underlying genetic capacity to produce two different phenological sets of leaves in the same individual and to have at least one of these active during a relatively cool portion of the year.

These observations suggest that leaf phenology may be evolutionarily rather non-labile within genera and that phylogenetic constraints may determine which lineages are represented within each temporal guild. Among the lineages in each guild, however, there also appears to be a remarkable degree of convergence in several leaf traits of ecological importance. To understand the fabric of evolution, one must understand both its ecological warp (the effects of environmentally imposed selective pressures) and its phylogenetic weft (the result of constraints inherent to different lineages). In this case, phylogenetic constraints appear to be most important in determining which lineages evolve a particular adaptation, not in determining whether the adaptation will arise. Considering phylogenetic constraints may thus not so much undermine analyses of adaptation (e.g. see Gould & Lewontin, 1979; Harper, 1982) as enrich the range and kind of conclusions that can be drawn from such analyses.

Limits of models based on phylogenetic constraints

However, even though both selective pressures and phylogenetic constraints are important in determining the pattern of evolution, I do not believe that we will ever be able to develop truly predictive evolutionary theories based on phylogenetic constraints. In principle, two different kinds of evolutionary 'predictions' can be made:

(1) *horizontal* (or within time-horizon) predictions, which attempts to relate the extant range of variation in traits within a lineage (or lineages) to the distribution of species bearing them; and

(2) *vertical* (or between time-horizon) predictions, which attempt to foretell or explain temporal, evolutionary shifts within lineages over micro- or macro-evolutionary time.

Horizontal predictions are based on analyses of the selective pressures on the expression of a given trait in different environments and on the assumption that selection operates simultaneously on several populations with different genetic backgrounds and inherent phylogenetic constraints, thus eliciting 'favorable traits favorably controlled' (Givnish, 1986a). Such predictions are made quite frequently and form a central feature of evolutionary ecology. Phylogenetic constraints are implicitly incorporated in the best of these predictions because the range of adaptations considered is not the greatest conceivable but that actually observed within the lineages in question (Givnish, 1986a). Nevertheless, once this retrospective information has been included, the 'predictions' made are (or at least can be) deductive and non-circular (Horn, 1979).

However, vertical predictions must be based explicitly on both selective pressures and phylogenetic constraints. Certain kinds of phylogenetic constraints – specifically those involving linkage, pleiotropy, epistasis and genetic variation – are, by their very nature, knowable only in retrospect, after careful investigation of the group in question, because they essentially are the result of accidents of history. They reflect the selective pressures operating at various times in the past, the bottlenecks faced by ancestral populations and the nature of genetic variation, linkage, pleiotropy and epistasis at various times. Prospective phylogenetic constraints are largely unknowable, at least beyond micro-evolutionary time scales: it is impossible to predict what qualitatively new adaptations, linkage patterns or epistatic or pleiotropic interactions may arise over long periods of time. Because phylogenetic constraints, whether prospective or retrospective, cannot be predicted or deduced based on general principles, it should generally be impossible to erect truly predictive theories based on phylogenetic constraints. This is an important conclusion, given the extensive press such constraint-based analyses have received in recent years (e.g. see Gould, 1984, 1986).

To illustrate this point, consider the excellent recent study by Alberch & Gale (1985) on the evolutionary pattern of toe reduction in salamanders and frogs. They concluded that the reduction or loss of distal digits in salamanders, and of peripheral (mainly proximal) digits in frogs, reflects the constraints imposed by phylogenetic differences in the temporal sequence of digit development. In both groups, the digits reduced or lost are those which are last to develop; their reduction or loss may be least likely to perturb development. On one level, this analysis provides a cogent explanation for a difference in evolutionary trends between lineages but, at a deeper level, it is ultimately unsatisfying. Why do *salamanders* show a proximodistal sequence of digit development, whereas *frogs* show a centrifugal sequence? The same kind of conclusion applies to the forest herbs discussed in this section. It is possible to make horizontal 'predictions' and explain convergence in several leaf traits within guilds on a deductive basis. It is also possible to note that the species found in each guild are those in genera that possess the corresponding leaf phenology. However, it is not possible now, nor is it likely ever to be possible, to explain why a specific genus has a given phenology.

ADAPTIVE RADIATION IN LEAF SHAPE AMONG VIOLETS OF EASTERN NORTH AMERICA

Givnish & Vermeij (1976) and Givnish (1986f) present a predictive model for the evolution of cordate (i.e. heart-shaped) or peltate (i.e. centrally attached) leaves, based on minimizing the total mass of stem and veins in leaves of a given area, length and effective size. A leaf should be deeply cordate or peltate and attached near its centroid to the petiole whenever its lamina is essentially perpendicular to the petiole. The point of attachment should move further from the centroid, resulting in a more shallowly cordate leaf, as the angle between petiole and blade decreases. Finally, leaf bases should be blunt or convex when the leaf and petiole lie in the same plane.

Givnish (1986f) used this model to analyze the pattern of adaptive radiation in leaf shape in violets (*Viola*) of eastern North America, which I review here. The 47 *Viola* species named by Fernald (1950) can be subdivided into nine ecological groups, based on leaf shape, leaf arrangement and underground morphology. Each ecological group, in turn, shows a characteristic range of habitats and represents a phylogenetic group. Taxonomic affinities of the ecological groups, based on chromosome number, potential for hybridization and flower colour are shown in Figure 8.

The largest group of species is characterized by stemless plants with deeply cordate leaves and rhizomatous rootstocks. The 13 blue-flowered species of group I are all closely related and mainly inhabit rich deciduous forests and thickets; a few also occur in meadows, bogs or gravelly shores (Fernald, 1950). The shady, productive habitat of these species favours horizontal leaves atop tall, roughly vertical petioles, thus favouring the deeply cordate leaf bases seen in this group. Two other groups bear similar, deeply cordate leaves and occur mainly in

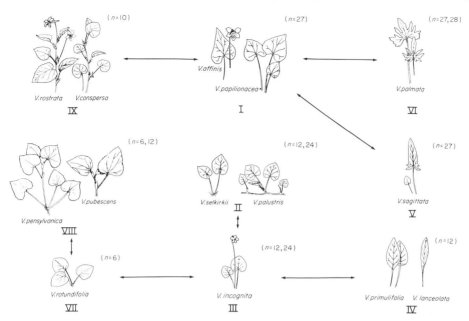

Fig. 8. Representative species of ecological groups of violets (*Viola* spp.) native to the northeastern United States, based on leaf form, leaf arrangement and underground morphology. Arrows indicate phylogenetic relationships inferred from chromosome number (in parentheses) and crossability (modified from Givnish, 1986f). Species in each groups are as follows: **I** (*V. affinis, cucullata, hirsutula, langloisii, latiuscula, missouriensis, nephrophylla, novae-angliae, papilionacea, pectinata, septentrionalis, sororia, villosa*); **II** (*V. selkirkii*); **III** (*V. blanda, incognita, pallens, palustris, renifolia*); **IV** (*V. lanceolata, primulifolia*); **V** (*V. emarginata, fimbriatula, sagittata*); **VI** (*V. brittoniana, chalcosperma, egglestoni, esculenta, lovelliana, palmata, pedata, pedatifida, septemloba, stoneana, triloba, viarum*); **VII** (*V. rotundifolia*); **VIII** (*V. canadensis, hastata, pensylvanica, pubescens, tripartita*); **IX** (*V. adunca, conspersa, rostrata, rugulosa*).

woodland habitats but differ in having stoloniferous rootstocks and either white (group II) or blue (group III) flowers.

At the opposite extreme in leaf shape, the two stoloniferous species (*V. lanceolata, V. primulifolia*) of group IV bear long, lanceolate leaves with blunt or convex leaf bases. Both occur in brightly lit, rather open habitats (such as damp meadows, bogs, open shores and thin woods) with their leaves held vertically and in the same plane as their petioles. Comparative photosynthetic studies by Curtis (1984) have shown that *Viola* spp. characteristic of open and wooded habitats have roughly the same, shade-adapted photosynthetic responses to light. Species growing in open habitats thus suffer no disadvantage by holding their leaves vertically and may thereby reduce transpirational costs and the potential for photobleaching and photoinhibition. Thus, *Viola* spp. of open habitats are expected to have vertical leaves, lack cordate bases and have a peripheral point of attachment, as observed.

Group V consists of stemless species with sagittate leaves that are essentially intermediate in form between the deeply cordate leaves of groups I to III and the lanceolate leaves of group IV. Not surprisingly, these occur mainly in habitats of intermediate openness, such as meadows, thickets, wood edges and open woods, which may favour slightly inclined leaves and a petiolar attachment near the leaf base. Group VI, another blue-flowered group of stemless rhizomatous violets apparently closely related to the sagittate- and cordate-leaved groups I and V, is characterized by deeply divided, palmatifid or pedatifid leaves. This group typically occupies nutrient-poor sites, which favour effectively narrow leaves (Givnish, 1979, 1984), such as peaty or sandy soils of the Coastal Plain, pine woods, prairies and calcareous soils (Fernald, 1950). The bizarre leaf form found in this group may partly reflect developmental constraints; the lobes form around a venation system typical of cordate-leaved *Viola* species.

Group VII consists of a single species (*V. rotundifolia*) with orbicular leaves (Fig. 8). This yellow-flowered species occurs in rich deciduous woods, often in microsites with sparse coverage by other herbs, as on shallow soil or moss-covered rocks. As expected, its horizontal leaves are at ground level. As a result, its horizontal petioles and leaves are coplanar and its leaves are only shallowly notched at the base.

Finally, groups VIII and IX are composed of stemmed violets that bear cauline leaves on an above-ground stem. Some species also bear basal leaves, like those of the stemless violets, that spring directly from the underground rootstock. Yellow- or white-flowered species in group VIII include *V. canadensis*, *V. hastata*, *V. pensylvanica* and *V. pubescens*. *V. tripartita* sometimes bears divided leaves. Typical blue-flowered species in group IX include *V. rostrata* and *V. striata*. All species in these groups are native to rich deciduous forests. As expected from the low irradiance in such forests, both basal and cauline summer-active leaves in groups VIII and IX tend to be held roughly horizontally. Basal leaves are invariably cordate (e.g. as in *V. pensylvanica* and *V. pubescens*), as expected from the horizontal posture of the leaves and the erect posture of their petioles. Cauline leaves, however, are borne on horizontal rather than vertical petioles in several species and should have blunter leaf bases for mechanical reasons, as well as to avoid self-shading (Givnish, 1979, 1984). Indeed, several stemmed species (e.g. *V. adunca*, *V. conspersa*, *V. hastata*, *V. pensylvanica*, *V. pubescens*, *V. rostrata*, *V. striata*) bear rhombic or hastate leaves with blunt or only shallowly notched leaf bases (Fernald, 1950; Ricketts, 1966).

Three conclusions can be drawn from this study. Firstly, the leaf shapes seen in each ecological group follow the trends expected on functional grounds: species of rich woodland habitats have deeply cordate, horizontal leaves on erect petioles; species of more open habitats have erect leaves with sagittate or completely blunt bases. Species of densely shaded habitats with low herb coverage have orbicular leaves on prostrate petioles; blunt leaf bases are also seen in cauline leaves. Violets of nutrient-poor sites have narrowly divided leaves.

Secondly, these trends in leaf shape are seen among both blue-flowered and yellow- or white- flowered violets, characterized by chromosomal base numbers of $n = 10$ and 27, and $n = 6$, respectively. Convergent patterns appear to occur in other groups of violets. For example, among yellow-flowered violets ($n = 12$, 24 or 48) of the western United States (Bokkhovskikh *et al.*, 1969), alpine or open-ground species such as *V. bakeri*, *V. nuttallii* and *V. venosa* have erect, lanceolate leaves with blunt or convex bases. Forest species such as *V. douglasii* have cordate leaves and species such as *V. sempervirens* and *V. orbiculata*, found in mossy, deeply shaded forests, have prostrate orbicular leaves (Klaber, 1976). The existence of parallel trends in leaf shape in separate lineages strengthens the conclusion that those trends are adaptive. The predictions made for violets also appear to apply generally to forest herbs (Givnish, 1986f). Genera with horizontal foliage on erect petioles bear peltate (*Diphylleia*, *Podophyllum*) or deeply cordate leaves (*Asarum, Sanguinaria*), or functionally equivalent leaf whorls (*Anemonella, Isotria, Medeola, Trillium*) or ternately compound leaves (*Actaea, Botrychium, Caulophyllum, Cimicifuga, Dicentra*). Rosette herbs with several leaves radiating from a common rootstock have more shallowly cordate leaves (*Hepatica, Heuchera, Tiarella*); species with recumbent foliage and petioles have ovate leaves (*Pyrola*).

Finally, the independent information on phylogeny provided by chromosome numbers allows some inferences to be made about the sequence of habitats invaded and the leaf shapes evolved during adaptive radiation in *Viola*. Given that the lowest chromosome numbers occur in forest species (at least in northeast North America) and that all species studied thus far have a shade-adapted photosynthetic response, it seems likely that open habitats were secondarily invaded. This suggests that cordate leaves may be phylogenetically primitive within *Viola* and lanceolate leaves advanced, although the latter may be primitive within the Violaceae as a whole. The primitive nature of cordate leaves may have imposed phylogenetic constraints on venation pattern and may ultimately be responsible for the evolution of palmatifid or pedatifid leaves in violets of nutrient-poor sites. The preceding inferences about phylogenetic constraints and adaptive radiation, of course, can only be made by combining functional analyses with comparative studies at ecological and phylogenetic levels, using measures of taxonomic relationship independent of the traits in question.

CONCLUSIONS

This paper has illustrated the value of comparative studies in generating and testing hypotheses regarding leaf adaptations, the relative importance of phylogenetic constraints and selective pressures and the rise of adaptations within lineages. Five general conclusions regarding the study of leaf adaptations and the role of comparative studies emerge.

(1) To understand the adaptive significance of variations in leaf form and physiology, we must consider the functional integration of leaves with other plant parts, particularly roots and mechanical support tissue. Many leaf traits that enhance photosynthesis have associated costs involving the uptake of water and nutrients, canopy support and ecological interactions with predators and mutualists. The balance between these costs and associated energetic benefits create trade-offs that underlie the economics of gas exchange, support and biotic interactions; these, in turn, help shape the evolution of many features of leaves and canopies. A major theme of research on photosynthetic adaptations over the next 10 years is likely to be an integrated approach to the study of above- and below-ground organs, because any completely quantitative theory for a trait influencing photosynthesis must incorporate the associated costs of transpiration and root function. Physiological ecologists – who heretofore have largely been leaf physiologists – must begin to collaborate more closely with root biologists to achieve significant advances in our understanding of photosynthetic adaptations (cf. Rorison, 1987).

(2) Similarly, studies of the allometry of support tissue as a function of plant height, leaf mass and canopy geometry should emerge as an important means for studying the significance of growth form. Such studies may usher in a new field of biomechanical ecology, whose interests would include not only those of traditional biomechanics (i.e. the properties that enable plants to withstand mechanical stresses in their environment) but also whether particular growth

forms confer a context-specific competitive advantage. This kind of approach has already led to new insights into the significance of leaf height, branching angle and the arching growth habit in forest herbs (Givnish, 1982, 1986f); branching angles in trees (Honda & Fisher, 1978, 1979; Borchert & Tomlinson, 1984; Fisher, 1986); optimal energy allocation between trunk and canopy (King, 1981); and the quantitative basis of the slope and intercept of the -3/2 self-thinning law (Givnish, 1986g). The opportunities for research on context-specific advantages of bio-mechanical adaptations are today as exciting, in many ways, as those facing physiological ecology in the 1960s and 1970s.

(3) The power of the comparative approach is directly related to the numbers of actual comparisons made. The greater the number of lineages that can be shown to display a particular pattern of convergence, both within habitats and in trends across habitats, the more persuasive is the evidence that the observed pattern is the result of selection, rather than species- or lineage-specific peculiarities involving drift, founder effects, pleiotropy, epistasis or historical availability of genetic variability. Furthermore, the greater the number of interspecific comparisons made within and between lineages, the better circumscribed are the evolutionary possibilities and phylogenetic constraints inherent to each lineage. Comparative studies should thus have as their first priority the inclusion of massive numbers of appropriate comparisons, both ecological (within and between habitats) and phylogenetic (within and between lineages). Such comparisons have revealed well-marked ecological patterns in at least 23 aspects of leaf form, physiology and arrangement.

(4) Comparative studies and optimality analyses play complementary roles in ecological research and the study of adaptations. The precision of the predictions produced by optimality models, and their relationship to specific environmental factors, make them an excellent means of generating sharply defined, testable hypotheses. Comparative studies, designed in such a way as to exclude or control potentially confounding factors (Clutton-Brock & Harvey, 1979), provide powerful tests of such hypotheses. Interspecific comparisons can also play a crucial role in generation of hypotheses by suggesting the ecological importance of certain traits by highlighting particular energetic trade-offs and by delineating the range of evolutionary possibilities available within a range of lineages.

(5) It should be evident that comparative studies and optimality analysis are complemented, in turn, by related research on physiological ecology, biomechanics, genetics and demography. Physiological and biomechanical experiments are required to demonstrate the functional implications of variations in a given trait, a prerequisite for any optimality analysis. Optimality models provide insight into a trait's ecological significance by analyzing the impact of its functional effects on resource allocation and competitive ability in various environments. Finally, although comparative studies can verify the predictions of an optimality model and suggest the correctness of its assumptions, only populational studies can demonstrate the actual impact of variation in a given trait on individual competitive ability and fitness. Such populational studies are not sufficient by themselves, because fitness measurements alone can only show that a particular trait is favoured, not why. Furthermore, fitness measurements within a single population or species may reflect not only the general effects of the trait in question but also aspects of pleiotropy, epistasis, linkage or genetic variation peculiar to that population or species. Such studies should thus be conducted on several unrelated taxa and be complemented by comparative studies on the actual occurrence of the trait.

The most powerful approach to the study of plant adaptation and ecology should thus be an integrated one, embracing comparisons at the ecological, phylogenetic and biogeographic scales, as well as studies in physiological ecology, biomechanics, optimality analysis, ecological genetics and demography. This conclusion clearly contradicts that voiced recently by a leading plant population biologist:

... the search for generalities in ecology has been disappointing – more so in plant than in animal ecology. The few generalities that have emerged come from studies of single species.
... It is from the work of many individuals working scattered over a variety of parts of the world, but concentrating their attention over long periods on the behaviour of individual plants, that the development of ecology as a generalizing and predictive science may be possible.
... The detailed analysis of proximal ecological events is the only means by which we can reasonably hope to inform our guesses about the ultimate causes of the ways in which organisms behave (Harper, 1982).

Were it not expressed by a scientist of such stature, this unidimensional view could be ignored as dismissive of alternative avenues of ecological research and the documented value of comparative studies. The findings of the Virginia study recounted in this paper vividly illustrate the unique power of an integrated approach involving comparisons at ecological, phylogenetic and biogeographic levels. The new insights produced by that study – bearing on the significance of several aspects of leaf shape and growth form in forest herbs, the importance of selection *vs* phylogenetic constraints in determining the ecological characteristics and taxonomic affinities of species in each temporal guild, and the basis for trends in temporal community structure and species richnes – could never have been obtained solely from the single-species studies of proximal ecological events advocated by Harper. As Grime (1984) has argued, Harper's view entails (1) a disregard of existing and potential research gains from studies of units which are larger and of greater antiquity and informational value than present-day single-species populations; (2) a neglect of opportunities for fruitful synthesis of evidence from complementary avenues of research; and (3) a delay in the development of general models needed to understand species interactions and vegetation processes. It is sincerely to be hoped that the complementary, rather than competitive, roles of broad-scale comparisons and intensive, single-species studies continue to be recognized in plant ecology, as they are throughout biology.

ACKNOWLEDGEMENTS

I wish to thank Donald Waller and Susan Knight for their helpful comments and suggestions on a draft of this paper, John Madsen for permission to cite results from his unpublished Ph.D. thesis and Philip Grime for stimulating conversations and generous hospitality. I also wish to thank the Natural Environment Research Council for providing travel funds.

REFERENCES

ALBERCH, P. & GALE, E. A. (1985). A developmental analysis of an evolutionary trend: digital reduction in amphibians. *Evolution*, **39**, 8–23.
ANDERSON, J. A. R. (1961). The structure and development of the peat swamps of Sarawak and Brunei. *Journal of Tropical Geography*, **18**, 7–16.
ASHTON, P. S. (1964). Ecological studies in the mixed dipterocarp forests of Brunei State. *Oxford Forestry Memoirs*, **25**, 1–75.
ASPREY, G. F. & LOVELESS, A. R. (1958). The dry evergreen formations of Jamaica. II. The raised coral beaches of the north coast. *Journal of Ecology*, **46**, 547–570.
BAILEY, I. W. & SINNOTT, E. W. (1916). The climatic distribution of certain types of angiosperm leaves. *American Journal of Botany*, **3**, 24–39.

BARLOW, B. A. & WIENS, D. (1977). Host–parasite resemblance in Australian mistletoes: the case for cryptic mimicry. *Evolution*, **31**, 69–84.

BEADLE, N. C. W. (1966). Soil phosphate and its role in molding segments of the Australian flora and vegetation, with special emphasis to xeromorphy and scleromorphy. *Ecology*, **47**, 992–1007.

BEARD, J. S. (1944). Climax vegetation in tropical America. *Ecology*, **25**, 127–158.

BEARD, J. S. (1955). The classification of tropical American vegetation-types. *Ecology*, **36**, 89–100.

BEWS, J. W. (1925). *Plant Forms and their Evolution in South Africa*. Longman, London.

BJÖRKMAN, O. (1968). Carboxydismutase activity in shade-adapted and sun-adapted species of higher plants. *Physiologia Plantarum*, **21**, 1–10.

BJÖRKMAN, O. (1981). Photosynthetic responses to different quantum flux densities. In: *Encyclopedia of Plant Physiology*, vol. 12A (Ed. by O. L. Lange, P. S. Nobel, C. B. Osmond & H. Ziegler), pp. 57–107. Springer-Verlag, Berlin.

BJÖRKMAN, O., BOARDMAN, N. K., ANDERSON, J. N., THORNE, S. W., GOODCHILD, D. J. & PYLIOTIS, N. A. (1972). Effect of light intensity during growth of *Atriplex patula* on the capacity of photosynthetic reactions, chloroplast components and structure. *Carnegie Institute of Washington Yearbook*, **1971**, pp. 115–135.

BOLKHOVSKIKH, Z., GRIF, V., MATVEJEVA, T. & ZAKHARYEVA, M. (1969). *Chromosome Numbers of Flowering Plants*. V. L. Komarov Botanical Institute, Academy of Sciences of the USSR, Leningrad.

BONE, R. A., LEE, D. W. & NORMAN, J. M. (1984). Leaf epidermal cells function as lenses in tropical rainforest shade plants. *Applied Optics*, **24**, 1408–1412.

BORCHERT, R. & TOMLINSON, P. B. (1984). Architecture and crown geometry in *Tabebuia rosea* (Bignoniaceae). *American Journal of Botany*, **71**, 958–969.

BROWN, W. H. (1919). *Vegetation of the Philippine Mountains*. Philippine Bureau of Science, Department of Agriculture & Natural Resources, Manila.

BRÜNIG, E. F. (1970). Stand structure, physiognomy and environmental factors in some lowland forests in Sarawak. *Tropical Ecology*, **11**, 26–43.

BRÜNIG, E. F. (1983). Vegetation structure and growth. In: *Tropical Rain Forest Ecosystems: Structure and Function* (Ed. by F. B. Golley), pp. 49–75. Springer-Verlag, Berlin.

BUCKLEY, R. C., CORBETT, R. T. & GRUBB, P. J. (1980). Are the xeromorphic trees of tropical upper montane rain forests drought resistant? *Biotropica*, **12**, 124–136.

BÜSGEN, M. & MÜNCH, E. (1929). *The Structure and Life of Forest Trees*. John Wiley & Sons, New York.

CAIN, S. A., DE OLIVIERA CASTRO, C., PIRES, J. M. & DE SILVA, N. T. (1956). Application of some phytosociological techniques to Brazilian rain forest. *American Journal of Botany*, **43**, 911–941.

CHABOT, B. F. & HICKS, D. F. (1982). The ecology of leaf life spans. *Annual Review of Ecology and Systematics*, **13**, 229–259.

CHEW, F. S. & RODMAN, J. E. (1979). Plant resources for chemical defense. In: *Herbivores: Their Interaction with Secondary Plant Metabolites* (Ed. by G. A. Rosenthal & D. H. Janzen), pp. 271–307. Academic Press, New York.

CHIARIELLO, N. (1984). Leaf energy balance in the wet lowland tropics. In: *Physiological Ecology of Plants of the Wet Tropics* (Ed. by E. Medina, H. A. Mooney & C. Vásquez-Yánes), pp. 85–98. Dr Junk, The Hague.

CLUTTON-BROCK, T. H. & HARVEY, P. H. (1979). Comparison and adaptation. *Proceedings of the Royal Society of London, Series B*, **205**, 547–565.

COLEY, P. D. (1983). Herbivory and defensive characteristics of tree species in a lowland tropical forest. *Ecological Monographs*, **53**, 209–233.

COWAN, I. R. (1977). Stomatal behaviour and environment. *Advances in Botanical Research*, **4**, 1176–1227.

COWAN, I. R. (1982). Regulation of water use in relation to carbon gain in higher plants. In: *Encyclopedia of Plant Physiology*, vol. 12A (Ed. by O. L. Lange, P. S. Nobel, C. B. Osmond & H. Ziegler), pp. 589–613. Springer-Verlag, Berlin.

COWAN, I. R. (1986). Economics of carbon fixation in higher plants. In: *On the Economy of Plant Form and Function* (Ed. by T. J. Givnish), pp. 133–170. Cambridge University Press, Cambridge.

COWAN, I. R. & FARQUHAR, G. D. (1977). Stomatal function in relation to leaf metabolism and environment. *Symposia of the Society for Experimental Biology*, **31**, 471–505.

COWLING, R. M. & CAMPBELL, B. M. (1980). Convergence in vegetation structure in the Mediterranean communities of California, Chile, and South Africa. *Vegetatio*, **43**, 191–198.

CURTIS, W. (1984). Photosynthetic light response in the genus *Viola*. *Canadian Journal of Botany*, **62**, 1273–1278.

DEAN, J. M. & SMITH, A. P. (1979). Behavioral and morphological adaptations of a tropical plant to high rainfall. *Biotropica*, **10**, 152–154.

DOLPH, G. E. & DILCHER, D. L. (1980a). Variation in leaf size with respect to climate in Costa Rica. *Biotropica*, **12**, 91–99.

DOLPH, G. E. & DILCHER, D. L. (1980b). Variation in leaf size with respect to climate in the tropics of the western hemisphere. *Bulletin of the Torrey Botanical Club*, **107**, 154–162.

EHLERINGER, J. (1981). Leaf absorptances of Mohave and Sonoran desert plants. *Oecologia*, **49**, 366–370.

EHLERINGER, J. & FORSETH, I. (1980). Solar tracking by plants. *Science*, **210**, 1094–1098.

EHLERINGER, J. R. & MOONEY, H. A. (1978). Leaf hairs: effects on physiological activity and adaptive value to a desert shrub. *Oecologia*, **37**, 183–200.

EHLERINGER, J. R. & WERK, K. S. (1986). Modifications of solar-radiation absorption patterns and implications for carbon gain at the leaf level. In: *On the Economy of Plant Form and Function* (Ed. by T. J. Givnish), pp. 57–82. Cambridge University Press, Cambridge.

EHLERINGER, J. R., MOONEY, H. A., GULMON, S. L. & RUNDEL, P. W. (1981). Parallel evolution of leaf pubescence in *Encelia* in coastal deserts of North and South America. *Oecologia*, **49**, 38–41.

EHLERINGER, J. R., ULLMANN, I., LANGE, O. L., FARQUHAR, G. D., COWAN, I. R., SCHULZE, E.-D. & ZIEGLER, H. (1986). Mistletoes: a hypothesis concerning morphological and chemical avoidance of herbivory. *Oecologia*, **70**, 234–237.

FERNALD, M. L. (1950). *Gray's Manual of Botany*. Van Nostrand, New York.

FIELD, C. (1981). Leaf-age effects on the carbon gain of individual leaves in relation to microsite. In: *Components of Productivity of Mediterranean-climate Regions* (Ed. by N. S. Margaris & H. A. Mooney), pp. 41–50. Dr Junk, The Hague.

FIELD, C. & MOONEY, H. A. (1986). The photosynthesis–nitrogen relationship in wild plants. In: *On the Economy of Plant Form and Function* (Ed. by T. J. Givnish), pp. 25–55. Cambridge University Press, Cambridge.

FISHER, J. B. (1986). Branching patterns and angles in trees. In: *On the Economy of Plant Form and Function* (Ed. by T. J. Givnish), pp. 493–523. Cambridge University Press, Cambridge.

GATES, D. M. (1965). Energy, plants, and ecology. *Ecology*, **46**, 1–13.

GATES, D. M. (1980). *Biophysical Ecology*. Springer-Verlag, New York.

GATES, D. M. & PAPIAN, L. E. (1971). *An Atlas of Leaf Energy Budgets*. Academic Press, New York.

GELLER, G. N. & SMITH, W. K. (1980). Leaf and environmental parameters influencing transpiration: theory and field measurements. *Oecologia*, **46**, 308–313.

GILBERT, L. E. (1975). Ecological consequences of a coevolved mutualism between butterflies and plants. In: *Coevolution of Animals and Plants* (Ed. by L. E. Gilbert & P. H. Raven), pp. 210–240. University of Texas Press, Texas.

GIVNISH, T. J. (1978a). Ecological aspects of plant morphology: leaf form in relation to environment. *Acta Biotheoretica*, **27**, 83–142.

GIVNISH, T. J. (1978b). On the adaptive significance of compound leaves, with particular reference to tropical trees. In: *Tropical Trees as Living Systems* (Ed. by P. B. Tomlinson & M. H. Zimmermann), pp. 351–380. Cambridge University Press, Cambridge.

GIVNISH, T. J. (1979). On the adaptive significance of leaf form. In: *Topics in Plant Population Biology* (Ed. by O. T. Solbrig, S. Jain, G. B. Johnson & P. H. Raven), pp. 375–407. Columbia University Press, New York.

GIVNISH, T. J. (1982). On the adaptive significance of leaf height in forest herbs. *American Naturalist*, **120**, 353–381.

GIVNISH, T. J. (1984). Leaf and canopy adaptations in tropical forests. In: *Physiological Ecology of Plants of the Wet Tropics* (Ed. by E. Medina, H. A. Mooney & C. Vásquez-Yánes), pp. 51–84. Dr Junk, The Hague.

GIVNISH, T. J. (1986a). On the use of optimality arguments. In: *On the Economy of Plant Form and Function* (Ed. by T. J. Givnish), pp. 3–9. Cambridge University Press, Cambridge.

GIVNISH, T. J. (1986b). Economics of gas exchange. In: *On the Economy of Plant Form and Function* (Ed. by T. J. Givnish), pp. 11–24. Cambridge University Press, Cambridge.

GIVNISH, T. J. (1986c). Optimal stomatal conductance, allocation of energy between leaves and roots, and the marginal cost of transpiration. In: *On the Economy of Plant Form and Function* (Ed. by T. J. Givnish), pp. 171–213. Cambridge University Press, Cambridge.

GIVNISH, T. J. (1986d). Economics of support. In: *On the Economy of Plant Form and Function* (Ed. by T. J. Givnish), pp. 413–420. Cambridge University Press, Cambridge.

GIVNISH, T. J. (1986e). Economics of biotic interactions. In: *On the Economy of Plant Form and Function* (Ed. by T. J. Givnish), pp. 667–680. Cambridge University Press, Cambridge.

GIVNISH, T. J. (1986f). Biomechanical constraints on crown geometry in forest herbs. In: *On the Economy of Plant Form and Function* (Ed. by T. J. Givnish), pp. 525–583. Cambridge University Press, Cambridge.

GIVNISH, T. J. (1986g). Biomechanical constraints on self-thinning in plant populations. *Journal of Theoretical Biology*, **119**, 139–146.

GIVNISH, T. J. & VERMEIJ, G. J. (1976). Sizes and shapes of liane leaves. *American Naturalist*, **100**, 743–778.

GOULD, S. J. (1984). Morphological channeling by structural constraint: convergence in styles of dwarfing and gigantism in *Cerion*, with a description of two new fossil species and a report on the discovery of the largest *Cerion*. *Paleobiology*, **10**, 172–194.

GOULD, S. J. (1986). Of kiwi eggs and the Liberty Bell. *Natural History*, **95(11)**, 20–29.

GOULD, S. J. & LEWONTIN, R. C. (1979). The spandrels of San Marco and the Panglossian paradigm: a critique of the adaptationist programme. *Proceedings of the Royal Society of London, Series B*, **205**, 581–598.

GRIME, J. P. (1984). The ecology of species, families and communities of the contemporary British flora. *New Phytologist*, **98**, 15–33.

GRUBB, P. J. (1977). Control of forest growth and distribution on wet tropical mountains. *Annual Review of Ecology and Systematics*, **8**, 83–107.

GRUBB, P. J., LLOYD, J. R., PENNINGTON, T. D. & WHITMORE, T. C. (1963). A comparison of montane and lowland rain forest in Ecuador. I. Forest structure, physiognomy, and floristics. *Journal of Ecology*, **51**, 567–602.

GULMON, S. L. & CHU, C. C. (1981). The effect of light and nitrogen on photosynthesis, leaf characteristics, and dry matter allocation in the chaparral shrub. *Diplacus aurantiacus. Oecologia*, **49**, 207–212.

HABERLANDT, G. (1912). *Physiological Plant Anatomy*. Macmillan, London.

HARPER, J. L. (1982). After description. In: *The Plant Community as a Working Mechanism* (Ed. by E. I. Newman), pp. 11–25. Blackwell Scientific Publications, Oxford.

HICKEY, L. J. & WOLFE, J. A. (1975). The bases of angiosperm phylogeny: vegetative morphology. *Annals of the Missouri Botanical Garden*, **62**, 538–589.

HONDA, H. & FISHER, J. B. (1978). Tree branch angle: maximizing effective leaf area. *Science*, **199**, 888–890.

HONDA, H. & FISHER, J. B. (1979). Ratio of tree branch lengths: the equitable distribution of leaf clusters on branches. *Proceedings of the National Academy of Sciences, USA*, **76**, 3875–3879.

HORN, H. S. (1971). *The Adaptive Geometry of Trees*. Princeton University Press, Princeton.

HORN, H. S. (1979). Adaptation from the perspective of optimality. In: *Topics in Plant Population Biology* (Ed. by O. T. Solbrig, S. Jain, G. B. Johnson & P. H. Raven), pp. 48–61. Columbia University Press, New York.

HOWARD, R. A. (1969). The ecology of an elfin forest in Puerto Rico. 8. Studies of stem growth and form and of leaf structure. *Journal of the Arnold Arboretum*, **50**, 225–267.

JANZEN, D. H. (1974). Tropical blackwater rivers, animals, and the evolution of mast fruiting in the Dipterocarpaceae. *Biotropica*, **6**, 69–103.

KING, D. (1981). Tree dimensions: maximizing the rate of height growth in dense stands. *Oecologia*, **51**, 351–356.

KLABER, D. (1976). *Violets of the United States*. A. S. Barnes & Company, Cranbury, New Jersey.

KOEHL, M. A. R. (1982). The interaction of moving water and sessile organisms. *Scientific American*, **247**, 124–134.

KOEHL, M. A. R. (1986). Seaweeds in moving water: form and mechanical function. In: *On the Economy of Plant Form and Function* (Ed. by T. J. Givnish), pp. 603–634. Cambridge University Press, Cambridge.

LEE, D. W. (1977). On iridescent plants. *Gardens Bulletin of Singapore*, **30**, 21–29.

LEE, D. W. (1986). Unusual strategies of light absorption in rain-forest herbs. In: *On the Economy of Plant Form and Function* (Ed. by T. J. Givnish), pp. 105–131. Cambridge University Press, Cambridge.

LEE, D. W. & LOWRY, J. B. (1975). Physical basis and ecological significance of iridescence in blue plants. *Nature*, **254**, 50–51.

LEE, D. W., LOWRY, J. B. & STONE, B. C. (1979). Abaxial anthocyanin layer in leaves of tropical rain forest plants: enhancement of light capture in deep shade. *Biotropica*, **11**, 70–79.

LEIGH, E. G. (1972). The golden section and spiral leaf arrangement. In: *Growth by Intussusception* (Ed. by E. S. Deevey), pp. 163–176. Archon Press, Hamden, Colorado.

LINCOLN, D. E., NEWTON, T. S., EHRLICH, P. R. & WILLIAMS, K. S. (1982). Coevolution of the checkerspot butterfly *Euphydryas chalcedona* and its larval food plant *Diplacus aurantiacus*: larval response to protein and leaf resin. *Oecologia*, **52**, 216–223.

LOVELESS, A. R. & ASPREY, G. F. (1957). The dry evergreen formations of Jamaica. I. The limestone hills of the south coast. *Journal of Ecology*, **45**, 799–822.

MADSEN, J. (1986). *The production and physiological ecology of the submerged aquatic macrophyte community in Badfish Creek, Wisconsin*. Ph.D. dissertation, University of Wisconsin-Madison, USA.

McKEY, D. (1979). The distribution of secondary compounds within plants. In: *Herbivores: Their Interaction with Secondary Plant Metabolites* (Ed. by G. A. Rosenthal & D. H. Janzen), pp. 55–133. Academic Press, New York.

McKEY, D. B., WATERMAN, P. G., MSI, C. H., GARTLAN, J. S. & STRUHSAKER, T. T. (1978). Phenolic content of vegetation in two African rain forests: ecological implications. *Science*, **202**, 61–63.

MEDINA, E. (1984). Nutrient balance and physiological processes at the leaf level. In: *Physiological Ecology of Plants of the Wet Tropics* (Ed. by E. Medina, H. A. Mooney & C. Vásquez-Yánes), pp. 139–154. Dr Junk, The Hague.

MOONEY, H. A. & DUNN, E. L. (1970). Photosynthetic systems of mediterranean-climate shrubs and trees of California and Chile. *American Naturalist*, **104**, 447–453.

MOONEY, H. A. & GULMON, S. L. (1979). Environmental and evolutionary constraints on the photosynthetic characteristics of higher plants. In: *Topics in Plant Population Biology* (Ed. by O. T. Solbrig, S. Jain, G. B. Johnson & P. H. Raven), pp. 316–337. Columbia University Press, New York.

MOONEY, H. A. & GULMON, S. L. (1982). Constraints of leaf structure and function in reference to herbivory. *BioScience*, **32**, 198–206.

MOONEY, H. A., FIELD, C., GULMON, S. L. & BAZZAZ, F. A. (1981). Photosynthetic capacity in relation to leaf position in desert versus old-field annuals. *Oecologia*, **50**, 109–112.

NOBEL, P. S. (1986). Form and orientation in relation to PAR interception by cacti and agaves. In: *On the Economy of Plant Form and Function* (Ed. by T. J. Givnish), pp. 83–103. Cambridge University Press, Cambridge.

ORIANS, G. H. & SOLBRIG, O. T. (1977). A cost-income model of leaves and roots with special reference to arid and semi-arid areas. *American Naturalist*, **111**, 677–690.

OSMOND, C. B. (1978). Crassulacean acid metabolism: a curiosity in context. *Annual Review of Plant Physiology*, **29**, 379–414.

PARKHURST, D. F. (1978). Adaptive significance of stomatal location on one or both surfaces of leaves. *Journal of Ecology*, **66**, 367–383.

PARKHURST, D. F. (1986). Internal leaf structure: a three-dimensional perspective. In: *On the Economy of Plant Form and Function* (Ed. by T. J. Givnish), pp. 215–249. Cambridge University Press, Cambridge.

PEACE, W. J. H. & MacDONALD, Q. D. (1981). An investigation of the leaf anatomy, foliar mineral levels, and water relations of trees of a Sarawak forest. *Biotropica*, **13**, 100–119.

PEARCY, R. W. & EHLERINGER, J. (1984). Comparative ecophysiology of C$_3$ and C$_4$ plants. *Plant Cell and Environment*, **7**, 1–13.

PRICE, P. W., BOUTON, C. E., GROSS, P., McPHERON, B. A., THOMPSON, J. N. & WEIS, A. E. (1980). Interactions among three trophic levels: influence of plants on interactions between insect herbivores and natural enemies. *Annual Review of Ecology and Systematics*, **11**, 41–66.

RADFORD, A. E. (1976). *Natural Areas of the Southeastern United States: Field Data and Information.* Department of Botany, University of North Carolina, Chapel Hill.

RAUNKIAER, C. (1934). *The Life Forms of Plants and Statistical Plant Geography.* Clarendon Press, London.

RAUSHER, M. D. (1978). Search image for leaf shape in a butterfly. *Science*, **200**, 1071–1073.

RAUSHER, M. D. (1980). Host abundance, juvenile survival, and oviposition preference in *Battus philenor*. *Evolution*, **34**, 343–355.

RICHARDS, P. W. (1952). *The Tropical Rain Forest: An Ecological Study.* Cambridge University Press, Cambridge.

RICKETTS, H. W. (1966). *Wild Flowers of the United States*, vol. I. McGraw-Hill, New York.

RORISON, I. H. (1987). Mineral nutrition in time and space. In: *Frontiers of Comparative Plant Ecology* (Ed. by I. H. Rorison, J. P. Grime, R. Hunt, G. A. F. Hendry & D. H. Lewis), *New Phytologist*, **106** (Suppl.), 79–92. Academic Press, London & New York.

SARMIENTO, G. (1972). Ecological and floristic convergence between seasonal plant formations of tropical and subtropical South American. *Journal of Ecology*, **40**, 367–410.

SCHIMPER, A. F. W. (1898). *Pflanzengeographie auf physiologischer Grundlage.* Fischer, Jena.

SCHULZE, E.-D. & HALL, A. E. (1982). Stomatal responses, water loss and CO$_2$ assimilation rates of plants in contrasting environments. In: *Encylopedia of Plant Physiology*, vol. 12B (Ed. by O. L. Lange, P. S. Nobel, C. B. Osmond & H. Ziegler), pp. 181–230. Springer-Verlag, Berlin.

SCULTHORPE, C. D. (1967). *The Biology of Aquatic Vascular Plants.* Edward Arnold, London.

SHIELDS, L. M. (1950). Leaf xeromorphy as related to physiological and structural influences. *Botanical Review*, **16**, 399–447.

SKOG, L. E. (1976). A study of the tribe Gesnerieae, with a revision of *Gesneria* (Gesnerieae: Gesneriaceae). *Smithsonian Contributions to Botany*, **29**, 1–182.

SMALL, E. (1972). Water relations of plants in raised *Sphagnum* bogs. *Ecology*, **53**, 726–728.

SOBRADO, M. A. & MEDINA, E. (1980). General morphology, anatomical structure, and nutrient content of sclerophyllous leaves of the 'bana' vegetation of Amazonia. *Oecologia*, **45**, 371–378.

STENECK, R. S. (1982). A limpet–coralline alga association: adaptations and defenses between a selective herbivore and its prey. *Ecology*, **63**, 507–522.

STOWE, L. G. & BROWN, J. L. (1981). A geographic perspective on the ecology of compound leaves. *Evolution*, **35**, 818–821.

SUGDEN, A. M. (1982). The vegetation of the Serrania de Macuira, Guajira Columbia: a contrast of arid lowlands and an isolated cloud forest. *Journal of the Arnold Arboretum*, **63**, 1–30.

TALBERT, C. M. & HOLCH, A. E. (1957). A study of the lobing of sun and shade leaves. *Ecology*, **38**, 655–658.

TANNER, E. V. J. & KAPOS, V. (1982). Leaf structure of Jamaican upper montane rain forest. *Biotropica*, **14**, 16–24.

TAYLOR, S. E. (1972). Optimal leaf form. In: *Perspectives in Biophysical Ecology* (Ed. by D. M. Gates & R. B. Schmerl), pp. 73–86. Springer-Verlag, New York.

WALTER, H. (1973). *Vegetation of the Earth.* Springer-Verlag, New York.

WEBB, L. J. (1968). Environmental relationships of the structural types of Australian rain forest vegetation. *Ecology*, **49**, 296–311.

WHITMORE, T. C. (1984). *Tropical Rain Forests of the Far East*, 2nd Edn. Clarendon Press, Oxford.

WILLIAMSON, G. B. (1981). Drip tips and splash erosion. *Biotropica*, **13**, 228–231.

WOLFE, J. A. (1978). Temperature parameters of humid to mesic forests of eastern Asia and relation to forests of other regions of the Northern Hemisphere and Australasia. *United States Geological Service Professional Paper No. 1106*. United States Government Printing Office, Washington, DC.

Wood, J. G. (1934). The physiology of xerophytism in Australian plants. The stomatal frequencies, transpiration and osmotic pressures of sclerophyllous and tomentose-succulent leaved plants. *Journal of Ecology*, **22**, 65–85.

New Phytol. (1987) **106** (Suppl.), 161–175

PHOTOSYNTHESIS AND CARBON ECONOMY OF PLANTS

By C. B. OSMOND

Department of Environmental Biology, Research School of Biological Sciences, Australian National University, Box 475, Canberra 2601, Australia and Biological Sciences Center, Desert Research Institute, University of Nevada, Box 60220, Reno, NV 89506, USA

SUMMARY

The extent to which studies of photosynthesis and carbon economy are, or should be, at the forefront of comparative plant ecology is discussed from the points of view of carbon pathways, water-use efficiency, irradiance and nutrients, especially nitrogen. The ways in which new integrative and reductionist techniques can be applied to photosynthesis in an ecological control are illustrated and their value assessed. The study of photosynthetic physiology of plants at the edges of their geographical range is urged as a means of understanding both their responses to stress and disturbance, and their competitive interactions.

Key words: Photosynthesis, carbon economy, irradiance, water-use efficiency, C_3, C_4 and CAM.

INTRODUCTION

Because the vast majority of plants, for most of their life-cycle, derive their energy and carbon from photosynthesis, it is inevitable that studies of photosynthesis and carbon economy should be at the forefront of comparative plant ecology. In the last quarter century, great strides have been made in photosynthetic physiology and biochemistry as a consequence of comparative, ecologically oriented enquiry. In nearly all instances, these advances have stimulated reductionist research rather than provided ecological insights. The principal reasons for this are that it has been easier to attract funding for short-term research projects on the one hand, and difficult to devise a framework for integration on the other. There are exceptions to this pessimistic assessment, however, and some of these are discussed below.

It has been argued frequently that two kinds of physiological understanding, of tolerance limits to environment, and of production and reproduction within these tolerance limits, are important in plant ecology (Scott, 1974; Osmond, Björkman & Anderson, 1980). Much of the ecologically relevant research in photosynthesis of the last decades has been directed towards small-scale to large-scale integration of CO_2 acquisition, from leaf to community, as an input to models of carbon economy and plant productivity. There has been relatively little analysis of the relationship between production and reproduction and, thus far, few attempts to evaluate the physiological significance of relationships between tolerance limits and productivity functions. Both of these shortcomings contribute to the difficulties inherent in linking photosynthesis and carbon economy during stress and competition in the trinity of vegetation strategies identified by Grime (1979). Thus, although a physiological ecologist of photosynthetic inclination may take immediate exception to statements such as, 'The productivity of a plant is directly related to the rate of photosynthetic carbon assimilation ...' (Barber & Baker, 1985; p. ix), he cannot contribute much to amplify statements such as, 'In

0028-646X/87/05S161+15 $03.00/0

Chenopodium rubrum, however, the seed is small and large size may be attained only by a sustained period of rapid photosynthesis' (Grime, 1979; p. 59).

It is difficult to foresee that this situation will change drastically in the near future. The best that can be done is to accept with satisfaction the stimulus that ecological perspectives continue to provide for reductionist research while, at the same time, seeking a pragmatic framework in which we can reduce an immense amount of data on photosynthetic activities into an ecologically meaningful context. I believe that experience shows it is often unrewarding to try and extract ecological insights from seasonal accounting of leaf photosynthetic activities. That is, detailed studies of response functions within the tolerance limits, no matter how we massage the data, tend not to provide the answers we seek. It is possible that more understanding of the photosynthetic tolerance limits themselves, with respect to major limiting factors, and of the stochastic characteristics of these factors in the environment, may better advance our quest. Almost certainly, such studies will provide a better physiological understanding of vegetation strategies in terms of disturbance, stress and competition. Here, I shall review, quite superficially, several new integrative and reductionist techniques which seem to be leading us along this path.

CARBON ECONOMY, CARBON PATHWAY AND WATER

One of the most spectacular outcomes of comparative, ecologically motivated photosynthetic research of the last decades has been the recognition of different pathways of photosynthetic carbon acquisition in vascular plants, both in terrestrial and aquatic environments. The biochemical and physiological attributes of these carbon pathways are reduced to their simplest expression in Figure 1 and amplified

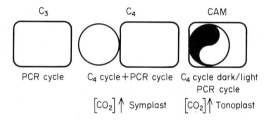

Fig. 1. Schematic representation of the structural and functional basis of different pathways of photosynthetic carbon metabolism. All are based on the photosynthetic carbon reduction (PCR) cycle. In C_4 plants, a C_4 cycle in mesophyll cells provides a CO_2 concentrating mechanism (via symplastic transport of C_4 acid and decarboxylation) to adjacent bundle sheath cells containing the PCR cycle and decarboxylation enzymes. In CAM plants, the C_4 cycle is found in the same cells as the PCR cycle and is temporally regulated, functioning as a C_4 carboxylation system in the dark and a decarboxylation system in the light. The CO_2 concentrating mechanism is based on tonoplastic transport of malic acid stored in the vacuole (after Osmond, 1984).

in the legend. It seems that embellishments of the fundamental photosynthetic carbon reduction cycle of C_3 plants, which characterize C_4 plants and CAM plants, are all to do with mechanisms for concentrating CO_2 in the vicinity of ribulose-1,5-bisphosphate carboxylase (Rubisco). This inhibits the oxygenase activity of the bifunctional enzyme and, by minimizing photorespiration, has a fundamental effect on carbon economy of C_4 plants in the very first steps of the acquisition of carbon. In CAM plants, the CO_2 concentrating function is based on quite different temporal, physiological and biochemical functions and is effective for only part of the light period. Until recently we have tended to underestimate the significance of the conservation of respiratory carbon during nocturnal acidification in the carbon economy of CAM plants.

From the outset, biochemists and physiologists were quick to claim 'manifold consequences in ecological relationships which may arise' (Black, 1971) from differences among species in the pathways of carbon acquisition. We have pointed out (Osmond, Winter & Ziegler, 1982) that it is difficult to sustain the functional significance of these different CO_2 fixation pathways in biochemical, physiological and ecological contexts in terms of carbon economy and productivity, as originally envisaged by Black. To a large extent, the higher productivity sometimes associated with C_4 plants reflects the longer, warmer growing seasons of tropical grasses (Gifford, 1974; Loomis & Gerakis, 1975; Monteith, 1978). The eco-logically interesting distributions of C_3 and C_4 grasses with respect to latitude and altitude are dominated by night temperature in the growing season (Teeri & Stowe, 1976; Tieszen *et al.*, 1979; Boutton, Harrison & Smith, 1980; Hattersley, 1983). That is, the effects of chilling temperature on development of the photosynthetic apparatus (Slack, Roughan & Bassett, 1974) during establishment seem more important than potential productivity.

It is clear that no generalizations with respect to the temperature responses in C_3, C_4 or CAM plants are possible; C_3 and C_4 species from cool and hot habitats show similar response functions and similar potential for acclimation (Osmond *et al.*, 1980). Temperature acclimation is a much vaunted property of plants but its significance for carbon economy has scarcely been tested. In an exemplary analysis, Lange *et al.* (1978) found that, in spite of acclimation of the thermal optimum for photosynthesis, branchlets of *Hammada scoparia* (C_4) were too cold for 66% of the growing season in the plant's natural habitat in the Negev Desert. *Prunus armenica* (C_3), an introduced species which was also capable of about a 7 °C shift in its temperature optimum, was marginally better suited, being close to the random expectation of 33% too cold, 33% just right and 33% too hot!

Perhaps the most comprehensive of all studies of carbon pathway and carbon economy so far undertaken is that of Caldwell *et al.* (1977), who compared *Ceratoides lanata* (C_3) and *Atriplex confertifolia* (C_4) in the Great Basin Desert. This ideal comparison of closely related species of similar life form established that carbon fixation by the C_4 plants, which continued into the hotter months of the year, was only marginally greater than that of the C_3 plants but that the higher water-use efficiency of C_4 photosynthesis prevailed throughout the growing seasons. The carbon economies of the two systems were remarkably similar overall but the slightly higher input of carbon in *Atriplex*, which was predominantly allocated below ground, seemed to permit deeper exploration of water reserves and prolonged carbon exchange. Schulze *et al.* (1980) found that, in *Hammada scoparia* (C_4), both the normalized maximum rate of photosynthesis and normal-ized daily carbon gain declined more slowly with water stress throughout the season than in four species of C_3 plants in the Negev Desert. The significance of higher intrinsic water-use efficiency of C_4 plants in this response is not clear and there is no evidence that photosynthesis in C_4 or CAM plants is intrinsically less sensitive to water stress than in C_3 plants.

Another example of pragmatic accounting of carbon economy in relation to the carbon pathway of photosynthesis is given in recent analyses of productivity in CAM plants (Nobel, 1985). Several factors simplify the analysis. In desert succulents, CO_2 fixation is largely confined to the dark (and is equivalent to nocturnal acidification), practically all the biomass is photosynthetically active and respiratory carbon is conserved. Nobel uses the relationships between CO_2 uptake and plant water status, temperature and PAR, determined in the laboratory, to construct an environmental productivity index. Using field enviornmental data, this index correlated well with the unfolding of new leaves in *Agave* and thus can be used to estimate productivity. Uncomplicated by significant CO_2 and H_2O vapour exchanges in the light, modelling of the water-use efficiency of CAM plants in desert habitats permits some spectacular conclusions, especially as to the water cost of flowering (Nobel, 1976). The rate of net nocturnal CO_2 fixation in stem succulents is sometimes inadequate to account for the acidification observed (Osmond, Nott & Firth, 1979), especially when these CAM plants experience soil water deficits. The conclusion that internal refixation of respiratory CO_2 makes a significant contribution to nocturnal acidification also seems to apply in recent studies of epiphytic *Tillandsia* (Martin & Adams, 1986), epiphytic bromeliads (Griffiths *et al.*, 1986) and aquatic CAM plants (Boston & Adams, 1986). It also makes a small contribution to the carbon economy of *Stylites* (Keeley, Osmond & Raven, 1984).

Baskin & Baskin (1978) proposed that C_4 photosynthesis was less important than other features in determining growth rate and competitive ability in weeds. There have not been many tests of the relationships between photosynthetic pathway and competitive relationships among plants of similar life form and reproductive pattern. Pearcy, Tumosa & Williams (1981) compared *Chenopodium album* C_3 and *Amaranthus retroflexus* (C_4) in competition at different temperatures and found the competitive outcome reflected photosynthetic performance as a function of temperature. However, the outcome was largely due to differences in relative growth rate prior to canopy closure, i.e. before competition began. Such experiments are tedious but necessary to establish the ways in which photosynthetic potential, water-use efficiency and assimilate partitioning may be translated into competitive advantage. Although the differences in stable carbon isotope ratio between C_3 and C_4 plants have been exploited to estimate the contributions to the biomass by these plants in natural communities (Tieszen *et al.*, 1979), they have not been widely applied in laboratory studies of competition. The potential of the method is considerable, especially because it permits partitioning of root biomass as to source and integration of carbon acquisition over the growing season.

Table 1. *Below-ground competition between* Triticum aestivum *(C_3) and* Echinchloa crus-galli *var.* frumentacea *(C_4) as indicated by the $\delta^{13}C$ values of root biomass (unpublished data of S. C. Wong, Z. Roksandic & C. B. Osmond)*

Growth conditions	Mixing ratios $\delta^{13}C$ value, ‰)			
	$3 \times C_3$	$2 \times C_3$; $1 \times 6_4$	$1 \times C_3$; $2 \times C_4$	$3 \times C_4$
Low light (500 μmol photons m^{-2} s^{-1})				
12·0 mM NO$_3^-$	−26·6	−23·5 (−21·4)	−16·9 (−16·1)	−11·1
2·0 mM NO$_3^-$	−29·9	−28·2 (−24·7)	−24·4 (−19·5)	−14·2
High light (2000 μmol photons m^{-2} s^{-1})				
12·0 mM NO$_3^-$	−24·5	−15·7 (−20·9)	−12·0 (−14·3)	−8·4
2·0 mM NO$_3^-$	−28·9	−27·6 (−23·4)	−20·4 (−17·2)	−12·3

Predicted $\delta^{13}C$ values, assuming proportional contributions of carbon according to mixing ratios, are shown in parentheses.

We have validated the method in studies of competition using *Triticum* (C_3 and *Echinochloa* (C_4) by mechanically separating shoot biomass and then estimating proportions due to C_3 and C_4 in total biomass from the $\delta^{13}C$ value. Table 1 shows data for roots. If the measured value is less negative than the predicted value, the C_4 plant shows a competitive advantage; if the measured value is more negative than predicted, the C_3 plant shows an advantage. In this experiment, the C_3 plant seems to compete effectively below ground, especially in low light, low nitrogen treatments. The converse is true in high light, high nitrogen treatments in which the C_4 plant shows advantage below ground. There is considerable scope for application of this technique in studies of C_3/C_4 competition in relation to water stress and elevated atmospheric CO_2 concentration.

Until relatively recently, it was thought that the $\delta^{13}C$ value of C_3 and C_4 plants was fixed by the biochemistry of CO_2 assimilation. The normal distribution of $\delta^{13}C$ values about two nodes was ascribed to variations in source CO_2 in different studies and to analytical unreliabilities in early measurements. Variation in source CO_2 almost certainly contributes to the variations in controls shown in Table 1. However, following the incisive analysis by O'Leary (1981) and improved models of C_3 photosynthesis (Farquhar, O'Leary & Berry, 1982), it has become clear that, at low stomatal conductance, diffusional processes can significantly alter the $\delta^{13}C$ value. Farquhar *et al.* (1982) showed that the $\delta^{13}C$ value of C_3 plants is related to the ratio of intercellular to ambient CO_2 partial pressure (p_i/p_a), the diffusional discrimination (a) and the biochemical discrimination (b_3) owing to Rubisco, as follows:

$$\delta^{13}C \ (C_3) = \delta^{13}C \ (air) - a - (b_3 - a) \ p_i/p_a.$$

Subsequent studies (Farquhar & Richards, 1984) demonstrate a good correlation between water-use efficiency and carbon isotope discrimination of wheat and other species. In C_3 plants, high water-use efficiency is associated with less negative $\delta^{13}C$ values. Whereas water-use efficiency was previously thought to be invariate within C_3 or C_4 plants (at a particular leaf to air vapour pressure difference), it is now a target for assessment in relation to carbon economy and productivity under water-limited conditions.

The potential ecological implications of this insight are considerable. We have been interested in the relative contributions of leaf and stem photosynthesis to the carbon economy of a biennial shrub, *Eriogonum inflatum* (C_3), of the Great Basin Desert. In this species, the rosette of leaves formed in early spring senesces as stems develop and the stems inflate to provide 66 to 77 % of the total photosynthetically active surface integrated over the life of the plant. Both in the field and in the laboratory, stems maintain lower conductance than leaves as a function of tissue to air vapour pressure difference. In the field, mean daily conductance of stems remains much less than that of leaves throughout the life-cycle. We have observed significant differences between the $\delta^{13}C$ values of leaves and stems (Table 2) which are consistent with the predictions of Farquhar *et al.* (1982). The implication that carbon acquisition by photosynthetic stems during the hot, dry summer takes place with lower p_i seems to fit with the low stomatal conductance and potentially higher water-use efficiency of such photosynthesis.

This new dimension and application for $\delta^{13}C$ value in the carbon and water budgets of C_3 plants seems unlikely to be extended to C_4 plants, because of analytical limitations and leakiness of the CO_2 concentrating mechanism of the bundle sheath. In CAM, the effect of low stomatal conductance during CO_2 fixation in the light of $\delta^{13}C$ value is in the same direction as an increase in the

Table 2. *Differences in $\delta^{13}C$ value of leaves and stems of* Eriogonum inflatum, *indicative of higher water-use efficiency and lower stomatal conductance during C_3 photosynthesis in stems compared with leaves (unpublished data of S. D. Smith, C. B. Osmond & Z. Roksandic)*

Location	Leaves ($\delta^{13}C$ value, ‰)	Stems $\delta^{13}C$ value, ‰)
Death Valley		
Site 1	−27·0	−24·1
Site 2	−26·9	−24·3
Walker Lake	−27·7	−25·6

proportion of carbon acquired via dark CO_2 fixation. Both changes are likely to occur in CAM plants in response to reduced water availability. Some careful carbon accounting, with the right plants, will be required to unravel these problems. For example, only nocturnal acidification serves to distinguish between a succulent C_3 plant (*Threlkeldia halocnemoides*, $\delta^{13}C$ value $= -24.8$ ‰) collected in mid summer and an adjacent CAM plant (*Calandrinia polyandra*, $\delta^{13}C$ value $= -24.7$‰) collected in spring (Winter, Osmond & Pate, 1981).

Perhaps the ecological importance of differences in water-use efficiency has been overrated and many other aspects of water relations deserve attention. It ought to be possible, for example, to put turgor maintenance into context with diurnal patterns of photosynthesis and transpiration, but it remains difficult to sustain a case for improved carbon gain as a consequence of turgor maintenance owing to osmotic adjustment in the face of increasing water deficit. One interesting and relevant model system, nocturnal malic acid accumulation in CAM plants, suggests a significant role for osmotically driven water uptake in these plants (Ruess & Eller, 1985; Lüttge, 1986). Kuppers (1984b) found that the theory of optimal water use in relation to carbon gain seemed to apply in hedgerow species with low water-use efficiency (e.g. *Crataegus*), but not in those in which high water-use efficiency was correlated with low stomatal conductance and low hydraulic conductivity (e.g. *Ribes* and *Rubus*). He found that these latter species desiccated on dry days in exposed habitats in spite of their high water-use efficiency and showed low competitive ability.

CARBON ECONOMY AND LIGHT

Just as ecologically motivated research played a major role in elucidation of different carbon pathways of photosynthesis, so also have studies of plant response to light helped to open up new fields of research in membrane biochemistry and photosynthetic physiology. Many herbaceous plants are capable of photosynthetic acclimation to a 10- to 20-fold difference in irradiance during growth, characterized by a two- to five-fold increase in the rate of light- and CO_2-saturated photosynthesis and no change in the initial slope of the light response curve (quantum yield), as shown for spinach in Figure 2(a). Much attention has been given to the importance of maximum photosynthetic rate in relation to successional status of species. In rain forests, gap pioneers tend to have higher photosynthetic rates than canopy dominants (Bazzaz & Pickett, 1980; Oberbauer & Strain, 1984). A similar trend was observed in hedgerows (Kuppers, 1984a) but competitive ability seemed

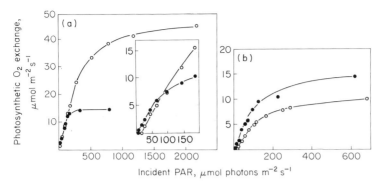

Fig. 2. Light response curves of photosynthesis measured by O_2 exchange at CO_2 saturation in leaves of (a) an herbaceous C_3 plant (*Spinacea*) showing conventional sun–shade acclimation and (b) of a climbing CAM plant (*Hoya*) native to shaded habitats but which grows well in bright light without acclimation. Both were grown in full sunlight in a glass house (1500 to 2000 μmol photons $m^{-2} s^{-1}$, ○) or under shade cloth (100 to 225 μmol photons $m^{-2} s^{-1}$, ●). Data of Walker & Osmond (1986) and Adams, Osmond & Sharkey (1987)

inversely related to photosynthetic capacity. Ability to acclimate to shade, however, was an important feature in establishment and maintenance of species composition in the system.

The significance for carbon economy of the adjustment in photosynthesis from shade to sun and *vice versa* as shown in Figure 2(a) has been illustrated by simple models of carbon gain in *Atriplex triangularis* (Osmond *et al.*, 1980). A central feature in the carbon economy of the light acclimation response in herbaceous plants is the higher light compensation point and dark respiration of sun-grown plants. It seems likely that sun species would be excluded from shade because of their profligate respiratory activity, and shade species might not compete well in the sun because of their low rate of light-saturated photosynthesis.

Plants which develop in shaded habitats are sensitive to light-dependent damage to the photosynthetic apparatus (photoinhibition) if they are exposed to bright light [see response of *Hoya* in Figure 2(b) discussed more fully below]. Although the mechanisms involved are currently contentious, many studies suggest that damage to primary photochemical process in the reaction centre of photosystem II is involved (Powles & Björkman, 1982; Powles, 1984). The hypothesis that transfer of excitation energy from the antennae system to the reaction centre at rates in excess of the capacity for energy transduction by electron transport system and carbon metabolism results in damaging secondary reactions in the reaction centre seems plausible but remains to be tested (Osmond, 1981). Plants which develop in bright light are also susceptible to photoinhibition if their ability to transduce photochemical energy is impaired by factors such as environmental stress or inadequate nutrition. Many studies now show that light exaggerates the effects of environmental stress on the photosynthetic apparatus.

Detailed biochemical analyses of plants grown under different light regimes and analysis of species such as *Alocasia macrorrhiza* from deeply shaded habitats have led to the view that acclimation to sun or shade is based on changes in the distribution of membrane and stromal components of the photosynthetic apparatus (Björkman, 1981). Figure 3 records the extent of these changes, based on average values for four herbaceous species. While much attention has been paid to these detailed state descriptions of photosynthetic membranes in shade-grown and

C. B. OSMOND

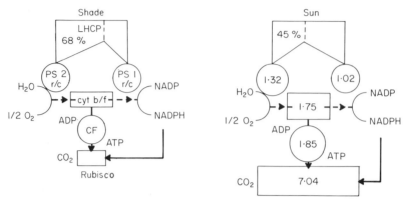

Fig. 3. Schematic representation of the changes in composition of some important components of the photosynthetic apparatus during sun–shade acclimation of herbaceous plants. Compared with shade-grown plants, the proportions of total light-harvesting chlorophyll–protein complexes (LHCP) associated with photosystem 2 (PS2) decreases from 68 % to 45 %. On a chlorophyll basis, the number of PS2 reaction centres (r/c) increases ($\times 1.32$) and that of photosystem 1 (PS1) reaction centres is unchanged ($\times 1.02$). Larger changes (on a chlorophyll basis) are observed in the cytochrome b/f complex in the electron transport chain ($\times 1.75$), in coupling factor (CF) associated with photophosphorylation ($\times 1.85$) (data cited by Anderson & Osmond, 1987) and in the amount (g m^{-2}) of ribulose–1,5–bisphosphate carboxylase (Rubisco, $\times 7.04$) (J. R. Seemann, T. D. Sharkey & C. B. Osmond, unpublished data).

sun-grown leaves (Fig. 3), little is known of the dynamics of acclimation. For example, composition and distribution of light-harvesting pigment–protein complexes, which presumably accompany the changes in chloroplast ultrastructure, must be co-ordinated with repair of the damage done to (or by) primary photochemistry in reaction centres. As acclimation to high light proceeds, one would expect the photosynthetic apparatus to become less sensitive to photo-inhibition.

Björkman (1981) would probably be among the first to concede that his conclusion 'that photoinhibition does not normally occur in plants under the light regimes encountered in their natural environments' needs qualification in the light of recent experiments. A theme emerging from these studies is that leaves of plants with intrinsically low rates of light- and CO_2-saturated photosynthesis show little acclimation of photosynthetic rate between shade and sun, and reduced quantum yield in sunlight. It is too early to reach any firm conclusions but it seems likely that, in bright light, the photosynthetic apparatus of these plants effects some sort of compromise between limited capacity for acclimation while experiencing substantial photoinhibition. In its natural habitats, *Alocasia* is often found in bright light at the margins of rain forests in comfortable coexistence with tropical grasses such as *Panicum maximum* (C_4) and it has become a weed of significant proportions in roadside habitats in Florida (Herndon, 1985). When growing in full sunlight, tropical rain forest understorey species, epiphytes or shade-tolerant seedlings of canopy trees do not acclimate as shown in Figure 2(a). In full sunlight, *Alocasia* shows about a two-fold increase in light- and CO_2-saturated photosynthetic rate but substantial reduction in quantum yield (Pearcy, 1986). In the emergent canopy of a rain forest, north-facing, sun-exposed leaves of *Castanospermum australe* show inhibition of quantum yield and of maximum photosynthesis compared to shaded, south-facing leaves (Pearcy, 1986). The climbing CAM plant, *Hoya australis*, flowers in full sunlight on rocky beaches, in spite of inhibited quantum yield and photosynthesis [see Fig. 2(b) above and Table 4 below]. A variety of responses, intermediate between those in Figures 2(a) and (b), is found among other rain forest plants (Langenheim *et al.*, 1984; Pearcy, 1986; Winter, Osmond & Hubick, 1986).

It could be that selection for tolerance of extremely low light in the seedling stage of shade-tolerant canopy dominants such as *Agathis* and *Castanospermum* is incompatible with the adjustments we have come to regard as normal photosynthetic acclimation. Leaves of these trees are long lived and longevity may bring with it constraints on rate and amplitude of metabolic adjustment. Thus, one could regard the steady state of chronic photoinhibition and limited acclimation in *Castanospermum* as equivalent to a time-base expanded, transitional phase of the acclimation process observed in herbaceous species. For example, when *Solanum dulcamara* is grown in about 100 μmol photons m^{-2} s^{-1} and then transferred to 10- to 20-fold brighter light, leaves suffer photoinhibition and, after 2 d, are indistinguishable in their light response from sun leaves of *Hoya* [Fig. 2(b)]. Two to 3 weeks later, these same leaves of *Solanum* have fully acclimated (Ferrar & Osmond, 1986), providing a before and after comparison similar to that of shade-grown and sun-grown spinach [Fig. 2(a)].

We are fortunate that new instrumentation is available which permits evaluation of the dynamics of these responses in ecologically interesting plants. The leaf disc electrode system, developed for other purposes in the Research Institute for Photosynthesis at the University of Sheffield, has proved to be a most effective means for screening maximum photosynthetic rate and estimating quantum yields [Fig. 2(a), (b); Walker & Osmond, 1986]. Reduced quantum yield is indicative of impaired efficiency of energy transduction in photosynthesis, potentially involving lesions in primary photochemistry, electron transport or changed demands of carbon assimilation. The latter source of variation can be overcome by using ultra-saturation of CO_2 in the leaf disc O_2 electrode. Changes in electron transport processes in relation to primary photochemistry and carbon metabolism are indicated by changes in oscillatory fluorescence at room temperature (Walker & Osmond, 1986). Changes in primary photochemistry are indicated by reductions in the ratio of variable to maximum fluorescence measured *in vivo* at the temperature of liquid N_2 (77 K fluorescence; Powles & Björkman, 1982). Björkman & Demmig (1987) and Adams *et al.* (1987) have established a good correlation between reduction in quantum yield measured in the leaf disc electrode and reduction in 77 K fluorescence when leaves of shade-grown plants are photoinhibited after exposure to bright light.

These methods can be applied to assess the cumulative and interactive effects of stress, as well as the dynamics of disturbance. For example, although prolonged drought only marginally reduces the water potential of *Opuntia basilaris*, these plants nevertheless show reduction in quantum yield and 77 K fluorescence in bright light and moderate temperatures (Table 3). In the field in Death Valley, high temperature exaggerates the effect of water stress on loss of photosynthetic integrity but only on the south face of cladodes exposed to the sun (Adams, Smith & Osmond, 1986). Before the leaf disc electrode became available, it was difficult to make meaningful estimates of quantum yield in CAM plants. The unexpectedly high values of quantum yield in CAM plants (Adams, Nishida & Osmond, 1986; Winter *et al.*, 1986) may explain why they are so common as epiphytes in tropical forests. The dimensions of shade and sun acclimation in these plants have yet to be established. However, like the rain forest, understorey and canopy elements,

Table 3. *Quantum yield and 77 K fluorescence properties of* Opuntia basilaris *from Death Valley (unpublished data of W. W. Adams III & C. B. Osmond)*

Treatment	Quantum yield* (mol O_2 mol^{-1} photons)	77 K fluorescence (F_v/F_m)
Glasshouse		
28 °C, well watered, −1·5 mPa	0·101	0·81
28 °C, no water 212 d, −2·3 mPa	0·050	0·63
Field, late summer		
> 52 °C, −2·2 mPA, north face	0·057	0·65
> 52 °C, −2·2 mPa, south face	0·031	0·36

* Measured at CO_2 saturation in leaf disc electrode.

Table 4. *Photosynthetic properties of a rain forest climber* (Hoya australis) *and epiphytes/lithophytes* (Pyrrosia confluens *and* Dendrobium speciosum) *growing in bright light or deep shade in natural habitats. All are CAM plants (unpublished data of W. W. Adams III & C. B. Osmond)*

Species, location and habitat	Quantum yield* (mol O_2 mol^{-1} photons)	Maximum† photosynthesis (μmol O_2 m^{-2} s^{-1})	Chlorophyll (μg g^{-1} f. wt)
Hoya australis			
Coastal headland			
Shade	0·105	10·0	568
Sun	0·050	10·5	141
Pyrrosia confluens			
Coastal headland			
Shade	0·088	17·0	258
Sun	0·037	15·0	396
Dendrobium speciosum			
Dry rain forest			
Shade	0·101	26·0	163
Sun	0·050	11·0	88

* Measured at CO_2 saturation in leaf disc electrode.
† Measured at CO_2 and light saturation.

epiphytes seem to accommodate bright light with reduced quantum yield and without acclimation of the light- and CO_2-saturated rate of photosynthesis (Table 4).

One of the most interesting of all problems in photosynthesis and carbon economy is that of utilization of sun flecks by understorey vegetation in rain forests where 55 to 80 % of the PAR is received in this form (Chazdon & Fetcher, 1984). Chazdon & Pearcy (1986a, b) have made comprehensive analyses of photosynthesis in response to this supremely stochastic environment and identified the importance of induction status, maintained by the background irradiance and post-illumination CO_2 fixation [Fig. 4(a), (b)]. This is one environment in which the induction phenomena examined at length in biochemical studies seem to play a major role in carbon economy. It is also an environment in which integration of photosynthesis and growth can be pragmatically simplified to a relationship between relative growth rate and duration per day of sun flecks [Fig. 4(c)]. Whether the organization

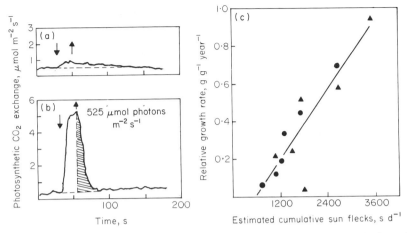

Fig. 4. Integration of photosynthetic carbon metabolism and carbon economy during utilization of sun flecks. The first or widely spaced sun flecks (a) are not very effective in *Alocasia* but, after induction in a period of low light, or after several sun flecks (b), a 20 s sun fleck can achieve nearly 50 % more than the same period of steady state photosynthesis. This is partly due to the substantial post-illumination CO_2 fixation [(b), shaded area]. Long-term estimates of growth rate in two Hawaiian understorey trees (c) show good correlations with potential cumulative sun flecks estimated from hemispherical photographs. Redrawn from Chazdon & Pearcy (1986b) and Pearcy (1983). (a) *Alocasia*, non-induced; (b) *Alocasia*, induced; (c) ●, *Claoxylon*; ▲, *Euphorbia*.

of the photosynthetic apparatus of rain forest understorey plants which optimizes use of sun flecks precludes acclimation to sustained high irradiance remains to be determined.

Ecological success, it seems, often depends on more than a thriving carbon economy and a well-tuned photosynthetic apparatus. From casual observation, *S. dulcamara* does not often flower in deep shade but reproduces successfully by vegetative means. The fronds of *Pyrrosia* exposed to the sun and the bleached runners of *Hoya* bear many more spores and flowers, respectively, than their shaded counterparts. It is possible that shaded *Solanum* and sun-exposed CAM epiphytes and climbers are near the limits of their distribution. We have learned much in physiological ecology by comparing successful, different species from extreme, contrasting habitats. If we are to understand the importance of disturbance, competition and stress in the dynamics of vegetation we might do well to examine species at the margins of their distribution rather than at the centre where, well within the tolerance limits, it may be difficult to discover what is important.

CARBON ECONOMY AND NUTRITION

In the last volume of the series, *Physiological Plant Ecology*, in the *Encyclopedia of Plant Physiology* dealing with ecosystem processes, contributors identified functional nutrient economy as one of the most important complexes of physiological processes deserving our attention. Although ecology has long been given to the accounting of nutrient capital and distribution, plant physiology has yet to provide the tools to translate these inventories into biological activity. On reflection, this may be because the best-funded areas of plant research in agriculture and crop physiology have been premised on the elimination of nutrition as a limiting factor. The fundamental links between leaf nitrogen economics and

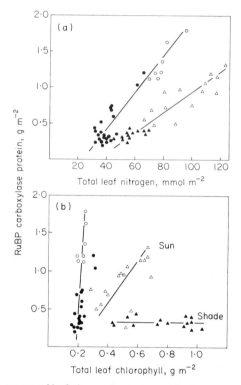

Fig. 5. Quantitative aspects of leaf nitrogen budgets in an herbaceous species from sunny habitats (*Phaseolus*, circles) and a shrub from deeply shaded habitats (*Alocasia*, triangles). Plants were grown in the shade [*Phaseolus* 300 to 500 μmol photons m^{-2} s^{-1} (●), *Alocasia* 20 μmol photons m^{-2} s^{-1} (▲)] or full sunlight (○, △) with abundant nitrogen nutrition (J. R. Seemann, T. D. Sharkey, & C. B. Osmond, unpublished date).

the machinery of photosynthesis on the one hand and between leaf phosphate pools and the currency of photosynthesis on the other are now enjoying long-deserved attention (Field & Mooney, 1985; Sharkey, 1985).

Leaf nitrogen budget and, indeed, the nitrogen budget of the biosphere is dominated by Rubisco, the primary carboxylase of photosynthetic CO_2 fixation. The availability of methods based on use of antibodies and analogues of substrates for quantitative estimation of this protein is providing functional insights into photosynthesis and leaf nitrogen relationships. *Alocasia* achieves about half the rate of photosynthesis as *Phaseolus* at the same concentration of nitrogen in the leaf because if contains only about half the amount of Rubisco [Fig. 5(a)]. Similar methods for other important soluble proteins of the photosynthetic apparatus may soon be available but less direct, more equivocal methods are available for the insoluble membrane proteins of the thylakoid. However, as shown in Figure 5(b), the ratio of Rubisco protein to chlorophyll as a function of nitrogen in the leaf reveals something about the functional nutrient economy of photosynthesis in *Phaseolus* and *Alocasia*. In *Phaseolus*, amounts of carboxylase protein may change seven- to eight-fold in response to light conditions, without change in the amount of chlorophyll. Although *Alocasia* may show a three-fold increase in concentration of Rubisco when grown in high light with high nitrogen, it seems unable to do this without increased chlorophyll and presumably increased investment in chlorophyll protein complexes. Greater negotiability of chloroplastic membrane and stromal proteins may underlie acclimation to light in herbaceous species.

Whatever the fine detail, there is now a clear link between nitrogen nutrition and shade–sun acclimation, as well as sensitivity to, and recovery from, photo-inhibition. *S. dulcamara* will not acclimate from shade to sun at low nitrogen supply (Osmond, 1983) and comparable data have now been obtained with several other species. Low nitrogen-grown *Solanum* plants transferred from shade to sun suffer extensive photoinhibition and take longer to recover maximum quantum yield and light-saturated rate of photosynthesis than do high nitrogen plants (Ferrar & Osmond, 1986). Casual observation suggests that older leaves of low nitrogen plants senesce more rapidly on transfer from shade to sun, perhaps thereby making limited nitrogen resources available for acclimation and repair of photoinhibition in younger leaves. Greer, Berry & Björkman (1986) have estab-lished the requirements for temperature and protein synthesis for recovery from photoinhibition in leaves from shade-grown *Phaseolus vulgaris*. When this species is grown in full sun or (50%) shade, leaves of low nitrogen-treated plants suffer more extensive photoinhibition in CO_2-free air with 1% O_2 than do leaves supplied with high nitrogen (Seemann, Sharkey & Osmond, unpublished data).

CONCLUSIONS

I have tried to illustrate how new integrative and reductionist techniques can be applied to the investigation of photosynthesis in an ecological context. These approaches tend to draw our attention away from attempts to construct carbon budgets on the basis of CO_2 exchange functions in leaves or canopies. Although new and more portable instruments are now available which facilitate such accounting, it is my view that this may continue to be a relatively unproductive exercise. Forays into carbon economy will be necessary to validate some of the more interesting systems, it is true, but to validate rather than serve as prime inputs. I suspect that it is much more important to understand the functional relationships of the productive structures in vegetation and to understand these in a sufficiently pragmatic way that we can make sense of ecological processes. This means, I think, that we need understanding of lots of untidy, often counter-intuitive, relationships in photosynthetic physiology of plants at the margins of their distribution. We need to know much more about photosynthetic responses to interactive stress factors and disturbance, especially near the tolerance limits where they are translated into presence or absence of the individual. We need to know much more about the partitioning of photosynthate in different plants, especially into reproductive structures which underlie competitive interactions, and which make some attributes vital.

REFERENCES

Adams, W. W., Nishida, K. & Osmond, C. B. (1986). Quantum yields of CAM plants measured by photosynthetic O_2 evolution. *Plant Physiology*, **81**, 297–300.

Adams, W. W., Osmond, C. B. & Sharkey, T. D. (1987). Responses of two CAM species to different irradiances during growth and susceptibility to photoinhibition by high light. *Plant Physiology*. (In press.)

Adams, W. W., Smith, S. D. & Osmond, C. B. (1987). Photoinhibition of the CAM succulent *Opuntia basilaris* growing in Death Valley: evidence from 77 K fluorescence and quantum yield. *Oecologia*, **71**, 221–228.

ANDERSON, J. M. & OSMOND, C. B. (1987). Shade-sun responses: compromises between photoinhibition and acclimation. In: *Photoinhibition, Topics in Photosynthesis* (Ed. by D. J. Kyle, C. B. Osmond & C. J. Arntzen), Vol. 9. Elsevier, Amsterdam. (In press.)

BARBER, J. & BAKER, N. R. (Eds) (1985). *Photosynthetic Mechanisms and The Environment, Topics in Photosynthesis*, vol. 6. Elsevier, Amsterdam.

BASKIN, J. M. & BASKIN, C. C. (1978). A discussion of the growth and competitive ability of C_3 and C_4 plants. *Castanea*, **43**, 71–76.

BAZZAZ, F. A. & PICKETT, S. T. A. (1980). Physiological ecology of tropical succession: a comparative review. *Annual Review of Ecology and Systematics*, **11**, 287–310.

BJÖRKMAN, O. (1981). Responses to different quantum fluxes. In: *Physiological Plant Ecology I, Encyclopedia of Plant Physiology* (New Series), vol. 12A (Ed. by O. L. Lange, P. S Nobel, C. B. Osmond & H. Ziegler), pp. 57–107. Springer-Verlag, Heidelberg.

BJÖRKMAN, O. & DEMMIG, B. (1987). Photon yield of O_2 evolution and chlorophyll fluorescence characteristics at 77 K among vascular plants of diverse origins. *Planta*. (In press.)

BLACK, C. C. (1971). Ecological implications of dividing plants into groups with distinct photosynthetic production capacities. *Advances in Ecological Research*, **7**, 87–114.

BOSTON, H. & ADAMS, M. (1986). The contribution of crassulacean acid metabolism to the annual productivity of two aquatic vascular plants. *Oecologia*, **68**, 615–622.

BOUTTON, T. W., HARRISON, A. T. & SMITH, B. H. (1980). Distribution of biomass of species differing in photosynthetic pathway along an altitudinal transect in south eastern Wyoming grassland. *Oecologia*, **45**, 287–293.

CALDWELL, M. M., WHITE, R. S., MOORE, R. T. & CAMP, L. B. (1977). Carbon balance, productivity, and water use of cold-winter desert scrub communities dominated by C_3 and C_4 species. *Oecologia*, **29**, 275–300.

CHAZDON, R. L. & FETCHER, N. (1984). Light environments of tropical forests. In: *Physiological Ecology of Plants of the Wet Tropics* (Ed. by E. Medina, H. A. Mooney & C. Vázquez-Yánes), pp. 27–36. Junk, The Hague.

CHAZDON, R. L. & PEARCY, R. W. (1986a). Photosynthetic responses to light variation in rain forest species. I. Induction under constant and fluctuating light conditions. *Oecologia*, **69**, 517–523.

CHAZDON, R. L. & PEARCY, R. W. (1986b). Photosynthetic responses to light variation in rain forest species. II. Carbon gain and light utilization during sunflecks. *Oecologia*, **69**, 524–531.

FARQUHAR, G. D. & RICHARDS, R. A. (1984). Isotopic composition of plant carbon correlates with water-use efficiency of wheat genotypes. *Australian Journal of Plant Physiology*, **11**, 539–552.

FARQUHAR, G. D., O'LEARY, M. H. & BERRY, J. A. (1982). On the relationship between carbon isotope discrimination and the intercellular carbon dioxide concentration in leaves. *Australian Journal of Plant Physiology*, **9**, 121–137.

FERRAR, P. J. & OSMOND, C. B. (1986). Nitrogen supply as a factor influencing photoinhibition and photosynthetic acclimation after transfer of shade grown *Solanum dulcamara* to bright light. *Planta*, **168**, 563–570.

FIELD, C. & MOONEY, H. A. (1986). The photosynthesis–nitrogen relationship in wild plants. In: *On the Economy of Plant Form and Function* (Ed. by T. J. Givnish), pp. 25–55. Cambridge University Press, Cambridge.

GIFFORD, R. M. (1974). A comparison of potential photosynthesis, productivity and yield of plant species with differing photosynthetic metabolism. *Australian Journal of Plant Physiology*, **1**, 107–117.

GREER, D., BERRY, J. A. & BJÖRKMAN, O. (1986). Photoinhibition of photosynthesis in intact bean leaves; Role of light and temperature and requirement for chloroplast–protein synthesis during recovery. *Planta*, **168**, 253–260.

GRIFFITHS, H., LÜTTGE, U., STIMMEL, K.-H., CROOK, C. E., GRIFFITHS, N. M. & SMITH, J. A. C. (1986). Comparative ecophysiology of CAM and C_3 bromeliads. III. Environmental influences on CO_2 assimilation and transpiration. *Plant Cell and Environment*, **9**, 385–393.

GRIME, J. P. (1979). *Plant Strategies and Vegetation Processes*. Wiley, London.

HATTERSLEY, P. W. (1983). The distribution of C_3 and C_4 grasses in Australia in relation to climate. *Oecologia*, **57**, 113–128.

HERNDON, A. (1985). Naturalized Aroids. *Aroideana*, **8**, 44–46.

KEELEY, J. E., OSMOND, C. B. & RAVEN, J. A. (1984). *Stylites*, a vascular land plant without stomata absorbs CO_2 via its roots. *Nature*, **310**, 694–695.

KUPPERS, M. (1984a). Carbon relations and competition between woody species in a Central European hedgerow. I. Photosynthetic characteristics. *Oecologia*, **64**, 332–343.

KUPPERS, M. (1984b). Carbon relations and competition between woody species and a Central European hedgerow. II. Stomatal responses, water use, and hydraulic conductivity of the root/leaf pathway. *Oecologia*, **64**, 344–354.

LANGE, O. L., SCHULZE, E.-D., EVENARI, M., KAPPEN, L. & BUSCHBOM, U. (1978). The temperature-related photosynthetic capacity of plants under desert conditions. III. Ecological significance of the seasonal changes of the photosynthetic response to temperature. *Oecologia*, **34**, 89–100.

LANGENHEIM, J. H., OSMOND, C. B., BROOKS, A. & FERRAR, P. J. (1984). Photosynthetic responses to light in seedlings of selected Amazonian and Australian rainforest tree species. *Oecologia*, **63**, 215–224.

LOOMIS, R. S. & GERAKIS, P. A. (1975). Production of agricultural ecosystems. In: *Photosynthesis and*

Productivity in Different Environments, IBP vol. III (Ed. by J. P. Cooper), pp. 145–172. Cambridge University Press, Cambridge.

LÜTTGE, U. (1986). Nocturnal water storage in plants having Crassulacean acid metabolism. *Planta*, **168**, 287–289.

MARTIN, C. E. & ADAMS, W. W. (1986). Crassulacean acid metabolism, CO_2-recycling, and tissue desiccation in the Mexican ephiphyte *Tillandsia schiedeana* Steud. (Bromeliaceae). *Photosynthesis Research*. (In press.)

MONTEITH, J. L. (1978). Reassessment of maximum growth rates for C_3 and C_4 crops. *Experimental Agriculture*, **14**, 1–5.

NOBEL, P. S. (1976). Water relations and photosynthesis of a desert CAM plant, *Agave deserti*. *Plant Physiology*, **58**, 576–582.

NOBEL, P. S. (1985). PAR, water and temperature limitations on the productivity of cultivated *Agave fourcroydes* (henequen). *Journal of Applied Ecology*, **22**, 157–173.

OBERBAUER, S. F. & STRAIN, B. R. (1984) Photosynthesis and successional status of Costa Rican rainforest trees. *Photosynthesis Research*, **5**, 227–232.

O'LEARY, M. H. (1981). Carbon isotope fractionation in plants. *Phytochemistry*, **20**, 553–567.

OSMOND, C. B. (1981). Photorespiration and photoinhibition; some implications for the energetics of photosynthesis. *Biochimica et Biophysica Acta*, **639**, 11–98.

OSMOND, C. B. (1983). Interactions between irradiance, nitrogen nutrition, and water stress in the sun–shade responses of *Solanum dulcamara*. *Oecologia*, **57**, 316–321.

OSMOND, C. B. (1984). CAM: regulated photosynthetic metabolism for all seasons. In: *Advances in Photosynthesis Research*, vol. 3 (Ed. by C. Sybesma), pp. 557–564. Martinus Nijhoff/Junk, The Hague.

OSMOND, C. B., NOTT, D. L. & FIRTH, P. M. (1979). Carbon assimilation patterns and growth of the introduced CAM plant *Opuntia mermis* in Eastern Australia. *Oecologia*, **40**, 65–76.

OSMOND, C. B., BJÖRKMAN, O. & ANDERSON, D. J. (1980). *Physiological Processes in Plant Ecology. Ecological Studies*, vol. 36. Springer–Verlag, Heidelberg.

OSMOND, C. B., WINTER, K. & ZIEGLER, H. (1982). Functional significance of different pathways of CO_2 fixation in photosynthesis. In: *Physiological Plant Ecology II, Encyclopedia of Plant Physiology*, New Series, vol. 12B (Ed. by O. L. Lange, P. S. Nobel, C. B. Osmond & H. Ziegler), pp. 479–547. Springer–Verlag, Heidelberg.

PEARCY, R. W. (1983). The light environment and growth of C_3 and C_4 tree species in the understorey of a Hawaiian forest. *Oecologia*, **58**, 19–25.

PEARCY, R. W. (1986). Photosynthetic responses of tropical forest trees. In: *Proceedings, International Conference on Tropical Plant Ecophysiology* (Ed. by D. Doley, C. B. Osmond & W. Wongkaew). Biotrop Special Publication 24, Bogor. (In press.)

PEARCY, R. W., TUMOSA, N. & WILLIAMS, K. (1981). Relationships between growth, photosynthesis and competitive interactions for a C_3 and C_4 plant. *Oecologia*, **48**, 371–376.

POWLES, S. B. (1984) Photoinhibition of photosynthesis by visible light. *Annual Review of Plant Physiology*, **35**, 15–44.

POWLES, S. B. & BJÖRKMAN, O. (1982). Photoinhibition of photosynthesis: effect on chlorophyll *a* fluorescence at 77 K in intact leaves and chloroplast membranes of *Nerium oleander*. *Planta*, **156**, 97–107.

RUESS, B. R. & ELLER, B. M. (1985). The correlation between crassulacean acid metabolism and water uptake in *Senecio medley-woodii*. *Planta*, **166**, 57–66.

SCOTT, D. (1974). Description of relationships between plants and environment. In: *Vegetation and Environment. Handbook of Vegetation Science*, part 6 (Ed. by W. D. Billings & R. R. Strain), pp. 49–69. Junk, The Hague.

SCHULZE, E.-D., LANGE, O. L., EVENARI, M., KAPPEN, L. & BUSCHBOM, U. (1980). Long-term effects of drought on wild and cultivated plants in the Negev Desert. II. Diurnal patterns of net photosynthesis and daily carbon gain. *Oecologia*, **45**, 19–25.

SHARKEY, T. D. (1985). Photosynthesis in intact leaves of C_3 plants: physics, physiology and rate limitations. *Botanical Review*, **51**, 53–105.

SLACK, C. R., ROUGHAN, R. G. & BASSETT, H. C. M. (1974). Selective inhibition of mesophyll chloroplast development in some C_4 pathway species by low night temperature. *Planta*, **118**, 67–73.

TEERI, J. A. & STOWE, L. G. (1976). Climatic patterns and the distribution of C_4 grasses in North America. *Oecologia*, **23**, 1–12.

TIESZEN, L. L., SENIYIMBA, M. M., IMBAMBA, S. K. & TROUGHTON, J. H. (1979). The distribution of C_3 and C_4 grasses and carbon isotope discrimination along an altitudinal and moisture gradient in Kenya. *Oecologia*, **37**, 337–350.

WALKER, D. A. & OSMOND, C. B. (1986). Measurement of photosynthesis *in vivo* with a leaf disc electrode: correlations between light dependence of steady-state photosynthetic O_2 evolution and chlorophyll *a* fluorescence transients. *Proceedings of The Royal Society, Series B*, **227**, 267–280.

WINTER, K., OSMOND, C. B. & PATE, J. S. (1981). Coping with salinity. In: *The Biology of Australian Native Plants* (Ed. by J. S. Pate & A. J. McCoomb), pp. 83–113. University of Western Australia Press, Perth.

WINTER, K., OSMOND, C. B. & HUBICK, K. T. (1986). Crassulacean acid metabolism in the shade. Studies on an epiphytic fern *Pyrrosia longifolia* and other rain forest species from Australia. *Oecologia*, **68**, 224–250.

New Phytol. (1987) **106** (Suppl.), 177–200

VARIATION IN GENOMIC FORM IN PLANTS AND ITS ECOLOGICAL IMPLICATIONS

By MICHAEL D. BENNETT

Plant Breeding Institute, Maris Lane, Trumpington, Cambridge CB2 2LQ, UK

SUMMARY

The gross form of the nuclear genome varies greatly among plant species in both anatomy and genetic organization. Chromosome number (n) ranges from 2 to over 600, and ploidy from 1 to over 20. The amount of DNA in the unreplicated haplophase genome (the $1C$ value) differs by more than 2500-fold among angiosperms. Although it has been questioned since the 1930s whether such variation is of adaptive significance and whether it is related, perhaps causally, with environmental factors, no direct or causal links have yet been found. However, variation in DNA C-value has far-reaching biological consequences and can be of considerable adaptive and hence ecological significance. Strikingly precise interspecific relationships exist between DNA C-value and many diverse phenotypic characters at the cellular level, and DNA can affect the phenotype in two ways, firstly by expression of its genic content and, secondly, by the biophysical effects of its mass and volume, the latter defined as nucleotypic effects. Nucleotypic variation in DNA C-value sets absolute limits to both the minimum size and mass of the basic unit of plant anatomy (i.e. the cell) and the minumum time needed to produce a similar cell with newly synthesized organic molecules. Moreover, in complex multicellular vascular plants, such effects at successive cell cycles are additive, so that DNA C-value influences many characters, including growth rate, seed weight, minimum generation time and type of life-cycle. Thus, the nucleotype profoundly affects where, when and how plants grow. Selection for a particular genomic form acting on its spatial or temporal consequences may occur at various levels ranging from the cell to the whole organism and may operate throughout the life-cycle or at just one stage. DNA C-value is often indirectly related to environmental factors which determine time-limited environments via selection acting on the temporal phenotypic consequences of nucleotypic variation. However, in the case of radio-sensitivity, selection for a low DNA C-value may act directly on the nucleotype itself, as the size of the nuclear DNA target directly affects the ability of the plant to survive.

Key words: Plant genomic form, DNA C-value, DNA amount, cytoecology, nucleotype.

GENOMIC FORM – CONSTANCY AND VARIATION

Many great discoveries in biology this century have concerned the genome and its form – that the genetic material is DNA, that this has a double helical structure, that this form is important in semi-conservative replication and cell heredity, and that the genetic code is written as triplets of nucleotides. In these important respects, genomic form is the same in all eukaryotes. These discoveries emphasized aspects of constancy in genomic form, except for the nucleotide base sequences of genic DNA whose variation determines the phenotype. However, much cytological work also showed that the genome can vary greatly in both anatomy and genetic organization among taxa.

This review is concerned with variation in aspects of gross genomic form in nuclei, and the search for unifying hypotheses to rationalize it and to relate the variation with associated environmental characters. Interesting results are also becoming available for mitochondrial and chloroplast genomes (Dyer, 1984; Bendich, 1985; Palmer, 1985).

0028-646X/87/05S177 + 24 $03.00/0

INTERSPECIFIC VARIATION IN GROSS GENOMIC FORM

Variation in chromosome number and size

In plants and other eukaryotes, the genome is contained in chromosomes, each of which contains a single DNA double helix. The number of chromosomes in the unreplicated gametic genome (n) varies greatly. In bryophytes n ranges from 4 to 66 (Fritsch, 1982). Some algae have more than 500 chromosomes (Nicholas, 1980)

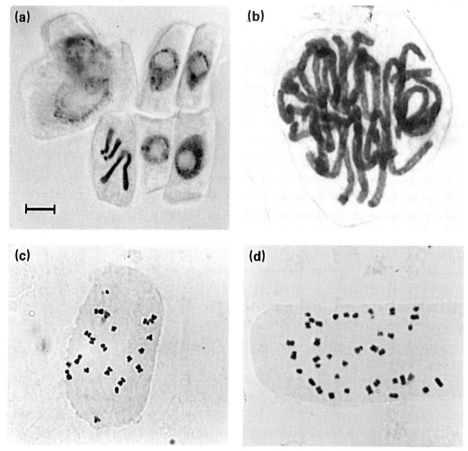

Fig. 1. A range of chromosome numbers and sizes in angiosperms seen in a comparison of the mitotic root-tip complements of (a) *Haplopappus gracilis* ($2n = 2x = 4$), $1C$ DNA = 2 pg; (b) *Fritillaria aurea* ($2n = 2x = 24$). $1C$ DNA = 81·9 pg; (c) *Brassica campestris* ($2n = 2x = 20$), $1C = 0·46$ pg and (d) *Brassica napus* ($2n = 4x = 38$), $1C = 0·86$ pg. (Bar = 5 μm.)

and in the fern *Ophioglossum reticulatum* L. n exceeds 600 (Abraham & Ninan, 1954). Two animal species are known where $n = 1$ (Crossland & Crozier, 1986) but no plant with only a single chromosome is known. In the angiosperms n ranges from two in dicotyledons [e.g. *Haplopappus gracilis* (Nutt.) A. Gray (Jackson, 1957) Fig. 1(a)] and monocotyledons [e.g. *Zingeria biebersteiniana* (Claus.) P. Smirnov (Bennett, Smith & Seal, 1986)] to at least 250 in a dicotyledonous species (Grant, 1982).

Chromosome size also varies greatly between species (Fig. 1). For example, mean chromosome length varied from 0·6 μm to more than 14 μm, while total chromosome length per diploid genome varied from 14·6 μm to more than 250 μm in a sample of 856 angiosperm species (Levin & Funderberg, 1979). [N.B. This excludes their listed mean (31·4 μm) and total (442·6 μm) chromosome lengths for *Phleum pratensis** which are about an order of magnitude too large.]

Variation in ploidy level

Variation in chromosome number (*n*) has varied causes (e.g. aneuploidy) but it mainly reflects changes in: (1) the number of genetically different chromosome types in the basic ancestral set (known as *x*), and/or (2) the number of such sets in the nuclear genome, i.e. in ploidy. [N.B. Species in which $2n = 2x$ are called diploids, while those with more than two basic sets, e.g. Fig. 1(d), are called polyploids.]

The incidence of polyploidy differs greatly between major plant groups (Lewis, 1980a), being uncommon in gymnosperms where fewer than 5% of species are polyploids, common in pteridophytes and very common in angiosperms. Lewis (1980a) and Goldblatt (1980) estimated that at least 58% of monocotyledons and 43% of dicotyledons are polyploids. The level of polyploidy also varies widely between species. For example, the genus *Rumex* has a polyploid series ranging from $2x$ to $20x$ (Ichikawa *et al.*, 1971).

DNA C-value

Nuclear DNA amount varies about 5000-fold among algae, 1300-fold among pteridophytes but only 15-fold among gymnosperms (Sparrow, Price & Underbrink, 1973; Cavalier-Smith, 1985a). Estimates of the amount of DNA in the unreplicated gametic genome (the $1C$ value) in angiosperms vary by more than 2500-fold, from 0·05 pg in *Cardamine amara* (S. R. Band, *pers. comm.*) to 127·4 pg in *Fritillaria assyriaca* Baker (Bennett & Smith, 1976). Thus, in angiosperms, DNA *C*-value is more variable than ploidy level or chromosome number.

In diploids and their newly formed polyploids DNA *C*-value is not independent of chromosome number and ploidy level – the former is directly proportional to the latter. While this relationship is seen in some naturally occurring polyploid series (e.g. *Hordeum*; Bennett & Smith, 1971), it is not seen in others (e.g. *Bulnesia*; Poggio & Hunziker, 1986). Thus, there is no general correlation between DNA *C*-value and either chromosome number or ploidy level. Indeed, comparisons of large samples of angiospermous species show that polyploids may have DNA *C*-values equal to, or lower than diploids (Bennett, 1972; Smith & Bennett, unpublished data).

Intraspecific Variation in Gross Genomic Form

Chromosome number ($2n$), ploidy level and DNA *C*-value are virtually constant for many species. Thus, these characters can have taxonomic value. Nevertheless, they are not immutable and many species display variation for one or more of these characters (Stebbins, 1971; Lewis, 1980b). For example, the greater celandine (*Chelidonium majus*) has a variable base number – $2n = 12$ in Britain and elsewhere but, in Japan, $2n = 10$ (Darlington & Wylie, 1955). Races with different ploidy

* Except where authorities are given, nomenclature follows Clapham, Tutin & Warburg (1981).

levels occur in many species. In Britain, the lesser celandine (*Ranunculus ficaria*) has diploid ($2n = 16$) and tetraploid ($2n = 32$) races and possible triploid hybrids ($2n = 24$) between them (Gill *et al.*, 1972).

Intraspecific variation in chromosome number and ploidy level approaches the same order as interspecific variation in angiosperms. Thus, in *Cardamine pratensis*. $2n = 16$, 24, 28, 30, 32 to 38, 40 to 46, 48, 52 to 64 and 67 to 96 (Clapham, Tutin & Warburg, 1962). Similarly, *Claytonia virginica* L. has base numbers of six, seven and eight, $2n$ ranges from 12 to *c.* 191, and a polyploid series ranging from $2x$ to $24x$ (Stebbins, 1971).

Perhaps because of the constancy in several aspects of genomic form (see above), the amount of DNA per species was at first regarded as a constant [e.g., Dounce (1955) wrote: '...the mean quantity of DNA per cell nucleus is constant within a species for normal resting diploid somatic cells...']. Indeed, in the term C-value coined by Swift 'C' stood for constant.

No significant variation in DNA C-value was detected in some species e.g. *Triticum monococcum* L. (Furuta, Nishikawa & Haji, 1978). However, it is now accepted that intraspecific variation in DNA C-value, despite a constant chromosome number, is not rare. Bennett (1985) listed 24 angiospermous examples including variation of 37% between lines of *Zea mays* L. (all with $2n = 20$) cultivated in Mexico or the USA (Laurie & Bennett, 1985), and about 100% between plants of *Poa annua* (all with $2n = 28$) from a field in Wales (Grime, 1983). The largest variation listed was 288% in *Collinsia verna* Nutt. ($2n = 14$), reported by Greenlee, Rai & Floyd (1984).

GROSS GENOMIC FORM AS PHENOTYPIC ADAPTATION

It has long been known that plants are adapted to their environments and that such adaptations can determine their distribution. As the large variation in chromosome size and number became clear, it was seen as a form of phenotypic variation at the cellular level. It seemed reasonable to ask whether such variation is environmental adaptation and whether it directly or indirectly affects plant distribution. If so, it might be possible to predict where a plant would grow best simply from a knowledge of its gross genomic anatomy. The idea was beguilingly simple and seemed easy to test.

Thus, since the 1930s biologists have sought relationships between variation in the number and morphology of chromosomes and plant form and distribution and environmental factors. Two main approaches have been followed. *Firstly*, enquiries in particular species, genera or families were made into the ecological relationships between different chromosome sizes or numbers, ploidy levels and later DNA C-values. *Secondly*, broader enquiries have analyzed the frequencies of species with different chromosome sizes or numbers, ploidy levels and DNA amounts, among local floras or plant communities, or between different major taxonomic (e.g. monocotyledons *vs* dicotyledons), life form (e.g. herbaceous *vs* woody), behavioural (e.g. annual *vs* perennial) or geographical (e.g. temperate *vs* tropical) groups.

CHROMOSOME SIZE AND NUMBER, AND PLANT DISTRIBUTION

Particular examples

In many particular examples, chromosomal variation is related to plant distribution or environment. Thus, different cytotypes often tend to occupy niches which differ in the availability of moisture, light, temperature or the special conditions found in cultivation. In *Cardamine pratensis*, low chromosome numbers appear to favour drier habits (Clapham *et al.*, 1962), while in *C. virginica*, diploids

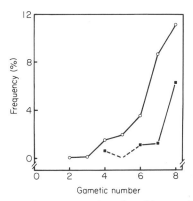

Fig. 2. The relative frequency of occurrence of species with gametic chromosome numbers (*n*) of eight or fewer in large samples of 2665 woody and 5287 herbaceous dicotyledons. ○, Herbaceous species; ■, woody species. [Data plotted from table 2 of Grant (1982).]

with $2n = 12$, 14 and 16 occur in different localized parts of the USA. Similarly, diploid *R. ficaria* is widespread in both sunny and shady places, while the distribution of the less common tetraploid is more local and chiefly in shade (Gill *et al.*, 1972). In *Empetrum nigrum*, the diploid ($2n = 26$) occurs at lower elevations in the temperate zone than does the tetraploid ($2n = 52$) form (Lewis, 1980b). In *Leucanthemum vulgare*, diploid ($2n = 18$) and tetraploid ($2n = 36$) cytotypes occur wild, while decaploid ($2n = 90$) and dodecaploid ($2n = 108$) forms are cultivated.

Wider surveys

Broader surveys also showed significant trends between chromosome number and morphology, and plant form, environment or distribution. The following are three noteworthy examples.

(i) *Chromosome number and life form.* Stebbins (1938) showed that the basic chromosome numbers (*x*) were significantly higher in woody than in herbaceous groups. Grant (1982) agreed with Stebbins' conclusions regarding *x*. The most common numbers are $n = 7$ to 9 in herbaceous dicotyledons but, in woody dicotyledons, they are $n = 11$ to 14 with a particularly high frequency of $n = 13$. Figure 2 shows the highly significant dearth of species with low gametic numbers ($2n = 8$ or less) in woody *vs* herbaceous dicots. The reason(s) for this difference are still unknown and the phenomenon merits further attention.

(ii) *Chromosome size and global distribution.* Avdulov (1931) noted that tropical grasses have uniformly small to medium sized chromosomes whereas most grasses of cool temperate regions have large chromosomes, and Stebbins (1966) noted that the same pattern might hold in other families. Recently, Levin &

Funderberg (1979) compared 892 angiospermous species from many families and showed that mean and total chromosome lengths per diploid genome were significantly lower in tropical than in temperate species. They concluded that 'in general genome size is larger in temperate than in tropical plants'.

(iii) *Ploidy level with global distribution.* Hagerup (1932) and Tischler (1934) showed that, among angiosperms, the frequency of polyploids in the flora increased with latitude in Europe. This was confirmed by Löve & Löve (1949) who showed that the frequency of polyploidy also increased with altitude (e.g. in alpine floras).

From early findings, it seemed that much variation in genomic form may have an ecological significance and that simple rules and broad generalizations might emerge from further work. Thus, Stebbins (1938) wrote with implicit optimism: '...there is a definite correlation between habit and growth and the cytological characters of angiosperms...evolution within the same genus or family of new species with different habits of growth has in general been accompanied by more visible changes in the chromosomes than has the evolution of new families or orders. A logical continuation of this study would be to determine to what extent such differences as those between the temperature, moisture and soil requirements of different plants are correlated with cytological differences.' Similarly, Löve & Löve (1949) stated that there is something of a direct relationship between latitude, temperature and the proportion of polyploids in a flora.

Later developments and conclusions

Later studies often confirmed the existence, and extended our knowledge of early trends but they also found many exceptions which, by breaking the 'rules', illustrate the fallacies of simple global calculations and generalized conclusions. For example, the trend of increasing frequency of polyploidy with increasing latitude and altitude was originally taken to indicate a general greater hardiness of polyploids for growth in harsh environments. However, as examples were found where, within a species, the polyploid form grows at a lower latitude or altitude than the diploid (Lewis, 1980b), this was shown to be incorrect. Depending on the species, diploids may be distributed towards higher latitudes and polyploids towards lower latitudes, or conversely.

A similar picture of exceptions to simple rules emerges from recent studies of chromosome size. Although Levin & Funderberg (1979) found a general difference between temperate and tropical species (see above), this was significantly expressed in some families but not others. They concluded that the overall temperate–tropical difference primarily is the result of the geographical replacement of families with different genome sizes rather than the product of parallel evolution within these taxa in response to similar environmental changes.

An overview of the field more than 50 years after its inception shows that variation in chromosomal form is significantly associated with plant form and distribution in many particular and broader comparisons and, indeed, it is often correlated with environmental factors. However, because of numerous exceptions to the trends, or opposite correlations for the same environmental factor in different species, the existence of any simple and general causal connections between chromosome number and size or ploidy level, and associated environmental factors is now largely discounted. For example, Ehrendorfer (1980) concluded; '...it is evident that there are no direct and general connections between polyploidy on the one hand and ecology, habitat or distribution of angiosperms on the other'. Instead of simple explanations involving environmental factors, more complex explanations are advanced which take account of the age of the flora and its spectra of growth forms.

Given the failure to demonstrate direct and causal connections between chromosome number and ploidy level, and environmental characters, interest in variation in gross genomic form focussed instead on the significance of the amount of DNA especially after it was shown that the $1C$ amount is very variable among cohesive groups of species of similar organizational complexity, such as the angiosperms. This puzzling phenomenon was epitomized by Thomas (1971) as the 'C-value paradox'. Given DNA's most fundamental role in determining phenotypic variation, might not DNA amount have a more obvious or direct adaptive significance than other characters of genomic form such as chromosome number?

DNA C-VALUE AND GEOGRAPHICAL DISTRIBUTION

Since 1976, several studies have reported interesting relationships between DNA amount and geographical distribution of angiosperms. In a comparison of cereal grains, pasture grasses and pulses, Bennett (1976) showed that cultivation of species with high DNA amounts per diploid ($2x$) genome tends to be localized in temperate latitudes or to seasons and regions at lower altitudes with temperate conditions. Moreover, man has tended to choose species for cultivation with increasingly lower DNA amounts at successively lower latitudes, so there is a positive cline for DNA amount and latitude. Comparison of DNA amount with sites of domestication led Bennett (1976) to conclude that this cline is a natural phenomenon which man has modified and exaggerated in agriculture.

These results support observations by Avdulov (1931) and Stebbins (1966) (see above) that species with large chromosomes tend to be concentrated in temperate latitudes. Given a close correlation between DNA C-value and chromosome volume (see below), such species are expected to have high DNA amounts. This expectation was confirmed by Levin & Funderberg (1979) who showed that the mean $4C$ DNA-value for temperate species (27·06 pg) was more than double that for tropical species (12·13 pg). (N.B. This difference stems from the greater range of DNA amounts and the higher frequency of species with large genomes in temperate as compared with tropical floras and not from any exclusion of species with small genomes from temperate floras.)

The above cline exists over most of the range of latitudes where agriculture is practised. Over this range, man chooses for cultivation cereals, pasture grasses and pulses with increasingly higher DNA amounts at successively higher latitudes, although obviously these have seemed suitable to him for other reasons. The prime factor suiting them to man's requirements was, and is, their relatively high yield of seed or leaf per unit area per unit time in each environment. Thus, there seems to be an important correlation between latitude and the DNA amount per $2x$ genome giving maximum crop yield, so that species with higher DNA amounts produce their maximum yields at higher latitudes than species with lower DNA amounts (Bennett, 1976).

This DNA-latitude cline for crops was particularly clear in comparisons of the northerly limits of cultivation in winter of cereals in the northern hemisphere. However, the cline was modified between winter and summer (Fig. 3). Thus, the species with high DNA amounts which show a pronounced DNA amount–latitude cline in winter are bunched at the north of their range in summer. Indeed, the cline appears to reverse its polarity in summer at the very high latitudes since, among cereals, barley ($4C = 22\cdot2$ pg of DNA) is grown nearer the pole in summer than rye ($33\cdot1$ pg). Similarly, peas ($4C = 19\cdot5$ pg) are grown nearer the pole in summer

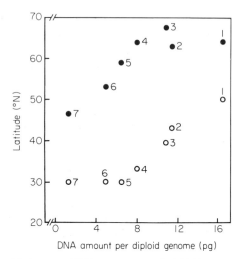

Fig. 3. The relationship between DNA amount per diploid ($2x$) genome and the northern limits of cultivation of several cereal grain species. Key to points: ○, for a transect from Hudson Bay to Key West in Florida (approximately 82° W) in winter; ●, for a transect from near Murmansk by the Arctic Ocean to Odessa by the Black Sea (approximately 32° E) in summer. 1, *Secale cereale*; 2, *Triticum aestivum*; 3, *Hordeum vulgare*; 4, *Avena sativa*; 5, *Zea mays*; 6, *Sorghum* spp.; 7, *Oryza sativa*.

than field beans ($53\cdot5$ pg) – the reverse of their northern limits of distribution in winter. This reversal at high latitudes in summer agrees with the general trend for DNA C-value and latitude noted for non-crop species at high latitudes. Thus, values for DNA amount and DNA amount per diploid ($2x$) genome for angiosperms from South Georgia and the Antarctic Peninsula were within the ranges known for angiosperms at temperate latitudes but, significantly, at their lower end (Bennett, Smith & Lewis Smith, 1982). They concluded that, above a certain high latitude, maximum DNA C-value per nucleus and per diploid genome ($2x$) decreases with increasing latitude (and with decreasing temperature). It was suggested, therefore, that at high latitudes, selection against species with high DNA amounts strongly increases towards the poles as the climate becomes progressively harsher.

Further support for this hypothesis came from Grime & Mowforth's (1982) survey of British angiosperms which showed a significantly higher mean genome size for 69 southern species ($2C = 13\cdot31$ pg) than for 80 widespread species ($7\cdot96$ pg), while the $2C$ DNA amount for eight northern species was even lower ($2C = 2\cdot25$ pg). Interestingly, published $4C$ DNA values for 25 *Allium* taxa range from $30\cdot4$ to 152 pg but that for *A. sibiricum* L., whose range extends inside the Arctic Circle, is the lowest. Even specific names which correctly describe geographical distributions may help to predict DNA C-values! Thus, a species named *antarctica*, *sibirica* or *alpina* etc. will probably not have a high DNA amount compared with the range for angiosperms as a whole.

The available data show that while species with low DNA amounts are ubiquitous, species with the highest DNA C-values are progressively excluded from increasingly harsh environments above a mid-latitude. Results for crop and non-crop species fit this view but the critical latitude above which this exclusion operates is probably higher for crops, owing to man's selection, than for non-crop species subject to natural selection alone. Whether species with high DNA amounts are also totally excluded from tropical floras so that the genomes of tropical floras are 'consistently small' as Grime (1983) stated seems unlikely. Thus, in *Sowerbaea juncea* Sm. from Queensland, Australia $4C = 105$ pg (Stewart & Barlow, 1976), while *Tradescantia guatemalensis* Vell. from Guatemala (Martinez & Ginzo, 1985) has $4C = 94\cdot8$ pg. Moreover, the large chromosomes in some tropical *Crinum* spp. [e.g. *C. latifolium* L. from Borneo (Jones & Smith, 1967)] and cycads indicate high DNA amounts in several diploid species. In the latter larger chromosomes 'proved to be a handicap in the preparation of slides for karyotypic studies' (Marchant, 1968). However, it seems clear that species with medium to very high DNA amounts are most common in latitudes with warm temperate climates. The greatest range of species DNA C-values probably occurs in these latitudes.

Interesting examples were recently noted of intraspecific variation in DNA amount significantly correlated with geographical distribution in both crop, e.g. *Z. mays*, and non-crop, e.g. *Gibasis karwinskyana* (Roem. et Schult.) Rohn, plants (see Bennett, 1985).

The characters on which natural and human selection operates to determine the broad relationships between DNA amount and geographical distribution are probably many and diverse, and are still largely unknown. While the findings just described strongly suggest that variation in DNA amount has adaptive significance related to environmental factors, such correlations neither establish the existence nor explain the nature of any causal links between DNA amount and any environmental factor. Before considering possible causal relationships between DNA amount and plant distribution and environment, it is necessary to describe some spatial and temporal consequences of interspecific variation in DNA amount at the level of the cell. It is the additive effect of these cellular consequences which may ultimately determine many of the relationships between DNA amount and environmental factors at the whole plant level.

CORRELATIONS BETWEEN DNA C-VALUE AND CELLULAR CHARACTERS

Background

Since 1960, considerable evidence has shown that variation in DNA amount correlates with a wide range of phenotypic characters and angiosperms have featured prominently in such work. For reviews, see Bennett (1973) and Cavalier-Smith (1985a, b). In particular, interspecific variation in DNA amount shows strikingly precise positive correlations with various characters of cell size and mass, and with the rate of cell development.

DNA C-value and cell size

DNA C-value is positively correlated with the total length and/or volume of chromosomes at metaphase of mitosis and/or meiosis (Rees *et al.*, 1966; Bennett *et al.*, 1983; Anderson *et al.*, 1985); the total volume of nucleolar material and of

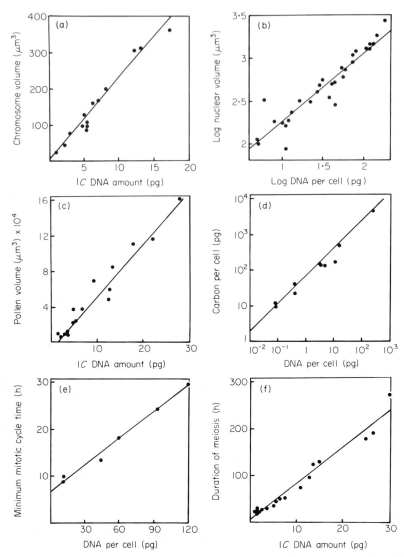

Fig. 4. Relationships between either 1C DNA amount, or mean DNA amount per cell and (a) total mitotic metaphase volume per cell in 14 species (from Bennett *et al*., 1983); (b) mean nuclear volume per cell in apical meristems in 30 herbaceous angiosperms [data from Baetcke *et al*. (1967) as plotted by Cavalier-Smith (1985b)]; (c) pollen grain volume at anther dehiscence in 16 species of wind-pollinated grasses (from Bennett, 1972); (d) total inorganic carbon per cell in 10 species of unicellular algae (from Holm-Hanson, 1969); (e) minimum duration of the mitotic cycle in root-tip cells of six angiosperms grown at 23 °C (redrawn from Van't Hof & Sparrow, 1963) and (f) duration of meiosis in 18 diploid angiosperms grown at 20 °C (from Bennett, 1977).

the interphase nucleus (Baetcke *et al*., 1967; Edwards & Endrizzi, 1975); cell area and volume (Lawrence, 1985); and chloroplast number per guard cell (Butterfass, 1983). For example, Figure 4(a) shows the relationship between DNA amount and total metaphase chromosome volume per roo-tip cell while Figure 4(b) compares DNA amount and nuclear volume in apical meristematic cells of angiosperms. Cell size and shape are not determined solely by DNA C-value and can vary considerably within and between organisms despite a constant nuclear DNA amount. Nevertheless, species with high DNA contents clearly show a general enlargement in cell size. Further evidence of the generality of this phenomenon is shown in Figure 7(a) below, which relates the DNA C-value to the linear dimensions of the leaf epidermal cells in 37 common British herbaceous angiosperms.

DNA C-value and cell weight or mass

Positive correlations also occur between DNA C-values and the total dry mass of metaphase chromosomes, nuclear RNA and protein content, the total dry mass of nucleoli and of the interphase nucleus (Sunderland & McLeish, 1961; Pegington & Rees, 1970), and cell weight (Martin, 1966). For example, Figure 4(d) shows the relationship between DNA content per cell and the total weight of carbon per cell for 10 species of unicellular algae.

DNA C-value and rate or duration of cell development

DNA C-value is positively correlated with the duration of mitotic cell cycles (Evans et al., 1972; Van't Hof, 1975), meiosis (Bennett, 1971, 1977) and pollen development. For example, Figure 4(e) shows the relationship between DNA amount and minimum mitotic cell cycle time in root-tip cells while Figure 4(f) compares DNA C-value and the duration of meiosis at 20 °C in angiosperms.

THE CAUSAL NATURE OF SUCH CORRELATIONS AT THE CELLULAR LEVEL

Clearly, DNA amount correlates closely with many important phenotypic characters at the cellular level and many of the relationships between DNA amount and various cellular characters are strikingly precise for biological phenomena (Fig. 4), being more reminiscent of physical or chemical relationships. As the genomic DNA is contained in, and forms a large proportion of the mass and volume of the chromosomes, large scale interspecific variation in DNA C-value is causally correlated with several biophysical characters at the chromosome level including minimum total volume and mass of chromosomes. Similarly, the chromosomes and the nucleus are contained in, and form a large proportion of, the minimum volume and mass of the nucleus and the cell, respectively, so DNA C-value is also causally correlated with other biophysical characters at the nuclear and cellular phenotypic level, including the minimum cell volume and mass. In other words, it is impossible to increase the DNA C-value greatly (say by 40-fold) without also increasing the minimum chromosome, nuclear and cell volume and mass. Figure 1(a) and (b), which compares typical diploid mitotic cells of *Fritillaria aurea* ($1C = 81.9$ pg) and *H. gracilis* ($1C = 2.0$ pg), shows the impossibility of containing the nucleus, the chromosomes, or even the DNA of the former in the small cells of the latter. Similar logic applies to variation in DNA C-amount in relation to the rate and duration of cell development. For biophysical reasons, it is impossible in practice to increase the DNA C-value greatly without also increasing the minimum time needed to produce a similar cell with newly synthesized molecules.

THE CONCEPT OF THE NUCLEOTYPE

As noted above, it is impossible to increase DNA C-value greatly without also

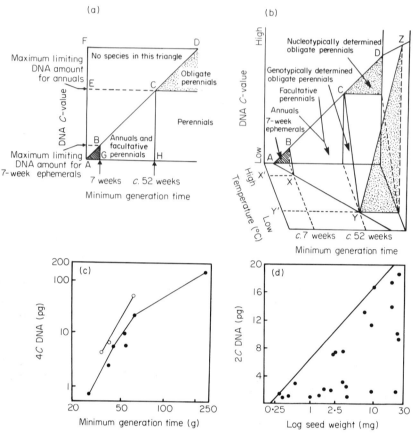

Fig. 5. Relationships between species DNA amount and minimum characters: (a) the model proposed by Bennett (1973) for a simple relationship between DNA C-value and minimum generation time (MGT) in a temperate environment. Points for the shortest MGT for a given DNA amount lie on the line A–D. All points lie on, or to the right of line A–D, with none in triangle ADF. Species with MGTs of 7 or fewer weeks lie in triangle ABG and hence all have low DNA C-values. Species whose DNA C-value exceeds the amount E have MGTs of 52 weeks or more, and are therefore nucleotypically determined, obligate perennials; (b) an extension of the model in (a) to allow for environmental variation in temperature [N.B. the diagram is slightly modified from figure 2 in Bennett et al. (1982)]. All points for angiosperm species plot below the plane trapezium (ADZY). With decreasing temperature, the maximum DNA amount permitting expression of, firstly, an ephemeral MGT (B–X) of 7 weeks or less and, secondly, an annual life-cycle (C–Y) decreases to the minimum DNA amount for angiosperms. Thus, no angiosperm can express an ephemeral life-cycle at temperatures below X′ while, at temperatures continuously below Y′, all angiosperms express an obligate perennial life-style; (c) the relationship between 4C DNA amount and the shortest MGT known for three polyploid (○) and seven diploid (●) angiosperms (Bennett & Smith, unpublished data); and (d) the significant relationship between 2C DNA amount and seed weight in 24 British legume species [modified from figure 3.2.A in Mowforth (1985)]. The distribution of points below the line suggests that a relationship between DNA C-value and minimum seed weight may exist in this cohesive sample of species.

increasing the *minimum* chromosome, nuclear and cell volumes. This fact is independent of the base sequence of the additional DNA and of whether it is genic or not. Thus, nuclear DNA influences the phenotype in two ways; firstly by the expression of its genic content and, secondly, by the physical effects of its mass and volume. The term 'nucleotype' (Bennett, 1971, 1972) defines those conditions of the nuclear DNA which affect the phenotype independently of its encoded informational content. Clearly, many of the correlations known between variation in DNA C-value and cell size, volume and mass and rate of development are largely or partly nucleotypic in origin and hence, to that extent, they are causal in nature.

Importantly, the nucleotype limits the range of phenotypic expression for some characters which can be achieved by gene action. For example, subject to genic control, nuclear volume can vary considerably during development despite a constant DNA content (Bennett, 1970). Nevertheless, such control can operate only at, or above, the limiting minimum nuclear volume determined by the DNA C-value. (N.B. The argument in the previous section was, therefore, concerned solely with the concept of minimum characters such as 'minimum nuclear volume' and 'minimum cell cycle time'.)

Considerable evidence shows that nucleotypic effects at the cellular level are additive and extend to the size, mass and rate of development of multicellular structures and organs. For example, Figure 4(c) shows a significant positive relationship between DNA C-value and the volume of the mature male gameto-phyte in a cohesive sample of wind-pollinated grasses; similar results are known for other families (Bennett, 1973; Lawrence, 1985).

Stebbins (1971) listed examples where reduction in the size of flowers and fruits is accompanied by reduction in the size and number of chromosomes (and hence in DNA amount). In this connection, it may be significant that the 4C DNA amounts of *Briza maxima*, *B. media* and *B. minima* are 43·2, 33·8 and 29·2 pg, respectively! Certainly, Lawrence (1985) reported positive correlations between DNA amount and both capitulum diameter of radiate species and the length of bisexual florets, as well as between plant height and 4C DNA amount for *Senecio* species. Significant positive relationships between DNA amount and seed weight have been found in the genera *Allium*, *Vicia* and *Crepis* (Bennett, 1972; Jones & Brown, 1976) and for both 24 British legumes [Fig. 5(d)] and 32 British grasses (Mowforth, 1985), but not in *Senecio* (Lawrence, 1985).

The clearest positive relationship between the duration of important life-cycle stages and DNA amount are known for animals (see next section and Horner & Macgregor, 1983). However, Mowforth (1985) noted a negative relationship between intraspecific variation in nuclear DNA amount and relative rate of dry matter production in *P. annua*, as might be expected if rate of cell division decreases with increasing DNA amount [Fig. 7(d) below]. As Grime (1983) noted, variance in these data is large, suggesting, as one might expect, that there are attributes of plant additional to those associated with DNA content which control relative growth rate.

THE RELATIONSHIP BETWEEN DNA C-VALUE AND MINIMUM GENERATION TIME

DNA C-value in plants is positively correlated with the two most fundamental characters (i.e. cell size and minimum cell doubling time) which interact to determine growth rate. Moreover, its effects apply to every cell and tissue, and act at all stages in the life history of the plant. Consequently, Bennett (1972) questioned whether variation in DNA C-value ultimately determines the maximum rate of development of the whole plant and hence its limiting minimum generation time. [N.B. Minimum generation time (MGT) is the duration of the minimum period from germination until production of the first mature seed.]

The original test (Bennett, 1972) compared four groups of herbaceous angiosperms: (1) ephemeral species which can complete a life-cycle in a very few weeks; (2) annuals which can complete a life-cycle in 52 weeks or less; (3) non-annuals which can set fertile seed within 52 weeks of germination (called facultative perennials) and (4) non-annuals which require more than 52 weeks to produce mature seed (called obligate perennials). The critical test was to see, firstly, whether the maximum DNA value for species in these classes increased as the longest MGT defining the classes increased, and, secondly, whether the range of life forms expressed by plants decreased with increasing DNA C-value.

The results showed that: (1) maximum DNA amount for ephemerals (MGT not greater than about 7 weeks) was much less than for annuals (MGT not greater than 52 weeks); (2) the maximum DNA amount for annuals (MGT not greater than 52 weeks) was much less than the maximum DNA amount for obligate perennials (maximum MGT much greater than 52 weeks); (3) the maximum DNA amount for facultative perennials (MGT not greater than 52 weeks, as in annuals) was similar to the maximum DNA amount for annuals, and much less than for obligate perennials; and (4) the range of types of life-cycle decreased with increasing DNA C-value. The MGTs for species with 4C DNA amount below 10·1 pg ranged from a month to many years but, above 10·1 pg, no 7-week ephemeral life-cycles and, above 82·8 pg, no annual life-cycles were found although many perennial species had DNA C-values in excess of this amount.

These results were interpreted as fitting a model [Fig. 5(a)] for a simple positive relationship between DNA C-value and the shortest MGT for species with each DNA amount. Later comparisons of DNA amounts for angiosperms also support this model, providing the original four groups are distinguished, e.g. in *Senecio* the maximum 4C DNA amount for ephemerals (14·9 pg) was lower than the maxima for annuals (42·9 pg) and facultative perennials (37·5 pg) (Lawrence, 1985). Similarly, analysis of data for 161 British species by Grime & Mowforth (1982) shows a maximum DNA C-value for ephemerals with a MGT of 7 weeks of less than 10 pg, higher but similar maximum DNA C-value for annuals and facultative perennials (59·6 pg and at least 45·2 pg for *Bromus erectus*, respectively) and a much higher maximum DNA C-value for obligate perennials (282·7 pg).

Published generation times are easy to find for many species but MGTs which closely approach the true minimum for species with their DNA amounts are hard to obtain. However, Figure 5(c) plots such 'record' generation times for 10 herbaceous angiosperms (ranging from 31 d in *Arabidopsis thaliana* to at least eight months in *Lilium longiflorum* Thunb.) against DNA amount. Clearly, the shortest MGT increases with increasing DNA C-value and the plot suggests that the MGT may be slightly shorter in polyploids than in diploids with the same nuclear DNA amount.

Together, these results support the conclusion (Bennett, 1972), 'that there is a relationship between nuclear DNA content and MGT in herbaceous higher plants. There is apparently a maximum limit to the mass of nuclear DNA for species which can complete development within a given time'.

DNA C-VALUE AND TYPE OF LIFE-CYCLE

The comparison mentioned above reveals a clear relationship between DNA C-value and the range of types of life-cycle displayed – the latter decreasing as the former increased. Species with low DNA amounts included ephemerals and long-lived perennials, while those with very high DNA amounts were exclusively obligate perennials. The data therefore show that nuclear DNA content can limit the types of life history which a species can display subject to genic control (Bennett, 1972; Smith & Bennett, 1975). For example, species with medium DNA amounts cannot be 7-week ephemerals, while those with very high DNA amounts cannot be ephemerals or even annuals [Fig. 5(a)].

The maximum DNA amounts for species which can display an ephemeral or an annual life-cycle are expected to vary between environments and to decrease with decreasing duration and/or temperature of the growing season (Bennett, 1972). Figure 5(b) which extends the original model for a temperate environment [Fig. 5(a)] to include the effects of temperature variation, predicts that, as the climate becomes cooler and the growing season shorter, the maximum DNA amount permitting, firstly, an ephemeral life-cycle (B–X) and, secondly, an annual life-cycle (C–Y) will decrease to the minimum DNA amount for angiosperms. As the environment becomes colder with increasing latitude, firstly, species displaying an ephemeral life-cycle and, secondly, an annual life-cycle are expected to become rare and then disappear from the angiospermous flora. These expectations are realized in Nature. Annual angiosperms are rare or absent in the tundra as they are from the native floras of South Georgia and the Antarctic Peninsula (Bennett *et al.*, 1982). Thus, in these environments, all species are nucleotypically determined, obligate perennials.

The fact that there is a maximum limit to the mass of DNA for species which can complete development in a given time and that this limit is positively correlated with MGT has considerable predictive value. For example, in any environment where, to survive, a plant must complete its life-cycle in 7 weeks or less, all species reproducing there are certain to have a very low DNA C-value. Such time-limited niches presumably exist in nature, e.g. where scouring water temporarily abates to uncover fertile silt in a river bed. The niche certainly exists in laboratories as scientists seek for angiospermous equivalents of *Drosophila* in order to maximize the number of experimental generations possible per year (Williams & Hill, 1986). Significantly, the species selected with MGTs of 4 to 6 weeks (e.g. *Arabidopsis thaliana* and *Brassica campestris* L.) all have very low DNA amounts. Similarly, any species with a very high DNA C-value ($4C > 150$ pg) will certainly not display an annual life-style nor will it establish and set seed in a single growing season when growing in a cool environment.

It is also important to note that a causal relationship involving two steps (Fig. 6) has been established linking DNA amount with any factor (such as temperature or water availability) which determines a time-limited environment. Firstly, a causal link has been established between variation in DNA C-value and variation in its temporal phenotypic consequences, such as the minimum time to produce mature seed. Secondly, such temporal consequences can determine the ability of a species to survive and reproduce in particular time-limited environments, so that minimum generation time can be of critical adaptive significance and the phenotypic character on which selection operates.

As shown above, selection for a particular genomic form (e.g. low DNA amount) can operate on its temporal consequences for the complete life-cycle. However, such selection can also operate on its temporal consequences at just one stage of the life-cycle. The clearest example concerns a study of 21 species of frog (Goin, Goin & Bachmann, 1968). The minimum duration of the tadpole stage was positively correlated with DNA C-value and ranged from 14 d in one species to

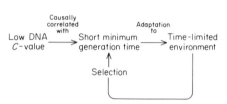

Fig. 6. A diagramatic illustration of a causal two-step relationship linking genome size and any environmental factor which determines a time-limited environment. Selection, acting on the phenotype for a short minimum generation time, results in co-selection for the genotypic character of low DNA C-value, with which it is causally correlated [see Fig. 5(c)].

about a year in others. The species were adapted to different environments for their larval stage, some colonizing transient puddles while others use permanent waters. The species whose tadpoles develop in transient puddles have a short tadpole stage and, because of the relationship just mentioned, a low DNA C-value. Goin *et al.* (1968) suggested that a high DNA amount and a low rate of development restrict some frogs to the vicinity of permanent water while a low DNA amount and a rapid rate of development allow others to invade arid environments. Here, the association between the genomic character (low DNA C-value) and the environmental factor (relative wetness) operates on the temporal consequences of variation in DNA amount at the tadpole stage in the life-cycle. Analogous examples presumably exist for plants. The progressive exclusion from the flora of angiosperms with high DNA amounts which occurs with increasing latitude in polar regions may result from selection acting at the establishment stage of the life-cycle. Bennett *et al.* (1982) suggested that selection acts against these species because the minimum duration of development essential for establishment and survival, determined by their high DNA C-value, exceeds the duration of the growing season in even the mildest years.

VARIATION IN DNA AMOUNT RELATED TO PHENOLOGY

The above examples show how DNA amount relates to geographical distribution and how the temporal consequences of a high DNA amount can exclude a species from certain local floras or prevent expression of certain life-styles. However, work at Sheffield has adopted another approach, investigating how variation in DNA amount relates to phenological characters, particularly timing of growth. Their interesting observations show how different species partition the available environment sequentially and suggest how different species can coexist within a plant community (Moore, 1985).

The first detailed study of genome size in relation to phenology was by Grime & Mowforth (1982) who compared DNA C-value with time of shoot expansion and temperature using 24 herbaceous species from the Sheffield region. Their results [Fig. 7(b)] indicated that, 'from early spring to mid-summer changes in the identity of the most active growing species are associated with a progressive reduction in genome size'. They suggested that the selective force determining this relationship between genome size and season of growth is in some way related to the capacity for growth at low temperatures and that it might arise from a differential effect of low temperature on cell division and cell expansion. Species which habitually grow by cell enlargement at low temperatures have relatively large cells and, for reasons already considered, such cells are expected to have large

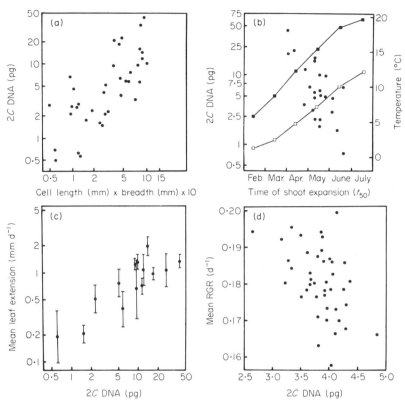

Fig. 7. Relationships between $2C$ DNA amount and leaf characters recently studied by Professor J. P. Grime and colleagues. (a) The relationship between DNA amount and the mean length × breadth of epidermal cells in mature leaves of a range of herbaceous species [modified from Grime (1983)]; (b) the relationship between DNA amount and the time of shoot expansion in 24 plant species commonly found in the Sheffield region [redrawn from Grime & Mowforth (1982)]. N.B. Temperature at Sheffield is expressed as the long-term averages for each month of daily minima (□) and maxima (■) in air temperature 1·5 m above the ground; (c) the relationship between DNA amount and the mean rate of leaf extension over the period 25 March to 5 April in 14 grassland species coexisting in the same turf [redrawn from Grime (1983)]. N.B. 95 % confidence limits are indicated by vertical lines; (d) the relationship between nuclear DNA content and mean relative growth rate of families of seedlings of *Poa annua* grown for 6 weeks in a controlled environment (30 °C day, 20 °C night) [redrawn from Grime (1983)].

DNA contents. They noted that in geophytes with very large genome sizes (e.g. *Fritillaria* and *Endymion*) which grow early in the spring, this growth occurs mainly by the expansion of cells formed during warm, relatively dry periods of the preceding year. They also suggested that, in species which follow the geophytes in the phenological succession, e.g. *Anthoxanthum odoratum*, development in spring may be sustained by the capacity for growth by cell expansion during intermittent cold periods.

Grime, Shacklock & Band (1985) tested the ecological implications of this idea that variation in DNA amount is positively correlated with the ability to grow at low temperature under natural field conditions in a north-facing area of damp limestone grassland in North Derbyshire. They compared the DNA C-values and rates of leaf elongation for all the major species while also monitoring the temperature. In the cold conditions of early spring, species with more than 10 pg of DNA grew two to four times as fast as those with less than 10 pg. Figure 7(c) shows such a relationship for mean rate of leaf extension plotted against $2C$ DNA amount. In warmer conditions by the beginning of May, growth rates in the two groups differed only slightly and, by June, no significant difference remained. These results clearly support the notion that variation in DNA C-value is positively correlated with the ability to grow at cool temperatures in the test environment.

The hypothesis that spring growth at low temperatures is superior in plants of higher nuclear DNA content may explain the geographical patterns which Bennett (1976) noted in crop plants in winter (see above). At high latitudes and elevations, the imperative in man's selection of crops has been to employ species which, through an early onset of growth, could maximize yield in the short growing season. It seems likely that the effect of this selection was unwittingly to introduce overwintering annual crops of high DNA content (Grime, 1983).

However, selection would be quite different in a summer annual growing only at warm temperatures. Grime's thesis that a small DNA content is characteristic of plants which grow by rapid and continuous cell division during favourable periods agrees with the explanation proposed by Rayburn et al. (1985) as to why DNA amount in Z. mays tends to decrease with increasing latitude of cultivation. This crop originated at low latitude and was taken north by man until environmental barriers (primarily a temperature-limited, shorter growing season) prevented normal maturation. Simultaneous selection for earlier maturation and maximum plant size and yield have involved selection for more cells which would result from the shorter mitotic cycle time that correlates with reduced DNA C-value.

If Grime's basic hypothesis of a relationship between DNA content and the time of shoot, and perhaps root growth has a wide application, then a simple anatomical examination of the spread of DNA amounts would provide considerable information about the nature of niche partitioning within a plant community and may show how certain vigorous community dominants can coexist. For example, the fact that one robust grass, *Brachypodium pinnatum*, has a nuclear DNA content of only 2 pg whereas another, *Bromus erectus*, has 22 pg per nucleus may help to explain how these species partition the environmental resources (Grime, 1983; Grime et al., 1985). They suggested that because of its higher DNA amount *B. erectus* can use its expanded cells as 'stored growth' more quickly to utilize the environment at times when *B. pinnatum* cannot, e.g. after the end of a drought. This interest is also relevant to mixed swards and intercropping in agriculture, and may also help to explain the success of mixtures of grasses with high DNA amounts with other grasses or clovers with low DNA amounts.

VARIATION IN DNA AMOUNT AND HARMFUL RADIATION

Background

The advent of the atomic age led biologists to try and predict the effects on plant life of environmental factors associated with nuclear war, notably of ionizing radiation from radioactive fallout. As the post-nuclear exchange scenario has developed, these predictions have evolved to include the effects of the so-called 'nuclear winter'. Both in the case of a nuclear winter (Grime, 1986) and in the case of ionizing radiation (Sparrow et al., 1965; Sparrow, Schwemmer & Bottini, 1971), variation in DNA amount has been singled out as important for predicting the likely response of different species.

Ionizing radiation

Much significant work leading to an understanding of the mechanisms determining radio-sensitivity in organisms used angiosperms because of their great range of genomic forms (Underbrink & Pond, 1976). Many studies report striking correlations between exposure to ionizing radiation and either nuclear DNA

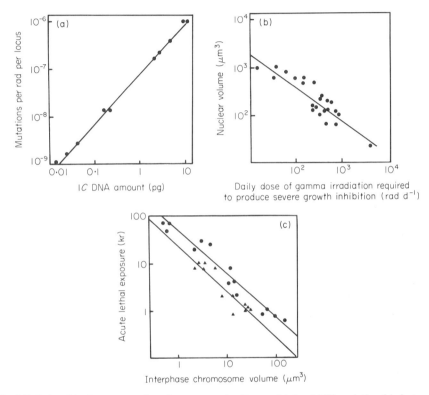

Fig. 8. Relationships between nuclear characters and radio-sensitivity. (a) The relationship between forward mutation rate per locus per rad and 1C DNA amount [modified from Abrahamson *et al.* (1973) and using the DNA amount for tomato (*Lycopersicon esculentum* Mill.) measured by Bennett & Smith (1976)]; (b) the relationship between nuclear volume and radio-sensitivity in 23 species of plants [redrawn from Sparrow & Miksche (1961)] and (c) the relationship between acute lethal exposure to ionizing radiation and interphase chromosome volume (ICV) for 15 herbaceous (▲) and 12 woody (●) plant species [redrawn from Sparrow *et al.* (1965)]. N.B. ICV = nuclear volume ÷ 2n. Thus, for species where 2n is constant, ICV is proportional to the mean DNA amount per chromosome, as DNA amount and nuclear volume are causally correlated [see Fig. 4(b)].

amount (Sparrow & Miksche, 1961) or some other character such as nuclear volume closely correlated with DNA amount. For example, Abrahamson *et al.* (1973) found a linear correlation between mutation per rad per locus and 1C DNA amount for different organisms including two angiosperms [Fig. 8(a)]. Similarly, highly significant relationships occur between (1) nuclear volume and radio-sensitivity in 23 angiosperms [Fig. 8(b)] and (2) acute lethal exposure to ionizing radiation and interphase chromosome volume (ICV) in 15 herbaceous and 12 woody plant species [Fig. 8(c)]. [N.B. The conceptual character, ICV, is obtained by dividing nuclear volume by the chromosome number (2n) and is therefore directly proportional to the mean DNA amount per chromosome.]

Nuclear volume and ICV are better indicators of ultimate biological effects than is the dose rate expressed in rad or roentgens. The response of plants to various types of ionizing radiation are predictable with considerable accuracy given only the DNA C-value, the chromosome number and the radiotaxon (Sparrow et al., 1965, 1971; Underbrink & Pond, 1976).

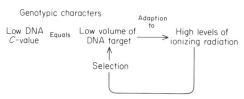

Fig. 9. A diagrammatic illustration of a causal one-step relationship linking genome size and the environmental factor of exposure to ionizing radiation. Selection for the ability to survive acts directly on a genotypic character, namely the minimum size of the DNA target, which is causally related to radio-sensitivity and is directly proportional to the DNA C-value (cf. Fig. 6).

Many factors affect radio-sensitivity (Underbrink & Pond, 1976) but it is generally held that (1) most radiation damage is mediated through lesions originating in the genomic DNA and (2) the volume of target DNA is a major factor determining radio-sensitivity. As noted above, minimum total chromosome volume per nucleus and minimum nucleus volume are both causally correlated with nuclear DNA amount and so the predictable relationships between radio-sensitivity and DNA target sizes are largely nucleotypic and, to that extent, are causal in origin.

The relationship between DNA amount and dosage of ionizing radiation differs from that linking DNA amount with a time-limited environment in one important respect. In the latter (Fig. 6), selection acts on a phenotypic character (MGT) to produce a two-step causal chain linking the genotypic and environmental variables. However, in the case of radio-sensitivity (where variation in the minimum volume of DNA amount directly affects the ability of the plant to survive) selection acts, at least in part, on a genotypic character, establishing a direct causal link between the genomic and environmental variables (Fig. 9).

The above research addressed the hopefully theoretical case of the effects of ionizing radiation after a nuclear exchange. However, recent events at Chernobyl (which aptly is the Ukrainian name for wormwood, Artemisia L.) shows that such knowledge may have practical relevance in predicting the effects on plant life of radiation from nuclear accidents. Given what is known about DNA amount and radio-sensitivity, it is worth considering whether similar relationships exist between DNA C-value and other naturally occurring types of radiation which although non-ionizing are potentially harmful.

A possible relation between DNA amount and UV-induced damage
Natural UV is an important cause of skin cancer. In man, the risk increases with UV dosage, perhaps not surprisingly as DNA absorbs strongly in the UV wavelengths. Moreover, UV radiation causes increased mutation rates and chromosome breakage in angiosperms. The amount of UV is generally higher in

tropical than in temperate latitudes and in summer rather than in winter. Variation in naturally occurring UV radiation and in DNA amount per chromosome might partly determine the localization of crop species with large DNA amounts to temperate latitudes and climates (Bennett, 1976). Increased UV in polar and some alpine environments may also be a cause of the exclusion of species with very large genomes and the increased incidence of polyploidy, and hence of reduced ICVs, in polar and some alpine floras. Selection for both low DNA *C*-value and low ICV may be especially strong in polar environments, when 24 h daylength allows no periods for repair of DNA in the dark.

Plants can adapt to minimize the harmful effects of UV, e.g. pigmentation increases with altitude in many species, while leaves of some alpine plants are coated with almost pure flavone with a very high near-UV absorption (Lautenschlager-Fleury, 1955). To be effective in nature, UV radiation need act at only one essential stage of the life-cycle. UV does not penetrate far into plant tissues but it would, for instance, penetrate pollen of wind-pollinated species. Perhaps deleterious effects of UV on pollen have contributed to selection for high frequencies of apomixis in alpine and polar environments, thus obviating the genetic contribution of a potentially damaged male gamete.

Investigations of the effects of UV on plant growth at different altitudes (e.g. Caldwell, 1968) suggested that damage caused by solar UV is mainly due to damage to the nucleic acids of cells, but whether nuclear, extra-nuclear DNA or both was the target locus was unexplored. Klein (1978) stated the case for new studies: 'That the near UV has profound effects on...plants in particular seems well established.... With a significant proportion of the solar spectrum at the earth's surface in the 280–390 nm range the effects of near-UV wavelengths on eco-systems...should be considered.... The known ability of near-UV to cause alterations in genomes...needs to be examined in vascular plants as a possible cause in ecotypic variation...'.

The effects of UV should be examined specifically in relation to major interspecific variation in DNA *C*-value and ICV. The need for such studies is increased by the recent depletion of atmospheric ozone especially over the poles (Lubinska, 1985; Tuck, 1986). Ozone prevents most harmful solar UV radiation from reaching the earth's surface so such changes could affect the global distributions of plants if variation in DNA *C*-value correlates with sensitivity to naturally occurring amounts of UV radiation.

CONCLUSIONS

Interspecific variation in DNA *C*-value correlates with a surprisingly wide range of important phenotypic and phenological characters at levels from the cell to the organism and many of its relationships are strikingly precise because of their biophysical nature. Large differences in DNA *C*-value have spatial and temporal phenotypic consequences which profoundly affect when, where or how plants grow both as to their global distribution and within a local community. Such work shows that variation in DNA *C*-value has considerable adaptive significance with far-reaching ecological implications and clearly illustrates the value of widely based, comparative studies. DNA *C*-values are known for less than 2000 of the estimated 240000 angiospermous species and for less than 2000 other plant species. No estimates are available for many geographical regions and for many types of plant community. Thus, a pressing need exists for new work both to extend the data base for this character and to increase our understanding of its ecological role.

It is too soon to abandon the search for unifying hypotheses to account for variation in chromosome number and ploidy level in relation to environmental factors. The progress achieved in understanding the ecological implications of variation in DNA C-value should provoke new work to study possible interactions between variation in DNA C-value and polyploidy in relation to environmental factors. It is now known that polyploidy may involve either no change in the DNA C-value per $2x$ genome, or a reduction in this character. The distribution of these two types of polyploids in relation to environmental factors such as temperature and duration of growing season should be examined.

Most new work should examine the plant in its external environment. However, one plant in its environment is part of its neighbour's environment while the cells and even the nuclear DNA of one species can form part of the environment of its endoparasites. It has been suggested that the genome of many plants contains a large proportion of parasitic DNA sequences. If the total DNA C-value comprises two kinds of genotype (host and parasite) which coexist, each forming the immediate environment of the other, it may be instructive to perceive the nucleotype as both genotype and environment. There may be an ecology of the genome itself and another new frontier to explore.

REFERENCES

ABRAHAM, A. & NINAN, C. A. (1954). The chromosomes of *Ophioglossum reticulatum* L. *Current Science*, **23**, 213–214.

ABRAHAMSON, S., BENDER, M. A., CONGER, A. D. & WOLFF, S. (1973). Uniformity of radiation-induced mutation rates among different species. *Nature*, **245**, 460–462.

ANDERSON, L. K., STACK, S. M., FOX, M. H. & CHUANSHAN, Z. (1985). The relationship between genome size and synaptonemal complex length in higher plants. *Experimental Cell Research*, **156**, 367–377.

AVDULOV, N. P. (1931). Karyo-systematische Untersuchungen der Familie Gramineen. *Bulletin of Applied Botany, of Genetics and Plant Breeding* (44th Supplement), **4**, 1–428.

BAETCKE, K. P., SPARROW, A. H., NAUMANN, C. H. & SCHWEMMER, S. S. (1967). The relationship of DNA content to nuclear and chromosome volumes and radiosensitivity (LD$_{50}$). *Proceedings of the National Academy of Sciences of the United States of America*, **58**, 533–540.

BENDICH, A. J. (1985). Plant mitochondrial DNA: unusual variation on a common theme. In: *Genetic Flux in Plants* (Ed. by B. Hohn & E. S. Dennis), pp. 111–138, Springer-Verlag, New York.

BENNETT, M. D. (1970). Natural variation in nuclear characters of meristems in *Vicia faba*. *Chromosoma*, **29**, 317–335.

BENNETT, M. D. (1971). The duration of meiosis. *Proceedings of the Royal Society of London, B*, **178**, 259–275.

BENNETT, M. D. (1972). Nuclear DNA content and minimum generation time. *Proceedings of the Royal Society of London, B*, **181**, 109–135.

BENNETT, M. D. (1973). Nuclear characters in plants. *Brookhaven Symposia in Biology*, **25**, 344–366.

BENNETT, M. D. (1976). DNA amount, latitude, and crop plant distribution. *Environmental and Experimental Botany*, **16**, 93–108.

BENNETT, M. D. (1977). The time and duration of meiosis. *Philosophical Transactions of the Royal Society of London, B*, **277**, 201–226.

BENNETT, M. D. (1985). Intraspecific variation in DNA amount and the nucleotypic dimension in plant genetics. In: *Plant Genetics* (Ed. by M. Freeling), pp. 283–302. A. R. Liss, New York.

BENNETT, M. D. & SMITH, J. B. (1971). The 4C nuclear DNA content of several *Hordeum* genotypes. *Canadian Journal of Genetics and Cytology*, **13**, 607–611.

BENNETT, M. D. & SMITH, J. B. (1976). Nuclear DNA amounts in angiosperms. *Philosophical Transactions of the Royal Society of London, B*, **274**, 227–274.

BENNETT, M. D., SMITH, J. B. & LEWIS SMITH, R. I. (1982). DNA amounts of angiosperms from the Antarctic and South Georgia. *Environmental and Experimental Botany*, **22**, 307–318.

BENNETT, M. D., HESLOP-HARRISON, J. S., SMITH, J. B. & WARD, J. P. (1983). DNA density in mitotic and meiotic metaphase chromosomes of plants and animals. *Journal of Cell Science*, **63**, 173–179.

BENNETT, M. D., SMITH, J. B. & SEAL, A. G. (1986). The karyotype of the grass *Zingeria biebersteiniana* ($2n = 4$) by light and electron microscopy. *Canadian Journal of Genetics and Cytology*, **28**, 554–562.

BUTTERFASS, T. (1983). A nucleotypic control of chloroplast reproduction. *Protoplasma*, **118**, 71–74.

CALDWELL, M. M. (1968). Solar ultraviolet radiation as an ecological factor for alpine plants. *Ecological Monographs*, **38**, 243–268.

CAVALIER-SMITH, T. (1985a). Eukaryote gene numbers, non-coding DNA and genome size. In: *The Evolution of Genome Size* (Ed. by T. Cavalier-Smith), pp. 69–103. John Wiley & Sons, Chichester.

CAVALIER-SMITH, T. (1985b). Cell volume and the evolution of eukaryotic genome size. In: *The Evolution of Genome Size* (Ed. by T. Cavalier-Smith), pp. 105–184. John Wiley & Sons, Chichester.

CLAPHAM, A. R., TUTIN, T. G. & WARBURG, E. F. (1962). *Flora of the British Isles*, 2nd Edn. Cambridge University Press, Cambridge.

CROSSLAND, M. W. J. & CROZIER, R. H. (1986). *Myrmecia pilosa*, an ant with only one pair of chromosomes. *Science*, **231**, 1278.

DARLINGTON, C. D. & WYLIE, A. P. (1955). *Chromosome Atlas of Flowering Plants*, 2nd Edn. George Allen & Unwin, London.

DOUNCE, A. L. (1955). The isolation and composition of cell nuclei and nucleoli. In: *The Nucleic Acid* (Ed. by E. Chargaff & J. N. Davidson), pp. 93–153. Academic Press, New York.

DYER, T. A. (1984). The chloroplast genome: its nature and role in development. In: *Topics in Photosynthesis. V. Chloroplast Biogenesis* (Ed. by N. R. Baker & J. Barber), pp. 23–69. Elsevier, Amsterdam.

EDWARDS, G. A. & ENDRIZZI, J. E. (1975). Cell size, nuclear size and DNA content relationships in *Gossypium*. *Canadian Journal of Genetics and Cytology*, **17**, 181–186.

EHRENDORFER, F. (1980). Polyploidy and distribution. In: *Polyploidy – Biological Relevance* (Ed. by W. H. Lewis), pp. 45–60. Plenum, New York.

EVANS, G. M., REES, H., SNELL, C. L. & SUN, S. (1972). The relationship between nuclear DNA amount and the duration of the mitotic cycle. *Chromosomes Today*, **3**, 24–31. Longmans, London.

FRITSCH, R. (1982). Index to plant chromosome numbers – bryophyta. *Regnum Vegetabile*, **108**, 1–268.

FURUTA, Y., NISHIKAWA, K. & HAJI, T. (1978). Uniformity of nuclear DNA amount in *Triticum monococcum* L. *Japanese Journal of Genetics*, **53**, 361–366.

GILL, J. J. B., JONES, B. M. G., MARCHANT, C. J., MCLEISH, J. & OCKENDON, D. J. (1972). The distribution of chromosome races of *Ranunculus ficaria* L. in the British Isles. *Annals of Botany*, **36**, 31–47.

GOIN, O. B., GOIN, C. J. & BACHMANN, K. (1968). DNA and amphibian life history. *Copeia*, **3**, 532–540.

GOLDBLATT, P. (1980). Polyploidy in angiosperms: monocotyledons. In: *Polyploidy – Biological Relevance* (Ed. by W. H. Lewis), pp. 219–239. Plenum, New York.

GRANT, V. (1982). Periodicities in the chromosome numbers of angiosperms. *Botanical Gazette*, **143**, 379–389.

GREENLEE, J. K., RAI, K. S. & FLOYD, A. D. (1984). Intraspecific variation in nuclear DNA content in *Collinsia verna* Nutt. (Scrophulariaceae). *Heredity*, **52**, 235–242.

GRIME, J. P. (1983). Prediction of weed and crop response to climate based upon measurements of nuclear DNA content. *Aspects of Applied Biology*, **4**, 87–98.

GRIME, J. P. (1986). Predictions of terrestrial vegetation responses to nuclear winter conditions. *International Journal of Environmental Studies*, **28**, 11–19.

GRIME, J. P. & MOWFORTH, M. A. (1982). Variation in genome size – an ecological interpretation. *Nature*, **299**, 151–153.

GRIME, J. P., SHACKLOCK, J. M. L. & BAND, S. R. (1985). Nuclear DNA contents, shoot phenology and species coexistence in a limestone grassland community. *New Phytologist*, **100**, 435–445.

HAGERUP, O. (1932). Über Polyploidie in Beziehung zu Klima, Ökologie und Physiologie. *Hereditas*, **16**, 19–40.

HOLM-HANSON, O. (1969). Algae: amounts of DNA and organic carbon in single cells. *Science*, **163**, 87–88.

HORNER, H. A. & MACGREGOR, H. C. (1983). C value and cell volume: their significance in the evolution and development of amphibians. *Journal of Cell Science*, **63**, 135–146.

ICHIKAWA, S., SPARROW, A. H., FRANKTON, C., NAUMAN, A. F., SMITH, E. B. & POND, V. (1971). Chromosome number, volume and nuclear volume relationships in a polyploid series ($2x$–$20x$) of the genus *Rumex*. *Canadian Journal of Genetics and Cytology*, **13**, 842–863.

JACKSON, R. C. (1957). New low chromosome number for plants. *Science*, **126**, 1115–1116.

JONES, R. N. & BROWN, L. M. (1976). Chromosome evolution and DNA variation in *Crepis*. *Heredity*, **36**, 91–104.

JONES, K. & SMITH, J. B. (1967). Chromosome evolution in the genus *Crinum*. *Caryologia*, **20**, 165–179.

KLEIN, R. M. (1978). Plants and near-ultraviolet radiation. *The Botanical Review*, **44**, 1–127.

LAURIE, D. A. & BENNETT, M. D. (1985). Nuclear DNA content in the genera *Zea* and *Sorghum*. Intergeneric, interspecific and intraspecific variation. *Heredity*, **55**, 307–313.

LAUTENSCHLAGER-FLEURY, D. (1955). Über die Ultraviolettdurchlässigkeit von Blattepidermen. *Bericht der Schweizerischen botanischen Gesellschaft*, **65**, 343–386.

LAWRENCE, M. E. (1985). *Senecio* L. (Asteraceae) in Australia: nuclear DNA amounts. *Australian Journal of Botany*, **33**, 221–232.

LEVIN, D. A. & FUNDERBERG, S. W. (1979). Genome size in angiosperms: temperate *versus* tropical species. *American Naturalist*, **114**, 784–795.

Lewis, W. H. (1980a). Polyploidy in angiosperms: dicotyledons. In: *Polyploidy – Biological Relevance* (Ed. by W. H. Lewis), pp. 241–273. Plenum, New York.

Lewis, W. H. (1980b). Polyploidy in species populations. In: *Polyploidy – Biological Relevance* (Ed. by W. H. Lewis), pp. 103–144, Plenum, New York.

Löve, A. & Löve, D. (1949). The geobotanical significance of polyploidy. I. Polyploidy and latitude. *Portugaliae acta Biologica (A), Special Volume* (Ed. by R. B. Goldschmidt), pp. 273–352.

Lubinska, A. (1985). Ozone depletion: Europe takes a cheerful view. *Nature*, **313**, 727.

Marchant, C. J. (1968). Chromosome patterns and nuclear phenomena in the cycad families *Stangeriaceae* and *Zamiaceae*. *Chromosoma*, **24**, 100–134.

Martin, P. G. (1966). Variation in the amounts of nucleic acids in the cells of different species of higher plants. *Experimental Cell Research*, **44**, 84–98.

Martinez, A. & Ginzo, H. D. (1985). DNA content in *Tradescantia*. *Canadian Journal of Genetics and Cytology*, **27**, 766–775.

Moore, P. D. (1985). Nuclear DNA content as a guide to plant growth rate. *Nature*, **318**, 412–413.

Mowforth, M. A. G. (1985). *Variation in nuclear DNA amounts in flowering plants: an ecological analysis.* Ph.D. thesis, University of Sheffield, UK.

Nichols, H. W. (1980). Polyploidy in algae. In: *Polyploidy – Biological Relevance* (Ed. by W. H. Lewis), pp. 151–161. Plenum, New York.

Palmer, J. D. (1985). Evolution of chloroplast and mitochondrial DNA in plants and algae. In: *Monographs in Evolutionary Biology: Molecular evolutionary genetics* (Ed. by R. J. MacIntyre), pp. 131–240. Plenum, New York.

Pegington, C. & Rees, H. (1970). Chromosome weights and measures in the Triticinae. *Heredity*, **25**, 195–205.

Poggio, L. & Hunziker, J. H. (1986). Nuclear DNA content variation in *Bulnesia*. *Journal of Heredity*, **77**, 43–48.

Rayburn, A. L., Price, H. J., Smith, J. D. & Gold, J. R. (1985). C-band heterochromatin and DNA content in *Zea mays*. *American Journal of Botany*, **72**, 1610–1617.

Rees, H., Cameron, F. M., Hazarika, M. H. & Jones, G. H. (1966). Nuclear variation between diploid angiosperms. *Nature*, **211**, 828–830.

Smith, J. B. & Bennett, M. D. (1975). DNA variation in the genus *Ranunculus*. *Heredity*, **35**, 231–239.

Sparrow, A. H. & Miksche, J. P. (1961). Correlation of nuclear volume and DNA content with higher plant tolerance to chronic radiation. *Science*, **134**, 282–283.

Sparrow, A. H., Sparrow, R. C., Thompson, K. H. & Schairer, L. A. (1965). The use of nuclear and chromosomal variables in determining and predicting radiosensitivities. *Radiation Research (Supplement)*, **5**, 101–132.

Sparrow, A. H., Schwemmer, S. S. & Bottini, P. J. (1971). The effects of external gamma radiation from radioactive fall-out on plants with special reference to crop production. *Radiation Botany*, **11**, 85–118.

Sparrow, A. H., Price, H. J. & Underbrink, A. G. (1973). A survey of DNA content per cell and per chromosome of prokaryotic and eukaryotic organisms: some evolutionary considerations. *Brookhaven Symposium in Biology*, **23**, 451–494.

Stebbins, G. L. (1938). Cytological characteristics associated with the different growth habits in the dicotyledons. *American Journal of Botany*, **25**, 180–198.

Stebbins, G. L. (1966). Chromosome variation and evolution. *Science*, **152**, 1463–1469.

Stebbins, G. L. (1971). *Chromosomal Evolution in Higher Plants.* Edward Arnold, London.

Stewart, D. A. & Barlow, B. A. (1976). Genomic differentiation and polyploidy in *Sowerbaea* (Liliaceae). *Australian Journal of Botany*, **24**, 349 367.

Sunderland, N. & McLeish, J. (1961). Nucleic acid content and concentration in root cells of higher plants. *Experimental Cell Research*, **24**, 541–554.

Thomas, C. A. (1971). The genetic organisation of the chromosomes. *Annual Review of Genetics*, **5**, 237–256.

Tischler, G. (1934). Die Bedeutung der Polyploidie für die Verbreitung der Angiospermen, erläutert an den Arten Schleswig-Holsteins, mit Ausblicken auf andere Florengebiete. *Botanische Jahrbücher für Systematik, Pflanzengeschichte und Pflanzengeographie*, **67**, 1–36.

Tuck, A. F. (1986). Depletion of antarctic ozone. *Nature*, **321**, 729–730.

Underbrink, A. G. & Pond, V. (1976). Cytological factors and their predictive role in comparative radiosensitivity: a general summary. *Current Topics in Radiation Research Quarterly*, **11**, 252–306.

Van't Hof, J. (1975). The duration of chromosomal DNA synthesis of the mitotic cycle, and of meiosis of higher plants. In: *Handbook of Genetics* (Ed. by R. C. King), pp. 363–377. Plenum Press, New York.

Van't Hof, J. & Sparrow, A. H. (1963). A relationship between DNA content, nuclear volume, and minimum cell cycle time. *Proceedings of the National Academy of Sciences of the United States of America*, **49**, 897–902.

Williams, P. H. & Hill, C. B. (1986). Rapid-cycling populations of *Brassica*. *Science*, **232**, 1385–1389.

New Phytol. (1987) **106** (Suppl.), 201–216

THE ECOLOGICAL SIGNIFICANCE OF FRUCTAN IN A CONTEMPORARY FLORA

By GEORGE HENDRY

Unit of Comparative Plant Ecology (NERC), Department of Botany, The University, Sheffield S10 2TN, UK

SUMMARY

From an extensive literature survey and from analytical data of 130 species from the Sheffield flora, the physiological and molecular attributes, and occurrence of fructan are considered. The exceptionally high concentrations of fructan in many native species and its exclusive vacuolar location are noted. In view of the contention that fructan may function in low temperature tolerance, estimates are given of their maximum vacuolar concentration. For most species examined, this is unlikely to make a significant contribution to low temperature tolerance under the field conditions of the Sheffield flora. The association of high concentrations in the shoots of several species with high nuclear 2C DNA values is recorded. It is this correlation which may be of significance to several early-season growing species. By maintaining supplies of fructose and sucrose from vacuoles in tissue undergoing expansion at low temperatures (a feature associated with high DNA values), such species obviate the need for transport of carbohydrate over distance as in starch-storing species. By shortening supply lines at critical cold periods, fructan-rich species may have a considerable advantage over starch-storing, small-celled, transport-dependent species. However, such an advantage is not commonly exploited by the contemporary Sheffield flora. It is suggested that the ecological significance of fructan may best be seen in floras of the past or in contemporary floras undergoing more severe environmental stresses.

Key words: Fructan, starch, sucrose, cold tolerance, osmoregulation.

INTRODUCTION

Starch is the principal or major reserve carbohydrate for the greater part of the world flora. However, a significant and discretely defined group of plants store not polymers of glucose but instead polymers of fructose. These fructan-containing species occur within a diverse range of families including the Gramineae and Compositae, and several families within the Liliales and in the Campanulales (Meier & Reid, 1982). This diversity and the occurrence of fructan among highly evolved families indicate that the genes for fructan metabolism in angiosperms may have arisen in response to one or a few selective pressures in the relatively recent past. Lewis (1984) has suggested that synthesis of this carbohydrate does not represent a minor pathway of declining evolutionary significance.

Most extant fructan-containing families first appear in the fossil record during the Oligocene and Miocene (Cronquist, 1981), that is some 10 to 30 million years ago, but the selective pressures which favoured accumulation of fructan in the Tertiary (or even Quaternary) period are no longer readily apparent. However, the nature of these pressures and the evolutionary significance of fructan can be examined in several ways using the contemporary flora. Two approaches have been used in this account. Firstly, the occurrence of fructan in species worldwide and within the Sheffield flora has been re-examined, both in terms of systematics and geographical distribution. Secondly, the known physiological and molecular

0028-646X/87/05S201 + 16 $03.00/0

characteristics of fructan have been reassessed from an ecological standpoint. The overall aim has been to assess the ecological significance of fructan accumulation with reference to the modern-day Sheffield flora and to draw wider conclusions.

In the first approach, an extensive literature search was made to reassess, in the light of modern-day criteria, the many reports of the occurrence of fructan, going back to the 19th century. These statements form the corpus from which chemotaxonomic summaries have been made in more recent years (e.g. Hegnauer, 1962–1973; Gibbs, 1974) and their citation in the modern literature may imply a contemporary validation. The second approach has been an analytical study of 130 native or long naturalized herbaceous and dwarf (< 1 m) shrubby species drawn from within 50 km of Sheffield, UK (latitude 53° N). The selected species include 103 dicots and 26 monocots. Some 64 families, about 80 % of those in the area, are represented. A diversity of ecological distributions and growth forms is also included. This survey was conducted once in winter (February to March) and once in summer (July to August) (G. A. F. Hendry, K. J. Brocklebank & J. G. Hodgson, unpublished data). In addition, data are drawn from a more detailed community-based study made every 3 to 4 weeks for 16 months (K. J. Brocklebank, unpublished data).

OCCURRENCE OF FRUCTAN

By establishing the identity of fructan-rich species and by correlating this with their habit, life history and distribution, clues to the function of fructan may be found. This worthy objective is, however, fraught with the presence of a substantial body of literature which today looks increasingly unreliable. In part, this may originate in the motives behind the original research. Many early reports going back to the 1870s appear in an area of literature concerned more with the quest for the elusive precursors of starch. The techniques employed were often cytological and based on Sachs' otherwise elegant studies of *Helianthus tuberosus* (Sachs, 1864). The techniques involved the infusion of alcohol into the plant tissue causing the precipitated fructan to become visible as birefringent spherocrystals in the vacuoles (Meier & Reid, 1982). In careful hands, the method may have had much to commend it. Unfortunately, it gave rise to many reports, now firmly embedded in the literature (see examples in Hegnauer, 1962–1973), which have subsequently failed to be confirmed using more modern analytical techniques. In this category come reports of fructan among the following orders: Violales, Malvales, Euphorbiales, Droserales, Theales, Myrtales, Polygalales, Primulales, among others (Hegnauer, 1962–1973; Gibbs, 1974). Most of these reports need modern-day confirmation. One of the few recent studies which records species in which fructan was *not* found is that of Pollard & Amuti (1981). However, even these authors have reported the sporadic occurrence of fructan in several unex-pected (previously unreported) orders, including the Dilleniales, Laurales and Cornales. Perhaps the best that can be done is to summarize the reports of fructan at the family level where the evidence is modern, repeatedly confirmed and has employed techniques now held to be reliable. This rather conservative approach does at least provide some systematic order. If small families with less than 50 genera worldwide are disregarded, fructans are widespread in just 10 families, including 1200 species from the economically important Gramineae (Table 1). The dominant family is the Compositae with perhaps 24 000 fructan-containing species. From Table 1, it is possible to predict that the number of species

Table 1. *The occurrence of fructan in vascular plants, based on modern, confirmed reports, using reliable analytical techniques, together with their distribution*

Order*	Principal*,† families	Estimated number of species	Distribution‡
Monocotyledones			
Cyperales	Gramineae	1200§	Pan temperate
Liliales‖	Haemodoraceae		Tropical, warm temperate
	Liliaceae		Cosmopolitan
	Iridaceae	6300	Richest in warm temperate
	Agavaceae		Subtropical, arid
Dicotyledones			
Asterales	Compositae	24000¶	Cosmopolitan outside tropical rain forest
Campanulales‖	Campanulaceae		N. temperate
	Stylidaceae	2500	S. temperate
	Goodeniaceae		S. temperate (esp. Australasian)
Lamiales	Boraginaceae	2000¶	Temperate and subtropical
	Total	36000	

  * Classification above family level follows that of Cronquist (1981), while that of Heywood (1978), Tutin *et al.* (1964–1980) and Clapham, Tutin & Warburg (1981) has been reserved for the families and species levels.
  † Excludes families with less than 50 genera.
  ‡ Based on Heywood (1978) and Tutin *et al.* (1964–1980).
  § Estimated value of subfamily Pooideae.
  ‖ Based on Pollard & Amuti (1981) or Pollard (1982).
  ¶ Assumes all species, annuals and perennials contain fructan.

worldwide which can be expected to contain fructan is about 36000, that is about 12 % of the angiospermous flora. Four of the five orders are considered to be highly evolved (Dahlgren, 1975; Cronquist, 1981). The Liliales, however, may not be so (Cronquist, 1981).

The geographical distribution of certain fructan-containing species, particularly among the Gramineae, has drawn comment. De Cugnac (1931) and Ojima & Isawa (1968) identified 'northern' grasses which accumulated fructan and 'southern' grasses from subtropical and tropical areas where starch or sucrose formed the principal reserve carbohydrates. The 'northern' types precisely coincide with the subtribe Pooinae (or subfamily Pooideae) which has a predominantly north temperate to subarctic distribution. This restriction to cool or cold climates has encouraged the proposition that the presence of fructan may confer cold tolerance (for example, see Pontis & del Campillo, 1985). However, if this attribute is to apply to all fructan-rich species, then a similar study of geographical distribution should be made of all fructan-containing families. Table 1 attempts to summarize the distribution of the principal families. Apart from the Gramineae and possibly the Campanulaceae, the majority of families are distributed in the warm temperate, subtropical and tropical areas, that is, regions today where cold tolerance is unlikely to be a significant advantage (other than in montane areas). The Liliaceae are considered to be particularly cosmopolitan (Heywood, 1978). The conclusion from Table 1 is that, whatever the advantages of fructan metabolism are, they are unlikely to be found by examining geographical distributions. The conditions or pressures which have determined a temperate distribution for many members of the Gramineae and Campanulaceae have apparently not applied to, or not favoured or restricted the distribution of, the other major fructan-containing families.

Within the British flora, the result of the survey of 130 native or naturalized species reveals just 20 species in which fructan forms a significant part of the reserve carbohydrates (Table 2), and 110 species in which fructan was neither detected chemically nor could be re-confirmed either by high performance, gas-liquid or paper chromatography (Table 3). There were few surprises, at least

Table 2. *Occurrence of fructan in a survey of the Sheffield flora, based on 130 native or long-naturalized herbaceous species*

Family*	Species*	Life history	Predominant climatic distribution†
Gramineae	*Aira praecox*	AV	NT
	Deschampsia flexuosa	P	NT/M
	Arrhenatherum elatius	P	NT/M
Iridaceae	*Iris pseudacorus*	P	NT/T
Amaryllidaceae	*Galanthus nivalis*	PV	T/WT
Liliaceae	*Allium ursinum*	PV	NT to WT
	A. vineale	P	NT to WT
	Hyacinthoides non-scripta	PV	T
Compositae	*Sonchus oleraceus*	A	T to STROP
	Tanacetum vulgare	P	NT to WT
	Tripleurospermum inodorum	A	SA to T
	Achillea millefolium	P	SA to T
	Bellis perennis	P (partly V)	NT to WT
	Tussilago farfara	PV	SA to WT
	Senecio vulgaris	A (partly V)	T
Campanulaceae	*Campanula rotundifolia*	P	SA to T
Boraginaceae	*Myosotis ramosissima*	AV	NT to WT
	M. secunda	P	NT
	Symphytum × uplandicum	P	NT
Monotropaceae	*Monotropa hypopytis*	P	NT/MAR

* Classification and nomenclature follows Clapham *et al.* (1981) in the text and in this and subsequent tables.
† Distribution follows Tutin *et al.* (1964–80).
A, annual; M, montane; MAR, maritime; NT, north temperate; P, perennial; SA, subarctic; STROP, subtropical; T, temperate; V, vernal/early spring flowering; WT, warm temperate.

in the taxonomic sense. The Iridaceae, regarded as a fructan-rich family (Meier & Reid, 1982), gave the expected high values for *Iris pseudacorus*. However, an ornamental species wildly grown in the area, *Crocus chrysanthus* (Herbert) Herbert, contained high concentrations of starch and glucose (February corm: starch 91 mg, glucose 10·96 mg g^{-1} fresh weight), while fructan was detected in only trace amounts. Pollock (1986) has reported similar findings. Among the non-fructan species (Table 3), several occur in families, such as the Liliales, which are generally classified as fructan-positive. The two grasses, *Phragmites australis* and *Molinea caerulea* (the latter a north temperate species) which did not contain detectable fructan, significantly perhaps are not members of the fructan-rich subfamily Pooideae.

Table 3. *Families from the Sheffield flora in which fructan was not detected*

Family	Number of species tested	Family	Number of species tested
Dicots		**Dicots** (*cont.*)	
Ranunculaceae	5	Pyrolaceae	1*
Nymphaeaceae	1	Empetraceae	1
Papaveraceae	2	Primulaceae	3
Fumariaceae	1	Gentianaceae	1
Cruciferae	5	Solanaceae	1
Resedaceae	1	Scrophulariaceae	5
Violaceae	1	Orobanchaceae	1
Polygalaceae	1	Labiatae	5
Hypericaceae	1	Plantaginaceae	1
Cistaceae	1	Rubiaceae	2
Caryophyllaceae	4	Caprifoliaceae	1
Portulaceae	1	Adoxaceae	1
Chenopodiaceae	3		
Malvaceae	1	Total species	91 out of 103 spp. tested
Linaceae	1		
Geraniaceae	1		
Oxalidaceae	1	**Monocots**	
Balsaminaceae	1	Alismataceae	1
Leguminosae	5	Hydrocharitaceae	1
Rosaceae	5	Potamogetonaceae	2
Crassulaceae	1	Liliaceae	1†
Saxifragaceae	3	Juncaceae	1
Onagraceae	1	Iridaceae	1‡
Haloragaceae	1	Dioscoreaceae	1
Hippuridaceae	1	Orchidaceae	1
Callitrichaceae	1	Araceae	2
Araliaceae	1	Lemnaceae	1
Umbelliferae	5	Sparganiaceae	1
Cucurbitaceae	1	Typhaceae	1
Euphorbiaceae	2	Cyperaceae	3
Polygonaceae	4	Gramineae	2§
Urticaceae	1		
Ericaceae	2	Total species	19 out of 27 spp. tested

* *Pyrola minor* with possible trace amounts of fructan.
† *Narthecium ossifragum.*
‡ *Crocus chrysanthus.*
§ *Phragmites australis* and *Molinea caerulea.*

Among the 20 fructan-containing species (Table 2), there was a range of life histories including annuals, perennials and vernals. The detection of substantial concentrations of fructan in annual species of the Compositae and Boraginaceae is not in agreement with Colin & Chollet (1939) and Bourdu (1957) who found fructan only in perennial species of these families using French material. The global natural distribution of the 20 species ranged from the subarctic to warm temperate with no indication that the British fructan flora was characteristic of exceptional climatic or edaphic areas. The habitats of the 20 species were diverse, ranging from stream margins to limestone grassland, woodland to disturbed areas. Indeed, all the major terrestrial habitats found in the Sheffield area and listed in Grime, Hodgson & Hunt (1987) contained at least one fructan-rich species.

The conclusion is that, despite the confinement of fructan-containing species to a small number of families, the species themselves have evolved a wide range of life histories and exploit a diverse selection of habitats and climates. This observation is based on the generalized evidence from the world flora as well as the more precise, but incomplete, evidence from a local flora.

CHARACTERISTICS OF FRUCTAN

In the search for the ecological significance of fructan accumulation, consideration should be given to the physiological, biochemical and chemical characteristics of the fructan molecule and to compare these with those of starch.

Table 4 lists the physiological conditions and developmental state favouring biosynthesis and accumulation of fructan. Those conditions which favour CO_2 fixation, such as long day lengths, but which do not promote growth, such as low temperatures, tend to favour its accumulation. Environmental stresses such as low temperature, low N fertilizer treatment and drought promote its accumulation as do biotic stresses such as fungal infection and removal of tiller or fruit (but not frequent defoliation) (Pollock, 1984). Although many studies of fructan physiology have been with cereal and pasture grasses, several of the conditions suppressing growth in non-graminaceous plants also result in accumulation. Accumulation, stimulated by low temperatures, is widespread in the Compositae and Liliaceae (Hendry & Brocklebank, unpublished data) and in the Boraginaceae (Bourdu, 1957). Conditions *promoting* growth, such as high temperatures or elevated supplies of N fertilizer, suppress accumulation or, in some species, promote the formation of starch (see Archbold, 1940; Pollock, 1986 and references therein).

Table 4. *Physiological conditions and developmental states favouring accumulation of fructan*

Conditions favouring CO_2 assimilation:
  Long daylength 1,2,5
Conditions supressing growth:
  Low temperatures 1,2,4,5
  Low supplies of N-fertilizer 1
  Drought 6
  Fungal infection 7
  Defoliation (infrequent) 1,2
Developmental state:
  After flowering 3,8
  Following tiller removal 1
  Following fruit removal 1

Sources: 1, Archbold (1940), Smith (1973), Pollock (1984); 2, Smith (1968); 3, Pollock & Jones (1979); 4, Bourdu (1957); 5, Hendry & Brocklebank, unpublished observation; 6, White (1973), Julander (1945); 7, Holligan, Chen & Lewis (1973); 8, Waite & Boyd 1953.

A second feature is the high concentration to which fructan accumulates in plants. This phenomenon has been recorded in a number of species [see, for example, in the Boraginaceae (Bourdu, 1957), Gramineae (Smith, 1967) and Liliaceae (Darbyshire & Henry, 1981)]. In isolation, the significance of these reports may be lost. Based on the survey of 130 British species, Table 5 lists the 12 species with the highest recorded reserve carbohydrate content. Although fructan was only recorded in 20 out of the 130 species examined, no less than nine fructan-containing species appear in Table 5. (The concentrations given are from material

Table 5. *Maximum concentration of reserve carbohydrates recorded in a survey of 130 species from the Sheffield flora*

Species	Tissue	Type	Carbohydrate Concn (mg/g f. wt)	Season	Starch concn in fructan-rich spp. (mg/g f. wt)
Allium vineale	so	Fructan	214	Summer*	0·21
Orchis mascula	so	Starch	163	Summer*	—
A. ursinum	so	Fructan	139	Summer*	0·32
Hyacinthoides non-scripta	so	Fructan	130	Summer*	0·22
Conopodium majus	so	Starch	115	Summer*	—
Crocus chrysanthus	so	Starch	92	Winter*	—
Bellis perennis	Root	Fructan	79	Summer	0·58
Galanthus nivalis	so	Fructan	77	Winter*	10·00
Campanula rotundifolia	Root	Fructan	75	Summer	0·33
Symphytum × uplandicum	so	Fructan	74	Summer	5·92
Iris pseudacorus	Root	Fructan	74	Summer	1·00
Tussilago farfara	Root	Fructan	73	Summer	0·41
Monotropa hypopytis	All	Fructan	72	Winter*	0·63

so, underground storage organ (bulb, corm, tuber, rhizome).

* Resting period for species.

Analyses were made according to the methods of Roe (1934) for fructan adjusted for glucose, fructose and sucrose determined enzymically by the methods of Klotsch & Bergmeyer (1963) and confirmed by HPLC, GLC and PC by the methods of Frehner, Keller & Wiemken (1984) or Lewis *et al.* (1972). Starch was determined also enzymically following, essentially, Haissig & Dickson (1979) and Bergmeyer & Bernt (1963).

collected either in July or in February; even higher concentrations may occur at other periods.) Only three of the 100 starch-rich species surveyed accumulate starch to anything approaching the concentrations of fructan. Most species in the list accumulate the highest concentrations of reserve carbohydrate in the summer. Interestingly, species which flower in late winter, *C. chrysanthus* (crocus) and *Galanthus nivalis* (snowdrop), store maximum concentrations of, respectively, starch and fructan in the winter months. A further feature of the list in Table 5 is that, with the exception of *G. nivalis* and *Symphytum × uplandicum*, fructan-rich species do not accumulate significant concentrations of starch in their vegetative parts. Indeed, in most cases cited in Table 5, starch is barely detectable. It follows that an outstanding feature of many British fructan-containing species is that fructans can be accumulated to very high concentrations and values exceeding 50% dry weight of tissue have been recorded for several members of the Compositae, Liliaceae and Gramineae (Edelman & Jefford, 1968; Darbyshire & Henry, 1981; Housley & Pollock, 1985). The significance of this will be considered below.

Knowledge of the molecular characteristics of fructan has improved in recent years. Certain of these are described in Table 6, together with those for starch for comparison. In summary, the fructan molecule when compared with starch is small, with comparatively little structural variation; it is relatively water-soluble and is synthesized, at significant rates, only in the presence of high concentrations of its substrate, sucrose. Degradation, usually associated with mobilization, appears to be initiated by only one enzyme, or two similar ones (Edelman & Jefford, 1968), with a high Michaelis constant. Starch, in contrast, is subject to breakdown

Table 6. *Molecular characteristics of fructan and starch*

Characteristic	Fructan	Starch
Polymer of	Fructose	Glucose
Ring structure	Flexible—furanose	Rigid—pyranose
Solubility in water	Relatively high	Low to $\pm$ insoluble
Degree of polymerization	Generally less than 30 (occasionally exceeds 250)	2000 to $>$ 200000
Relative molecular mass $\times 10^3$	0·5–50	300 to $>$ 36000
Structural variation	Two linear types One mixed/branched type	Numerous, some species specific
Initiating biosynthetic enzymes	Sucrose-sucrose fructosyl transferase	Two starch synthases and, possibly, others
K_m for initiation	*c*. 100 mM (sucrose)	*c*. 33 μM (ADP glucose)
Degrading enzymes	β-Fructofuranosidases	Various, depending on polymer size and/or on species
K_m for degradation	*c*. 100 mM	Various, generally $<$ 10 mM
Storage location	Vacuole (exclusively?)	Chloroplast, amyloplast, chromoplast, leucoplast

by a range of structurally unrelated enzymes, several with a high affinity for (parts of) the starch molecule (Duffus, 1984; Stitt, 1984). These unusual characteristics of fructan metabolism probably ensure that synthesis only occurs under conditions strongly favouring accumulation of sucrose and that degradation is most active in conditions where high concentrations of fructan have previously been built up (Pontis, 1970). These attributes suggest that fructan is unlikely to be subject to rapid turnover (defined as simultaneous synthesis and degradation) as might well be expected of a long-term storage carbohydrate. Farrar & Farrar (1985) have shown that fructans are turned over, in barley, but that much of this turnover is confined to fructan of a low degree of polymerization. Studies of turnover of fructan in *Dactylis glomerata* confirm that the time-scale (half-life) is considerably longer than turnover rates of starch in other grasses (Pollock, 1982a, b). Starch, at least that deposited in the chloroplast, can be turned over in a matter of hours (Dickson & Larwood, 1973; Chen-She, Lewis & Walker, 1975; Herold, 1984), being both an early product of photosynthesis (Walker & Herold, 1977) and possibly a short-term reservoir for sucrose metabolism (Stitt, 1984). There appears to be no uncontroverisal evidence that fructan also undergoes such rapid or short-term turnover. Finally, a potentially significant difference between starch and fructan lies in their respective subcellular locations. In the short term, the chloroplast acts as the principal or even the sole starch store (Stitt, 1984). Long-term major starch depositions are largely confined to amyloplasts at sites remote from areas of high photosynthetic activity. Fructan, however, is stored exclusively in the vacuole (Frehner, Keller & Wiemken, 1984), both within (potentially) photosynthetic tissue and in underground storage organs. It has been suggested that this vacuolar location may be significant if the function of fructan is in the regulation of osmotic potential (Pontis & del Campillo, 1985) or in low temperature tolerance (Edelman & Jefford, 1968).

A third cellular characteristic of fructan is its relation to cell size. On superficial examination, a number of fructan-containing species in the British flora appear to be soft, rather watery plants with large cells. Examples include members of the genus *Allium, Hyacinthoides non-scripta, Taraxacum* spp., *Tussilago farfara, Symphytum* spp. and *Monotropa hypopytis.* Data on cell size for most of the fructan species described here has not, unfortunately, been published. However, an alternative correlation with cell size is available from the studies on the amount (mass) of DNA in the nuclear genome (see for example, Bennett & Smith, 1976). In turn, in extensive comparative studies, a correlation has been found between genomic DNA mass and geographical distribution. Genomic masses of tropical species are comparatively small when compared with the wide range observed among temperate floras (Bennett, 1976; Levin & Funderburg, 1979). Recently, nuclear DNA content has been correlated with shoot phenology; early-spring growing species have a larger genomic DNA size than the late spring- and summer-growing species (Grime & Mowforth, 1982; Grime, Shacklock & Band, 1985). In addition, and of immediate relevance, large nuclear DNA size is associated with large cell size, consistent with the hypothesis (Grime & Mowforth, 1982) that, for such species, early-season growth is achieved by cell expansion rather than by division. Bennett (1987) provides a fuller account of these relationships.

From the data available for 15 of the 20 fructan-rich species covered in the survey (Bennett & Smith, 1976; Grime & Mowforth, 1982; Grime & Band, pers. comm.), Figure 1 shows the relationship between fructan concentration and

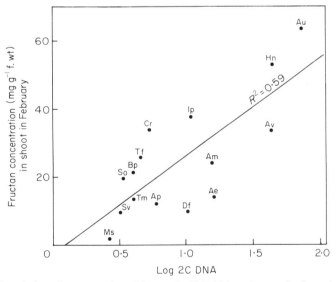

Fig. 1. Correlation of concentration of fructan in 15 British native species from the Sheffield flora with 2C DNA values. Initials are those of the generic and specific names listed in Table 2.

nuclear DNA mass. A positive correlation exists between species with high DNA values and the concentration of fructan in the principal, underground storage organ (using carbohydrate data derived, in all cases, from one mid-February field collection). For comparison, the DNA size of 31 of the 100 starch-rich species (Table 3) is also known and ranged from 0·1 to 20·6 pg (2C DNA) with a mean value of 3·35 ± 3·87. This compares with the 15 fructan-rich species which range from 2·7 to 71·4 pg (2C DNA) with a mean value of $16·0 \pm 20·1$. There was no correlation between starch concentration and DNA size.

There are at least two ways of interpreting these data. It may be that, in order to store large concentrations of fructan, large vacuoles are required which, in turn, are associated with large cells. It is perhaps as likely that large cells with large vacuoles have a high demand for a non-electrolyte such as fructan which both regulates the osmotic potential of the vacuole and, where climatically relevant, acts as a protectant against freezing. Starch-rich species, particularly non-wetland species, appear to have relatively low 2C DNA values, small cells and presumably small vacuoles. The significance of this is considered below.

PROBABLE AND IMPROBABLE FUNCTIONS OF FRUCTAN

Because fructan largely replaces starch in the vegetative parts of species in several families, attempts have often been made to establish the function of fructan. Three functions appear in the literature: as a reserve carbohydrate, as an osmoregulator and in chill or freeze tolerance. The first suggestion owes much to the French groups active 50 years ago (see, for example, Colin & Belval, 1922; de Cugnac, 1931) and does not today seem in dispute. In many species, particularly among the Compositae and Liliaceae, fructan is the *major* reserve carbohydrate. In other families, such as the Gramineae and Boraginaceae, it is the major carbohydrate during early vegetative growth. As a reserve carbohydrate, fructan may be considered as one alternative to starch.

The second and third proposals have persisted in the literature for some years, though with little supporting data. The origin of these proposals goes back at least to de Cugnac (1931) who, as noted earlier, classified grasses into, a 'northern type' which accumulated fructan and a 'southern type' (from warmer latitudes) which stored starch or sucrose. Similar observations were made by Ojima & Isawa (1968). Cooper (1964) suggested there was a link between the accumulation of soluble carbohydrates and cold-hardiness, a view supported by Eagles (1967) who proposed that the ecological significance of fructan, at least in *D. glomerata*, was in cold resistance. Similar possibilities were considered by Edelman & Jefford (1968) in their pioneering work on the mechanisms of fructan metabolism and the idea was given further support by Rutherford & Weston (1968) working also with Jerusalem artichoke. By 1981, Darbyshire & Henry were able to propose a role of fructan as an osmotic regulator during bulbing in onions and, in 1985, Gunn & Walton showed that a subantarctic winter-green grass may secure freeze resistance from fructan. Pontis & del Campillo (1985), reviewing recent progress, proposed that fructan might function as an 'osmotic buffer' in species that endure cold or dry periods where the (presumably depolymerized) fructan might increase resistance to freezing. Pollock, in his several commentaries, has however stated a number of reservations (Pollock, 1984, 1986). He has noted that the accumulation of fructan occurs, in some species, in the same storage organ as starch, that growth at low temperature occurs in species with no detectable fructan and that much of the evidence supporting the cryoprotection theory is hedged with qualifications and exceptions. To some extent, much of the following experimental evidence merely adds to Pollock's reservations.

If fructan, or fructose + sucrose, is to function within the vacuole in chill-freeze tolerance, then the vacuolar concentration must be high. To depress the freezing point of water by 1 °C, a concentration of 0·46 M sucrose is required (Weast, 1975).

Likewise, a 1 M solution of sucrose (or fructose) will depress the freezing point of water by 1·38 °C. To establish whether or not such concentrations are achieved in the vacuole, *in vivo*, a number of field-collected species were examined. However, to calculate the carbohydrate concentration in the vacuole, several assumptions have to be made. Ideally, estimates for vacuolar dimensions and relative volumes would come from artifact-free, stereological microscopy. Theoretical evidence and experimental data have been provided by several workers (Berlin *et al.*, 1982; Hajibagheri, Hall & Flowers, 1984) for mature shoot tissue and by Raven & Rubery (1982) for whole plants. This last paper, together with additional information (J. A. Raven, pers. comm.), may be summarized. The relative volumes occupied by the various fractions within the whole herbaceous plant are: gas space 0·05, cell wall 0·05, symplast 0·05, phloem and xylem conduits 0·06, vacuole 0·79. Within these compartments, the fractional volume of water is estimated as: phloem 0·75, symplast 0·67 to 0·75, cell wall 0·5 to 0·6, and xylem fluid and vacuolar sap > 0·95. Using these values, it is possible to estimate the vacuolar concentrations of various carbohydrates. Fructan is probably entirely confined to the vacuole (Frehner *et al.*, 1984), while sucrose, at least in some species, accumulates principally in the vacuole (Leigh *et al.*, 1979; ap Rees, 1984). For the purposes of this investigation, a vacuolar location is also assumed for the relatively low concentrations of glucose and fructose. The mean molecular weights of fructan were not determined in all cases during the survey and a low degree of polymerization (DP = 10) has been assumed. The data, described in Table 7, are based on eight fructan-rich species including three with exceptionally high concentrations in their underground storage organs (*H. non-scripta*, *Allium ursinum* and *A. vineale*) and three with relatively low values (*Myosotis secunda*, *Deschampsia flexuosa* and *Arrhenatherum elatius*). While fructan is predominantly stored underground in cold-hardy species, it is the emerging shoot which is likely to experience potentially injurious cold-air temperatures. Therefore, the values for the vacuolar concentrations of each carbohydrate have been calculated for the vacuoles of the shoot, rather than of the storage organ. Finally, the carbohydrate concentrations represent the maximum values detected within the winter period (October to March) for each species. Table 7 shows that seven of the eight species had a maximum (winter) vacuolar concentration of fructan + sucrose + glucose + fructose in the range 110 to 248 mM. Such a range of concentrations would depress the freezing point of vacuolar sap by about 0·2 to 0·5 °C. One species, however, *A. vineale*, accumulates exceptionally large amounts of fructan in the shoot during November and gives an estimated maximum vacuolar carbohydrate concentration of 459 mM (DP = 10), a value which, if in solution, would indeed depress the freezing point of water by 1 °C. However, separation by HPLC showed that the greater part of the fructan in this species during November had a DP exceeding 10, thereby significantly decreasing the estimated vacuolar molarity.

If, as has been suggested, depolymerization of fructan contributes to chill tolerance, then winter-induced increases in sucrose and fructose may be expected. With the sole exception of *M. secunda*, the vacuolar concentrations of sucrose and fructose in shoots during winter do exceed that found in summer, often three- to four-fold (data not shown). The highest concentrations of sucrose were for *Campanula rotundifolia* (January) at an estimated 70 μM and *H. non-scripta* (March) at 58 μM, concentrations which would depress the freezing point of water by 0·14 and 0·11 °C. Concentrations of free hexoses (assuming a vacuolar location) rarely exceed 70% that of sucrose. However, the greatest contribution to the

Table 7. *Estimated maximum winter concentrations of vacuolar carbohydrates in shoots*

Species	Carbohydrate (mM)				Total concn	Fructan (%)	Month selected
	Fructose	Glucose	Sucrose	Fructan			
Deschampsia flexuosa	5	5	18	140	168	83	December
Arrhenatherum elatius	18	20	43	110	191	58	January
Allium ursinum	32	32	15	31	110	28	March
A. vineale	8	14	39	459	520	88	November
Hyacinthoides non-scripta	7	11	47	183	248	73	February
Achillea millifolium	4	4	7	116	131	89	February
Campanula rotundifolia	4	7	16	201	228	88	December
Myosotis secunda	1	< 1	1	140	142	98	November

Relative volume of vacuole calculated from Raven & Rubery (1982) using maximum carbohydrate values detected in each species in months of October to March. DP of fructan assumed to be 10, location of fructose, glucose and sucrose assumed to be exclusively vacuolar. All field collections were made at 09.00 to 10.00 hours to minimize diurnal variations (Waite & Boyd, 1953).

See Table 3 for an outline of analytical methods.

vacuolar molarity of carbohydrate, in seven of the eight species, was from fructan itself and generally exceeded 70%. Concentrations ranged from 31 to > 450 mM (assuming a low DP of 10). There must be some doubt, however, about the solubility of such large amounts of fructan. *In vitro*, inulin from chicory root (Sigma Ltd, Poole) with a mean DP > 10 has a maximum solubility in distilled water of 80 mM at 25° (unpublished observation), although inulin from *Helianthus* (DP = 35) remains in solution at *c*. 100 mM at room temperature (C. J. Pollock, pers. comm.).

The exception to this pattern of high concentrations of fructan in shoots was *A. ursinum*. In this species, the highest recorded winter concentration of fructan was only 31 mM (Table 7). In this species, the molar concentration of fructose and glucose totalled 63 mM with a further 15·6 mM from sucrose. In total, these three carbohydrates would depress the freezing point of water by only 0·16 °C, a not very convincing example of carbohydrate-mediated cryoprotection in this early spring-growing species.

A further problem arises in explaining the variation in time when the maximum carbohydrate concentrations in the vacuoles are achieved. These range from November (*M. secunda*) to March (*A. ursinum*) and even in one community, limestone grassland where uniform air temperatures are likely to prevail, the period of maximum accumulation ranged from November (*A. vineale*) to February (*Achillea millifolium*). The period of maximum vacuolar concentration in these shoots varies not with the temperature of the environment but with the individual developmental history of each species.

The Ecological Significance of Fructan – And Conclusions

The vacuolar location of fructan, often in high concentration, provides a potentially large reservoir of carbohydrate during early to late spring, both to meet respiratory demands and to regulate chloroplast metabolism (Stitt, 1984). These attributes have been discussed more fully by Pollock (1986). Other workers, notably Wagner,

Keller & Weimken (1983), have shown that starch metabolism in the Gramineae is more sensitive to temperatures below about 5 °C than is that of fructan. Although the molecular basis for this difference is not known, nevertheless for some cold-season growing grasses at least, the advantage in utilizing fructan rather than starch may be significant. The advantage of this phenomenon to other families, however, remains untested but, within the Sheffield flora, does not appear to include the cold tolerance in which the vacuolar molarity of fructan is important. The estimates of the concentration of carbohydrates in the vacuole of the shoot, given in Table 7, are likely to be overestimates. The data were selected from the highest recorded concentrations in winter regardless of the developmental state of the shoot. The calculations assumed an improbably low degree of polymerization for fructan and an unlikely vacuolar location for all of the glucose and fructose. Yet, by making the most favourable interpretation of these data, there is no strongly convincing evidence that fructan makes any significant contribution to low temperature tolerance in the species considered under the field conditions prevailing. The benefit gained from depressing the freezing point of water by 0·1 to 0·5 °C could be of some modest advantage in the field but at latitude 53° N, the low temperatures of winter or early spring are unlikely to be sufficiently stressful to give fructan-rich species a significant advantage over starch-containing plants. Only in the case of *A. vineale* was there any indication that freeze tolerance, through fructan metabolism, was likely to be of any appreciable consequence. For those species, at the latitude of Sheffield, which have rather lower concentrations of fructan, the particular value of fructan over starch may now be lost.

There is, however, one further attribute of fructan which may be of significance to those species undergoing cold season growth. Cells which expand (rather than divide) during early spring will physically carry a full complement of reserve carbohydrate within the same cell. Thus, for example, in the elongating shoots of *G. nivalis*, *A. ursinum* and *H. non-scripta* the need for carbohydrate transport over a distance may be circumvented. By shortening supply lines at critical cold periods, such species may have a considerable advantage over starch-storing, small-celled, transport-dependent species. To take advantage of the vacuolar location of fructan, such species would require relatively large concentrations of the carbohydrate together with large vacuoles in which to store it (and retain it in solution). The large-celled, exceptionally fructan-rich, early spring-growing species, such as *A. ursinum* and *H. non-scripta*, may be among the few contemporary species in the Sheffield flora to gain a significant advantage from fructan. This advantage is only exploited because of their early season growth, during which changes in osmotic potential brought about by depolymerization of fructan may also facilitate the uptake of water necessary for cells to expand.

There are however several species, including *A. millefolium*, *A. vineale* and *C. rotundifolia*, which maintain high concentrations of fructan but do not undergo early spring growth (although their growth may be prolonged into late autumn). The peculiar attributes of fructan do not appear to be exploited by such species. Although there may be attributes of fructan yet to be discovered, it is possible that, for many species in the contemporary Sheffield flora, the environmental pressures encouraging early spring growth (such as summer shade or low temperature survival) are neither widespread nor severe enough to make fructan an attribute of high ecological significance. In other floras and at other times, this may be or may have been different.

ACKNOWLEDGEMENTS

Certain aspects of the experimental data reported here are based on the work of Jane Brocklebank, whose permission to discuss these, in advance of the publication of her thesis, is gratefully acknowledged. A debt is also due to Dr John Hodgson for his advice and care over selecting species from within the Sheffield flora. Several helpful discussions with Professors David Lewis and John Raven, and Dr Chris Pollock are also acknowledged with thanks.

REFERENCES

AP REES, T. (1984). Sucrose metabolism. In: *Storage Carbohydrates in Vascular Plants* (Ed. by D. H. Lewis), pp. 53–74. Cambridge University Press, Cambridge.

ARCHBOLD, H. K. (1940). Fructosans in the monocotyledons: a review. *New Phytologist*, **39**, 185–219.

BENNETT, M. D. (1976). DNA amounts, latitude and crop plant distribution. *Environmental & Experimental Botany*, **16**, 93–108.

BENNETT, M. D. (1987). Variation in genomic form in plants and its ecological implications. In: *Frontiers of Comparative Plant Ecology* (Ed. by I. H. Rorison, J. P. Grime, R. Hunt, G. A. F. Hendry & D. H. Lewis), *New Phytologist*, **106** (Suppl.), 177–200. Academic Press, New York & London.

BENNETT, M. D. & SMITH, J. F. (1976). Nuclear DNA in angiosperms. *Philosophical Transactions of the Royal Society of London. Series B*, **274**, 227–274.

BERGMEYER, H. U. & BERNT, E. (1963). D-glucose – determination with glucose oxidase and peroxidase. In: *Methods of Enzymatic Analysis* (Ed. by H. U. Bergmeyer), pp. 123–130. Verlag Chemie, Weinheim.

BERLIN, J., QUISENBERRY, J. E., BAILEY, F., WOODWORTH, M. & MCMICHAEL, B. L. (1982). Effect of water stress on cotton leaves. *Plant Physiology*, **70**, 238–243.

BOURDU, R. (1957). Contribution a l'étude du métabolisme glucidique des Boraginacées. *Revue générale de botanique*, **64**, 153–192, 197–260.

CHEN-SHE, S.-H., LEWIS, D. H. & WALKER, D. A. (1975). Stimulation of photosynthetic starch formation by sequestration of cytoplasmic orthophosphate. *New Phytologist*, **74**, 383–392.

CLAPHAM, A. R., TUTIN, T. G. & WARBURG, E. F. (1981). *Excursion Flora of the British Isles*, 3rd Edn. Cambridge University Press, Cambridge.

COLIN, H. & BELVAL, H. (1922). Les hydrocarbones solubles du grain de blé au cours du développement. *Comptes rendus de l'Académie des Sciences*, **177**, 343–346.

COLIN, H. & CHOLLET, M.-M. (1939). L'inulogénèse chez les plantes annulles. *Comptes rendus de l'Académie des Science*, **208**, 549–552.

COOPER, J. P. (1964). Climatic variation in foliage grasses. Leaf development in climatic races of *Lolium* and *Dactylis*. *Journal of Applied Ecology*, **1**, 45–61.

CRONQUIST, A. (1981). *An Integrated System of Classification of Flowering Plants*. Columbia University Press, New York.

DAHLGREN, R. (1975). A system of classification of the angiosperms to be used to demonstrate the distribution of characters. *Botaniska Notiser*, **128**, 119–147.

DARBYSHIRE, B. & HENRY, R. J. (1981). Difference in fructan content and synthesis in some *Allium* species. *New Phytologist*, **87**, 249–256.

DE CUGNAC, A. (1931). Les glucides des Graminées. Importance du fructoholosides. *Bulletin de la Société de chimie biologique*, **13**, 125–132.

DICKINSON, R. E. & LARWOOD, P. R. (1973). Incorporation of ^{14}C photosynthate into major chemical fractions of source and sink leaves in cottonwood. *Plant Physiology*, **57**, 185–193.

DUFFUS, C. M. (1984). Metabolism of reserve starch. In: *Storage Carbohydrates in Vascular Plants* (Ed. by D. H. Lewis), pp. 231–252. Cambridge University Press, Cambridge.

EAGLES, C. F. (1967). Variation in the soluble carbohydrate content of climatic races of *Dactylis glomerata* L. at different temperatures. *Annals of Botany*, **31**, 645–651.

EDELMAN, J. & JEFFORD, T. G. (1968). The mechanism of fructosan metabolism in higher plants as exemplified by *Helianthus tuberosus*. *New Phytologist*, **67**, 517–531.

FARRAR, S. C. & FARRAR, J. F. (1985). Carbon fluxes in leaf blades of barley. *New Phytologist*, **100**, 271–283.

FREHNER, M., KELLER, R. & WIEMKEN, A. (1984). Localisation of fructan metabolism in the vacuoles isolated from protoplasts of Jerusalem Artichoke tubers (*Helianthus tuberosus* L.). *Journal of Plant Physiology*, **116**, 197–208.

GIBBS, R. D. (1974). *Chemataxonomy of Flowering Plants*, **1–4**. McGill-Queens University Press, Montreal.

GRIME, J. P. & MOWFORTH, M. A. (1982). Variation in genome size – an ecological interpretation. *Nature*, **299**, 151–153.

GRIME, J. P., SHACKLOCK, J. M. L. & BAND, S. R. (1985). Nuclear DNA contents, shoot phenology and species coexistence in a limestone grassland community. *New Phytologist*, **100**, 435–445.

GRIME, J. P., HODGSON, J. G. & HUNT, R. (1987). *Comparative Plant Ecology: A Functional Approach to Common British Species*. George Allen & Unwin, London. (In press.)

GUNN, T. C. & WALTON, D. W. H. (1985). Storage carbohydrate production and overwintering strategy in a winter-green tussock grass on South Georgia (Sub-Antarctic). *Polar Biology*, **4**, 237–242.

HAISSIG, B. E. & DICKSON, R. E. (1979). Starch measurement in plant tissue using enzymatic hydrolysis. *Physiologia Plantarum*, **47**, 151–157.

HAJIBAGHERI, M. A., HALL, J. L. & FLOWERS, T. G. (1984). Stereological analysis of leaf cells of the halophyte *Sueda maritima* (L.) Dum. *Journal of Experimental Botany*, **35**, 1547–1557.

HEGNAUER, R. (1962–1973). *Chemotaxonomie der Pflanzen*, **1–6**. Birkhauser, Basel.

HEROLD, A. (1984). Biochemistry and physiology and synthesis of starch in leaves: autotrophic and heterotrophic chloroplasts. In: *Storage Carbohydrates in Vascular Plants* (Ed. by D. H. Lewis), pp. 181–204. Cambridge University Press, Cambridge.

HEYWOOD, V. H. (1978). *Flowering Plants of the World*. Oxford University Press, Oxford.

HOLLIGAN, P. M., CHEN, C. & LEWIS, D. H. (1973). Changes in the carbohydrate composition of leaves of *Tussilago farfara* during infection by *Puccinia poarum*. *New Phytologist*, **72**, 947–955.

HOUSLEY, T. L. & POLLOCK, C. J. (1985). Photosynthesis and carbohydrate metabolism in detached leaves of *Lolium temulentum* L. *New Phytologist*, **99**, 499–507.

JULANDER, O. (1945). Drought resistance in range and pasture grasses. *Plant Physiology*, **20**, 573–599.

KLOTSCH, H. & BERGMEYER, H. U. (1963). D-fructose. In: *Methods of Enzymatic Analysis* (Ed. by H. U. Bergmeyer), pp. 156–159. Verlag Chemie, Weinheim.

LEIGH, R. A., AP REES, T., FULLER, W. A. & BANFIELD, J. (1979). The location of acid invertase activity and sucrose in the vacuoles of storage roots of beetroot (*Beta vulgaris*). *Biochemical Journal*, **178**, 539–547.

LEVIN, D. A. & FUNDERBURG, D. W. (1979). Genome size in angiosperms: temperature versus tropical species. *American Naturalist*, **114**, 784–795.

LEWIS, D. H. (1984). Occurrence and distribution of storage carbohydrates in vascular plants. In: *Storage Carbohydrates in Vascular Plants* (Ed. by D. H. Lewis), pp. 1–52. Cambridge University Press, Cambridge.

LEWIS, D. H., CHEN, C., WOODS, G. & CULPIN, L. A. (1972). A modified *p*-anisidine technique for the detection and permanent recording of free and combined fructose on paper chromatograms. *Journal of Chromatography*, **74**, 369–373.

MEIER, H. & REID, J. S. G. (1982). Reserve polysaccharides other than starch in higher plants. In: *Encyclopedia of Plant Physiology*, vol. 13A. *Plant Carbohydrate I. Intracellular Carbohydrates* (Ed. by P. A. Loewus & W. Tanner), pp. 418–471. Springer-Verlag, Berlin.

OJIMA, K. & ISAWA, T. (1968). The variation of carbohydrates in various species of grasses and legumes. *Canadian Journal of Botany*, **46**, 1507–1511.

POLLARD, C. J. (1982). Fructose oligosaccharides in monocotyledons: a possible delimitation of the order Liliales. *Biochemical Systematics and Ecology*, **10**, 245–249.

POLLARD, C. J. & AMUTI, K. S. (1981). Fructose oliosaccharides, possible markers of phylogenetic relationships among dicotyledonous plant families. *Biochemical Systematics and Ecology*, **9**, 69–78.

POLLOCK, C. J. (1982a). Oligosaccharide intermediates of fructan synthesis in *Lolium temulentum*. *Phytochemistry*, **21**, 2461–2465.

POLLOCK, C. J. (1982b). Patterns of turnover of fructans in leaves of *Dactylis glomerata* L. *New Phytologist*, **90**, 645–650.

POLLOCK, C. J. (1984). Physiology and metabolism of sucrosyl-fructans. In: *Storage Carbohydrates in Vascular Plants* (Ed. by D. H. Lewis), pp. 97–113. Cambridge University Press, Cambridge.

POLLOCK, C. J. (1986). Fructans and the metabolism of sucrose in vascular plants. *New Phytologist*, **104**, 1–24.

POLLOCK, C. J. & JONES, T. (1979). Seasonal patterns of fructan metabolism in forage grasses. *New Phytologist*, **83**, 9–15.

PONTIS, H. G. (1970). The role of sucrose and fructosylsucrose in fructosan metabolism. *Physiologia plantarum*, **23**, 1089–1100.

PONTIS, H. G. & DEL CAMPILLO, E. (1985). Fructan. In: *Biochemistry of Storage Carbohydrates in Green Plants* (Ed. by P. M. Dey & R. A. Dixon), pp. 205–227. Academic Press, London.

RAVEN, J. A. & RUBERY, P. H. (1982). Co-ordination of development and hormone receptors, hormone action and hormone transport. In: *The Molecular Biology of Plant Development* (Ed. by H. Smith & D. Grierson), pp. 28–48. Blackwell Scientific Publications, Oxford.

ROE, J. H. (1934). A colorimetric method for the determination of fructose in blood and urine. *Journal of Biological Chemistry*, **107**, 15–22.

RUTHERFORD, P. P. & WESTON, E. W. (1968). Carbohydrate changes during cold storage of some inulin-containing roots and tubers. *Phytochemistry*, **7**, 175–180.

SACHS, J. (1864). Über die Sphaerokristalle des Inulins und dessen mikroskopische Nachweisung in den Zellen. *Botanische Zeitung*, **22**, 77–81, 85–89.

SMITH, D. (1967). Carbohydrates in grasses. II. Sugar and fructosan composition of the stem bases of bromegrass and timothy. *Crop Science*, **I**, 62–67.

SMITH, D. (1968). Carbohydrates in grasses. IV. Influence of temperature on the sugar and fructosan composition of timothy plant parts at anthesis. *Crop Science*, **8**, 331–334.

SMITH, D. (1973). The non-structural carbohydrates. In: *Chemistry and Biochemistry of Herbage* (Ed. by D. Smith, G. W. Butler & R. W. Bailey), **1**, 105–155. Academic Press, New York.

STITT, M. (1984). Degradation of starch in chloroplasts: a buffer to sucrose metabolism. In: *Storage Carbohydrates in Vascular Plants* (Ed. by D. H. Lewis), pp. 205–229. Cambridge University Press, Cambridge.

TUTIN, T. G., HEYWOOD, V. H., BURGES, N. A., MOORE, D. M., VALENTINE, D. H., WALTERS, S. M. & WEBB, D. A. (Eds) (1964–1980). *Flora Europaea*, **1–5**. Cambridge University Press, Cambridge.

WAGNER, A. W., KELLER, F. & WIEMKEN, A. (1983). Fructan metabolism in cereals: induction in leaves and compartmentation in protoplasts and vacuoles. *Zeitschrift für Pflanzenphysiologie*, **112**, 359–371.

WAITE, R. & BOYD, J. (1953). The water-soluble carbohydrates of grasses. *Journal of the Science of Food and Agriculture*, **4**, 197–204.

WALKER, D. A. & HEROLD, A. (1977). Can the chloroplast support photosynthesis unaided? In: *Photosynthetic Organelles: Structure and Function*, a special issue of *Plant and Cell Physiology* (Ed. by C. Miyachi, S. Katoh, Y. Fugita & K. Shibata), pp. 295–310. The Japanese Society of Plant Physiologists, Tokyo.

WEAST, R. C. (Ed.) (1975). *Handbook of Chemistry and Physics*, 56th Edn. CRC Press, Cleveland.

WHITE, L. M. (1973). Carbohydrate reserves of grasses: a review. *Journal of Range Management*, **26**, 13–18.

New Phytol. (1987) **106** (Suppl.), 217–233

TRANSPORT PROCESSES AND WATER RELATIONS

By JOHN A. RAVEN and LINDA L. HANDLEY

Department of Biological Sciences, University of Dundee, Dundee DD1 4HN, Scotland

SUMMARY

A cost–benefit analysis of transport and of water relations in plants of a number of life forms is presented. For planktophytes, it appears that an increase in size above $\sim 50\ \mu m$ diameter would restrict growth rate in oligotrophic environments owing to a greater increment of restriction by boundary layer of nutrient uptake than of maximum specific growth rate as organism size increases. For macroscopic aquatic haptophytes (mainly algae), even the most favourable combination of velocity of water flow and plant morphology cannot reduce these restrictions enough to make nutrient diffusion to the plant surface less limiting for growth in oligotrophic environments than is the case for planktophytes. Macroscopic aquatic rhizophytes, with their need for intraplant N- and P- flux from rhizoid or root to shoot, and for reduced C-flux from shoot to rhizoid or root, involves cytoplasmic streaming in giant-celled algae, and transport in the phloem and (probably) the xylem (using root pressure) in vascular plants. It is likely that the energetic running costs of nutrient transport (joules used per mol solute transported over a distance of 1 m) is higher for the giant-celled algae than for the vascular plants. The predominant (in terms of global biomass) terrestrial plants are the rhizophytic, homoiohydric desiccation-intolerant sporophytes of vascular plants. Such plants incur greater penalties, in terms of reduced specific growth rate under resource-saturated or resource-limited conditions, and of reduced resource use efficiency, with increased plant height. This is a result of diversion of resources to producing supporting tissue, and xylem and phloem, in larger amounts per unit biomass than is the case for smaller plants. By virtue of increased size, however, a plant can command a higher incident photon flux density as well as access to a greater depth of soil from which nutrients and water can be extracted. Other major categories of terrestrial rhizophytes are desiccation-tolerant homoiohydric sporophytes of vascular plants, endohydric but poikilo-hydric gametophytes of mosses, and ectohydric gametophytes of archegoniates and thalli of many algae and lichens. In the order in which they are listed, these plants have a smaller potential stature (none are more than 2 m high), and thus can command access to fewer resources than can taller plants with deeper root systems in the same community. However, the smaller plants have the potential for higher specific growth rates under resource-saturated and resource-limited conditions, and for higher resource use efficiencies, than do the homoiohydric, desiccation-intolerant plants of larger stature, since they have a smaller diversion of resources to supporting and long-distance transport tissues. The extreme of these trends is found in the haptophytic soil algae with no non-green cells. Quantitation of the differences between life forms requires much more investigation, as do differences within life forms with respect to the handling of water and solute transport. Investigating the possible selective significance of these differences will be an even more challenging task.

Key words: Transport processes, water relations, cost benefit analysis, planktophytes, hapto-phytes, rhizophytes.

INTRODUCTION

The comparisons which this article attempts to make are, perhaps, more wide-ranging than those which are attempted in some other contributions to this symposium in that aquatic and amphibious, and fossil plants are considered in addition to extant terrestrial plants. An underlying theme is that of the costs and benefits of transport and water handling in relation to the acquisition and retention

0028-646X/87/05S217 + 17 $03.00/0

of resources in the broader context of the survival and dispersal of the genotype. While such approaches have their limitations (Woolhouse, 1981), they do have value in delimiting what is possible within a given set of constraints of rate, efficiency and safety of net resource acquisition (Odum & Pinkerton, 1955; Raven, 1984a, b, 1985a, b, 1986a; cf. Watt, 1986). How these constraints influence inclusive fitness is not readily discerned, particularly if the implicit assumption of competition is rendered less significant by a frequency of disturbance which prevents the full expression of competitive exclusion (Reynolds, 1984; Harris, 1986). We feel that, this possibility notwithstanding, optimization of some combination of the rate, efficiency and safety of resource acquisition, manipulation, storage and protection is likely to be of selective significance in most situations. However, measuring inclusive fitness is a much more difficult problem to approach from the reductionist comparisons made here (Osmond, Björkman & Anderson, 1980); a more holistic approach is required (see Bradshaw, 1987).

In considering transport and water relations, emphasis is placed on transport occurring (by diffusion or mass flow) outside membranes. This is not intended to downgrade the role of membranes but reflects the greater ease of producing a coherent account in the context of comparative ecology. The discussion is organized around comparisons of different life forms (Raven, 1986a).

PLANKTOPHYTES

Planktophytes (*sensu* Luther, 1949; see Appendix) are the sole photosynthetic inhabitants of over half of the world's surface, i.e. water more than some 300 m deep: Littler *et al.*, 1985, 1986). Their habitats are generally characterized by low steady-state concentrations of resources such as nitrate, ammonium and phosphate. The small size of most planktophytes, i.e. less than some 50 μm diameter, means that the surface area per unit volume is large. Consideration of the volume-based elemental content of the organisms, their maximum specific growth rates and the unstirred layer thickness around the organisms (approximately equal to the radius of the organism under natural conditions) suggests that restrictions on the acquisition of dissolved nutrients owing to unstirred layer effects are relatively small at natural nutrient concentrations (Raven, 1987a, b). The movement of planktophytes relative to their bulk water medium (involving flagellar motility or sinking of non-motile, denser-than-water cells) does little to alter unstirred layer effects on nutrient acquisition. These movements do, however, serve to move organisms between parcels of water with different nutrient concentrations. This is most significant for planktophytes in stratified (as opposed to mixed) water bodies, where motile (flagellate) organisms can, apparently, optimize acquisition of light and dissolved nutrient (Raven & Richardson, 1984) via diel vertical migrations through gradients of availability of photons (more near the surface; only available in the daytime) and of dissolved nutrients (more at depth, available day and night).

It would appear that the small size of planktophytes may be mainly related to acquisition of resources. Larger phototrophs in the open ocean would have unstirred layer restrictions on nutrient acquisition which would have larger constraining effects on specific growth rate than occurs in real, smaller planktophytes, since the decrement in maximum specific growth rate with increasing volume of organism is not sufficient to offset the increasing restriction on nutrient resource acquisition by the larger unstirred layer thickness (Raven, 1987a, b).

Small size also favours the efficient use chromophore molecules in photosynthetic light harvesting (Kirk, 1983; Raven, 1984b, 1987a)

Among the costs of small size are those related to turgor and regulation of volume where increased surface area per unit volume can increase resource costs of 'osmoregulation' in both walled and wall-less cells from freshwater or marine habitats (Raven, 1982, 1984a, 1987a, b).

This brief discussion only serves to indicate in very broad terms aspects of the transport and water relations of planktophyte ecology. Many problems (e.g. that of the environmental factors which selectively favour different cell sizes and shapes within the general 'small' size of planktophytes which covers several orders of magnitude of cytoplasmic volume per cell) are essentially unresolved (Sournia, 1982; Raven, 1987a).

AQUATIC HAPTOPHYTES

Most haptophytes (*sensu* Luther, 1949; see Appendix) are, like essentially all planktophytes, algae (Raven, 1981a, 1984a, 1986a). Many of them are large differentiated plants. This can be interpreted in terms of shading out competitors, avoidance of grazing and the possibility of temporal averaging of annual variations in the availability of resources (Raven, 1981a, 1984a, 1986a).

The smaller surface area per unit volume for a larger organism does not generally offset the lower potential specific growth rate of the larger aquatic haptophytes. Accordingly, growth at a certain fraction of the maximum specific growth rate requires a larger nutrient solute influx on a unit surface area basis in the larger aquatic macrophytes, even when the reduced content of many nutrient elements per unit plant volume in larger aquatic plants is taken into account (Raven, 1984a). Unstirred layers are, then, more likely to restrict growth of haptophytes than of planktophytes at a given nutrient concentration in the bulk phase and water velocity relative to the organism. Haptophyte environments generally afford higher water velocities relative to the attached organism than the less than 1 mm s^{-1} achieved by sinking or swimming haptophytes. These higher velocities reduce unstirred layer thickness (and hence diffusive limitation on nutrient solute acquisition), especially if 'habitat choice' is combined with the presence of hairs, analagous to the rhizoids, root hairs and mycorrhizas of rhizophytes, which project through the 'bulk' unstirred layer around the plant body (Raven, 1981a, 1984a).

An example of 'habitat choice', combined with structural characteristics which reduce the 'bulk' unstirred layer thickness, is afforded by the freshwater red alga *Lemanea mamillosa* Kütz. (Raven, Griffiths & Beardall, 1982; MacFarlane & Raven, 1985; Raven *et al.*, 1987a, b). The diffusion boundary layer under natural conditions is only 10 to 20 μm thick for the haploid 'bristles' (tapering cylinders up to 300 mm long and about 1 mm in radius) with turbulence-generating 'knobbles' at intervals, living in a rapidly flowing (in excess of 1 m s^{-1}), cool stream. The unstirred layer thickness is equivalent to that around a planktophyte of 10 to 20 μm radius. Despite the best (evolutionary) endeavours of *L. mamillosa* in reducing boundary layer thickness, it appears that its growth is more limited by unstirred layers than is that of a planktophyte growing at the same bulk phase nutrient concentration, as the following calculation shows.

A steady-state growth rate of about 0·1 d^{-1} (8 × 10^{-7} s^{-1}) of *L. mamillosa* at 10 °C needs an influx of 0·135 μmol N (m^2 surface)$^{-1}$ s^{-1} (computed from a mean net C influx over a 24 h light–dark cycle of 1·35 μmol (m^2 surface)$^{-1}$ s^{-1}, and a C/N ratio of 10: MacFarlane & Raven, 1985; Raven *et al.*, 1987a, b, and unpublished data. A 20 μm radius planktophyte with 125 kg C (m^3 cell volume)$^{-1}$, a molar C/N ratio of 7, and a specific growth rate of 0·5 d^{-1} (4 × 10^{-6} s^{-1}) at 10 °C (see Raven, 1982, 1984a, 1987a, b, c) requires a mean N influx of 0·04 μmol N (m^2 surface)$^{-1}$ s^{-1}. Thus, the higher C/N ratio and lower specific growth rate of *Lemanea* does not offset its lower surface area per unit volume in terms of the required nutrient influx per unit surface area, so that equal boundary layer thicknesses for *Lemanea* and the planktophyte, and diffusive constraints on growth are greater at a given (limiting) nutrient concentration in the bulk phase. This sort of problem is more severe for haptophytes growing in slower-moving environments than does *Lemanea* and having even lower surface area per unit volume. More work is needed to analyze the restrictions on resource acquisition imposed by habitat and plant morphology.

Larger and more differentiated haptophytes have localized meristematic regions and resource-acquiring regions (including hairs). This implies resource transport within the plant, a process which generally occurs symplastically rather than apoplastically (see Appendix) [Raven (1984a, chapter 9) and Bauman & Jones (1986)]. Cyclosis in small cells in a symplastic network is not guaranteed to increase solute flux when the transport through plasmodesmata is diffusive; fluxes in such small-celled symplasts are generally limited to distances of a few millimetres. In many of the larger Phaeophyta, phloem-like tissue is present and involved in long-distance transport, presumably involving mass flow of solution through the cytosol of the cells and of the enlarged plasmodesmata (sieve pores) which link the cells (chapter 9 of Raven, 1984a).

A final point relates to the water relations of periodically emersed, and particularly of intertidal, haptophytes. These haptophytes are poikilohydric (see Appendix). Notwithstanding this, Raven *et al.* (1987a, b) have suggested that loss of 20 % of the plant water during an illuminated, emersed period of 6 h d^{-1} could be associated (with a 'C$_4$-like' loss of 300 g water per g dry weight gain) with a substantial fraction of the overall annual productivity of the 'top' plants of a population of *Ascophyllum nodosum* (L.) Le Jol. (cf. Tajiri & Aruga, 1984; Oates, 1985, 1986). Haptophytes, of course, must rely on their endogenous water for 'transpiration' when emersed; unlike rhizophytes, they have no connection to a supply of water from the sediment.

AQUATIC RHIZOPHYTES

While some aquatic rhizophytes (*sensu* Luther, 1949; see Appendix) are algae, many are secondarily aquatic bryophytes and tracheophytes. The greater concentration of available nutrients in the sediment than in the bulk water phase means that, even if the effective boundary layer thickness in the (unstirred) sediment is much higher than in the bulk phase, the sediment can be a preferred source of N, P, Fe and even, for vascular plants of the isoetid life form, of C (Raven, 1984a, 1987b).

The rhizophyte's environment is likely to be less well stirred than that of haptophytes in that it is depositing rather than eroding the substratum. Despite this, many rhizophyte shoots may have fewer 'design constraints' related to resource acquisition, since they are primarily photon- and C-acquiring moieties, with less involvement in acquisition of N and P present at very low concentrations in the bulk phase (Raven, 1981a; chapter 8 of Raven, 1984a). Examples are the relative absence of colourless hairs on, and the higher area-based maximum photosynthetic rates of, the shoots of rhizophytes.

The increased polarity of resource acquisition in aquatic rhizophytes relative to aquatic haptophytes is reflected in intraplant transport pathways which catalyze shoot to root/rhizoid transport of photosynthate and root/rhizoid to shoot transport of N, P, etc. Vascular aquatic rhizophytes apparently use the mass flow transport systems inherited from their terrestrial ancestors. The phloem moves photosynthate from leaves to non-green tissues, while the xylem (operating in a root pressure mode) may move N, P, etc. from roots to shoots (chapter 9 of Raven, 1984a). The larger algal rhizophytes use cytoplasmic streaming for the movement of photosynthate, N, P, etc.; plasmodesmatal diffusive constraints on the mass flow of solution within the algae are minimized by the organism's being composed of a single giant cell (many larger Chlorophyta: Ulvophyceae) or having giant cells symplastically connected by a few small cells (Chlorophyta: Charophyceae: Charales). Raven (1984a, chapter 9) discusses the relative energy costs (joules needed to move one mol of solute over a given distance of 1 m) of phloem-like transport and cytoplasmic streaming; the latter is energetically more expensive. Gas transport within the organism (e.g. of O_2 to parts in anoxic sediments) is much more readily explained in vascular rhizophytes (with intercellular gas spaces) than in algal rhizophytes (see Raven, 1981a, 1984a).

Water relations of aquatic (amphibious) rhizophytes when emersed have not been extensively studied; it is likely that the presence of roots/rhizoids in moist sediments could lead to ectohydric (see Appendix) water movement in non-vascular plants lacking xylem.

TERRESTRIAL RHIZOPHYTES

Introduction

Most of the larger terrestrial plants are rhizophytes; the great majority of these are vascular plants. As with the attached aquatics, we can find putative selective advantages of large size in attached terrestrial plants. Some of these are avoidance of shading by other plants, increased dispersal potential when air-borne propagules are released into turbulent air above the 'soil' boundary layer and tapping of water and nutrients from a greater depth (and hence volume) of soil (Raven, 1986a).

Terrestrial rhizophytes live in an even more polar environment with respect to resource acquisition than do aquatic rhizophytes, since nutrients (including water) are, with the exception of CO_2, only available from the soil. The acquisition of CO_2 from the gas phase implies substantial loss of water vapour to the atmosphere, thus increasing the demand for movement of soil-derived water to the shoot to hundreds of grams per gram dry weight gain for all but CAM (Crassulacean Acid Metabolism) plants. Terrestrial rhizophytes may be divided into four categories on the basis of three groups of characteristics of their water relations, i.e. the pathway of water transport to the shoot, the tolerance of desiccation of vegetative tissues and the extent to which the hydration state of the plant is independent of soil water supply and evaporative demand. The relevant terms (ectohydric/endo-hydric, desiccation intolerant/desiccation tolerant, poikilohydric/homoihydric) are defined in the Appendix and Table 1 presents a classification of terrestrial rhizophytes in relation to these criteria.

Table 1. *A classification of terrestrial rhizophytes in relation to aspects of water relations. Terms are defined in the Appendix*

Category	Examples
Homoiohydric, endohydric, desiccation-intolerant	Most vascular plant sporophytes ($\sim 99\%$ of pteridophytes, 100% of gymnosperms, 99.9% of monocotyledons, 99.9% of dicotyledons)
Homoiohydric, endohydric, desiccation-tolerant	A few vascular plant sporophytes ($\sim 1\%$ of pteridophytes, 0% of gymnosperms, 0.1% of monocotyledons, 0.01% of dicotyledons), all less than 1 to 2 m high
Poikilohydric, endohydric, desiccation-tolerant?	Some bryophyte gametophytes (e.g. *Funaria, Polytrichum, Dawsonia*)
Poikilohydric, ectohydric, desiccation-tolerant (or intolerant)	Most bryophyte gametophytes; all pteridophyte gametophytes; some algae; lichens

Water relations

The poikilohydric, ectohydric organisms include the terrestrial rhizophytic algae which probably represent the ecophysiological and structural state of the algal ancestors of the vascular plants (cf. the Silurian *Eohostimella*). These organisms are not able to regulate their water content as a function of soil water supply and atmospheric evaporative demand and are unable to grow to a great height (more than a few tens of millimetres). However, they do not bear the burden of having to produce the endohydric water transport system or the other adjuncts of homoiohydry, which might reduce the capacity for resource acquisition per unit biomass from a given incident photon flux density and a given nutrient concentration in the soil solution.

The poikilohydric, endohydric plants are represented by the extant gameto-phytes of various mosses. These plants have a water-repellent cuticle which discourages ectohydric water supply and a hydrome of dead conducting cells which facilitates endohydric water transport. This permits the plants to grow to a height of almost a metre. However, the plants lack stomata and thus are not homoiohydric, although the movements of leaf evaginations of the Polytrichaceae during water loss may serve some of the same functions (see Raven, 1977a, 1984c). In the fossil record of terrestrial plants, this stage may be represented by *Cooksonia* from the Upper Silurian and Lower Devonian strata of what is today Europe and Eastern North America, although some specimens have recently been shown to have stomata (Edwards, Fanning & Richardson, 1986). We note that the greater potential for growth in height of these plants permits them to have higher incident photon flux densities, and their rhizoids or roots to reach water and nutrient resources deeper in the soil, than is the case for ectohydric plants. However, the diversion of resources to form cuticle, hydrome and mechanical tissue limits the fraction of the biomass available to produce machinery for the acquisition of limiting supplies of resources and for high specific growth rates.

The homoiohydric plants are all endohydric. Those which are desiccation-tolerant in the vegetative state are limited to heights of 1 to 2 m (probably a function of problems with the refilling of long runs of xylem elements after desiccation), while desiccation-intolerant plants can be up to 100 m high (Gaff, 1981; Raven, 1986a). With the exception of the astomatous specimens of *Cooksonia*, it is likely that all fossil terrestrial vascular plant sporophytes, as well as their extant descendants, are potentially homoiohydric. An exception is the extant amphibious vascular plant *Stylites* (*Isoetes*) *andicola* Amstutz. This astomatous plant has a leaf cuticle which is very impermeable to gases, and even specimens which are very rarely submerged (i.e. are essentially completely terrestrial) obtain almost all of their CO_2 via their roots rather than from the air around their leaves (Keeley, Osmond & Raven, 1984; Sternberg *et al.*, 1985). *Stylites* is very probably secondarily astomatous and secondarily terrestrial, a conclusion based on the likely selective pressures of transpiratory water loss during CO_2 acquisition from the atmosphere favouring the presence of xylem (Raven, 1977a, 1984c). *Stylites* thus avoids gas exchange with the atmosphere and, while it can probably remain hydrated for a significant time in relatively dry soil with an unsaturated atmosphere around the leaves, it does not achieve this by means of a controllable variation in the resistance to gas exchange between the leaves and the surrounding atmosphere and so should probably not be considered to be truly homoiohydric. Another exception, this time in the shape of a homoiohydric terrestrial rhizophyte which is *not* a vascular plant, is the Devonian (Rhynie Chert) fossil *Aglaophyton major* (Kidston & Lang) D. S. Edwards (formerly *Rhynia major* Kidston & Lang). The free-living sporophytes (contrast extant bryophytes with sporophytes dependent on the gametophyte) of this plant have stomata, cuticle, intercellular gas spaces and tissue which apparently functioned in the apoplastic mass flow of water to transpirational termini (Edwards, 1986). However, the putative water-conducting tissue is analogous to the hydrome of extant bryophytes rather than to true xylem, and *Aglaophyton major* (unlike *Rhynia gwynne-vaughanii*) is *not* a vascular plant.

Desiccation-intolerant plants in particular are both important casters of shade on lower growing plants and exploiters of water and nutrients from deeper in the soil than smaller plants. However, even more than the endohydric poikilohydric plants, they have to divert resources into the structures and catalysts of water transport and its regulation (cuticle, stomata, intercellular gas spaces and xylem) and of mechanical support. The fractional diversion increases with height (see below), so that the capacity per unit total biomass to absorb photons from a given incident photon flux density, or to acquire nutrients from a given bulk phase solute concentration, is less than that of the poikilohydric plants. This diversion of resources may also restrict the specific growth rate of the organism. We note that the homoiohydric system also has running costs (e.g. of stomatal function) as well as capital costs (Cowan, 1986).

The diversion of resources into structural tissues and into the apparatus of water transport and its regulation is a larger fraction of the biomass in taller, free-standing plants than in shorter plants. A simple calculation illustrates this point for the xylem. We posit a C_3 herbaceous plant 1 m high (i.e. with 1 m of xylem between the water-absorbing zone of the roots and the transpirational termini). We consider the quantity of plant equivalent to 1 m² of leaf area. With a net CO_2 fixation rate (corrected for whole plant dark respiration over 24 h) of 10 μmol (m² leaf area)$^{-1}$ s^{-1} over a 12 h photoperiod and a water-use efficiency of 1 g dry weight gain per 500 g water transpired, then the transpiration rate in the light is 120×10^{-9} m³ water (m² leaf area)$^{-1}$ s^{-1}. This is the quantity of water which must be transported to the leaf in the xylem each second. We assume that the specific conductivity of the xylem is 25×10^{-9} m³ (m² total xylem transverse section area)$^{-1}$ s^{-1} (Pa m^{-1})$^{-1}$ or 25×10^{-9} m² s^{-1} Pa^{-1} (Huber, 1956; Heine, 1971; Milburn, 1979; Woodhouse & Nobel, 1982, taking into account the different areal bases used by different authorities). The water potential difference between leaves and roots is taken to be -500 kPa (leaves relative to roots), permitting some soil drying without implying a leaf water potential so negative as to inhibit photo-synthesis substantially. The required transverse sectional area of the xylem to support the flux of 120×10^{-9} m^3 water m^{-2} s^{-1} is then 9.6×10^{-6} m^2 so that the total xylem volume is 9.6×10^{-6} m^3 and the total xylem dry weight (if there is 200 kg m^{-3} xylem volume) is 1·92 g dry weight, both xylem volume and xylem dry weight being expressed per m^2 of leaf area. For a plant with a specific growth rate of 4×10^{-7} s^{-1} (doubling time of 20 d), we can posit [on the basis of the unit leaf rate (Evans, 1972) quoted above of 10 μmol m^{-2} s^{-1}] a plant dry weight of 380 g m^{-2} leaf area. This dry weight might reasonably be distributed as follows: leaves 180 g, absorbing parts of roots 100 g, rest 100 g. The fraction of the plant dry weight which is taken up by water-conducting xylem is, then, only 1·92/380 or 0·0053. This seems to be rather a low fraction, but it is of the right order of magnitude for a 1 m tall herb (Raven, 1977a; Raven & Rubery, 1982), especially since it makes no provision for either surplus conducting capacity which might maintain water flow after damage to the xylem, or for xylem elements which are involved in supporting the plant rather than in conducting water.

If we now double the height of the plant, the dry weight (excluding both conducting and non-conducting xylem elements) is now 480 g per m^2 leaf area, assuming (as before) 180 g dry weight of leaves and 100 g dry weight of absorbing roots per m^2 leaf area and doubling the dry weight of the rest of the plant to 200 g. The pro rata increase in the 'rest of the plant' fraction with the increase in distance between the water-absorbing zone of the roots and the transpirational termini of the leaves is justified by (*inter alia*) assuming that the phloem transectional area needed to move assimilates from 1 m^2 of leaf area with a given rate of photosynthesis does not vary with the length of the path, which in turn requires a 'non-decremental' mechanism of conduction such as is implicit in the relay hypothesis (see chapter 9 of Raven, 1984a). To estimate the requirements for conducting xylem in such a plant, we assume that the difference in water potential between leaf and root, the specific conductivity of the xylem, and the transpiration rate per m^2 of leaf area, are the same as for the 1 m tall plant. The gradient of water potential driving the transpiratory flux has fallen from 500 kPa m^{-1} to 250 kPa m^{-1}, so that the transectional area of conducting xylem needed to maintain the transpiratory flux is doubled to 19.2×10^{-6} m^2 which, with the doubled length of xylem, means a xylem volume of 38.4×10^{-6} m^3 per m^2 leaf area, and a xylem dry weight of 7·68 g per m^2 of leaf area. The conducting xylem thus adds 1·6 % to the dry weight of the plant; this is still a small value but the strictures mentioned at the end of the consideration of the 1 m high plant apply with more force in the case of the requirement for 'supporting' as well as 'conducting' xylem.

If we continue to double the height of the plant, we can consider the case of a tree which has 64 m between the absorbing zone of the roots and the transpiration sites in the leaves. If the assumptions as to water potential difference between the ends of the xylem, the xylem-specific conductivity and the transpiration rate per m^2 of leaf area are unaltered, then such a plant needs 7·68 kg of conducting xylem per m^2 of leaf area. Assuming (as before) that the dry weight per m^2 of leaf area consists of 180 g of leaf, 100 g of absorbing roots and 100 g per m plant height of non-xylem material, the non-xylem dry weight of the plant is 6·68 kg. This means that the conducting xylem (7·68 kg per m^2 leaf area) now weighs more than the non-xylem parts of the plant (6·68 kg per m^2 leaf area). This suggestion of 42·7 g dry weight of conducting xylem per g of foliage may be compared with the value of at least 18·6 g dry weight of above-ground conducting xylem per g of foliage of *Pseudotsuga menziessi* with an above-ground height of 53 m (Grier & Logan, 1977). While the transpiration rate per unit of foliage mass may be lower in *P. menziesii* than in our example, the specific conductivity of the xylem is also likely to be lower than our assumed value of 25×10^{-9} m² s⁻¹.

Our model plants with a total xylem length of 1 m, 2 m and 64 m show that, if the water potential difference between the leaf and root is to be held constant along with the transpiration rate per unit foliage biomass and the specific conductivity of the xylem, then a very substantial increase in the mass of conducting xylem per unit foliage must occur with plant height. With our assumed values for the parameters, a 1 m tall plant has 0·011 g of conducting xylem per g foliage, while a 64 m tall plant has 42·7 g of conducting xylem per g foliage. The requirement for conducting xylem in the taller plant is increased if we take into account the necessity for a static, gravitational water potential gradient of 10 kPa m⁻¹ (top of plant negative relative to bottom). This component of the water potential gradient is, of course, not available for moving water up the plant (see Milburn, 1979).

A simple test of the validity of our hypothesis that the ratio of dry weight of the conducting xylem to the dry weight of the foliage increases with plant height in a manner which keeps the water potential difference between leaves and roots constant for a given rate of transpiration per unit foliage and a given specific conductivity in the xylem comes from measurements of the ratio of transectional area of sapwood in the bole to foliage mass (or area) for specimens of varying heights (ages) of a given provenance of a tree species growing under the same conditions. The hypothesis predicts an increase in this ratio in direct proportion to the height within a given species. However, the data for both coniferous (Grier & Waring, 1974) and dicotyledonous (Waring *et al.*, 1977) woody plants show height independence of the ratio (see also Waring, Schroeder & Oren, 1982). The data suggest that, if the specific conductivity of the xylem and the transpiration rate per unit foliage biomass or area are invariant with tree height, then the water potential difference between leaves and roots during steady-state transpiration in a tree species must increase *more* than in direct proportion to height owing to the static, gravitational component. On the basis of present evidence, then, it would seem that trees with secondary thickening increase the ratio of dry weight of conducting xylem to that of foliage in direct proportion to the height of the tree, and that, unless xylem conductivity is higher and/or transpiration rate per unit foliage is lower to an extent which more than offsets the static water potential increase with height, the water potential difference between leaves and roots of a transpiring tree must increase more than in direct proportion to the height of the tree.

Our considerations of the dependence of the mass of conducting xylem per unit of photosynthetic (transpiring) apparatus as a function of plant size have not explicitly considered the other components of the homoiohydric apparatus or the quantitative role of xylem in mechanical support of the aerial parts of the plant. The non-xylem components of the homoiohydric water transport regulatory system (cuticle, stomata, gas spaces) will be rapidly dismissed as involving less input of resources per unit foliage mass than does the xylem of even a small plant and, furthermore, as having a resource requirement per unit foliage mass which is not necessarily a function of plant height. Costing of the involvement of xylem (conducting sapwood and non-conducting heartwood) in mechanical support of the plant involves consideration of (*inter alia*) whether secondary thickening permits the production of a conical trunk rather than the cylindrical trunk characteristic of trees with only primary growth and with the occurrence of the self-supporting as opposed to the liane habit (Givnish & Vermeij, 1976; Niklas, 1978; Niklas & O'Rourke, 1982; Givnish, 1984, 1986; Valentine, 1985; Raven 1986a).

The extent to which the differences in specific conductivity of the xylem between plants are related to a trade-off between safety and efficiency is not clear. Recent work suggests that high-conductivity xylem may not be more prone to cavitation on a unit-conducting element basis than is low-conductivity xylem (Tyree & Dixon, 1986; cf. Raven, 1986a). Over-provision of xylem conduction capacity may help to counter cavitation-induced impediments to transport, especially in plants with a low capacity to refill embolized elements and/or without the capacity to replace (by secondary thickening) non-functional xylem elements in non-elongating parts of the axis (see Raven, 1986a). An apparently unconsidered aspect of comparative xylem anatomy and physiology is possible variation in the rate constant for diffusion of gases into a cavitated, conducting element, since an embolized element containing gas at or near atmospheric pressure is less easy to refill than is a just embolized element containing a 'Torricellian vacuum' (cf. Zimmermann & Milburn, 1982; Sauter, 1984).

We now attempt to summarize the costs and benefits of the four categories in Table 1, and of increasing size of homoiohydric desiccation-intolerant plants. We can examine the benefits in terms of the increased supply of resources in the environments which the larger plants can occupy and the costs in terms of diversion of resources from their acquisition and from other growth-promoting structures and catalysts.

We have already seen that, in a relatively resource-rich environment, a taller plant can avoid shading by other plants and exploit deeper-lying water and nutrients in the soil. The larger homoiohydric plant has more of its resources committed to support and the machinery of water transport and thus has a restricted capacity for growth under resource-saturated conditions. In other words, a smaller fraction of biomass is devoted to catalysis of photosynthesis and nutrient uptake and metabolism, resulting (with constant specific reaction rate of catalysts in a lower specific growth rate of the plant. When availability of resources is less than that needed to yield the maximum specific growth rate, the larger plant is further disadvantaged. The larger plant has a smaller fraction of its biomass devoted to resource-harvesting machinery so that, at a given availability of resources (incident photon flux density or soil nutrient concentration), the rate at which the plant can acquire resources per unit biomass is lower than for a smaller plant. Furthermore, the increased diversion of resources to the production and maintenance of transporting and supporting machinery and structures means that less resources are available to make catalysts of 'basic' metabolism, so that the resource-use efficiency of the production of 'basic plant' (i.e. not structures and not xylem) biomass is reduced. This argument applies to the quantity of 'basic plant' produced per mol photon absorbed, per mol N absorbed, or per g water transpired, to mention but three of the frequently growth-limiting resources. The concepts in this paragraph have been developed in other (but related) contexts by Raven (1984a, b, 1985a, b, 1987a, b).

We thus see that larger, desiccation-intolerant, homoiohydric plants may live in more resource-rich portions of a community but have a restricted capacity for growth (specific growth rate) when resources are saturating or limiting, and reduced resource-use efficiency of production of 'basic' biomass relative to smaller homoiohydric plants. These concepts of resource diversion may be extended to the other three categories in Table 1 and, as the most extreme case, to unicellular, haptophytic soil algae (cf. Raven, 1985a). The smaller desiccation-intolerant homoiohydric plants may have less resource diversion from 'basic' biomass than do desiccation-tolerant homoiohydric plants if the latter have additional resource costs related to tolerance of desiccation in the vegetative state (cf. Gaff, 1981). The endohydric yet poikilohydric *Polytrichum* and *Dawsonia* may have similar resource diversion to structures and to water transport as do homoiohydric plants of a similar size: the endohydric bryophyte gametophytes have xylem-like hydrome tissue and cuticle but lack stomata and intercellular gas spaces. The ectohydric, poikilohydric plants have less resource diversion from 'basic' biomass than any of the other categories and thus should have higher specific growth rates at high and low resource availability and higher resource-use efficiency. However, the resource availability for such plants, occupying the few millimetres on either side of the soil–atmosphere interface, is limited relative to that of larger plants in a multistratum community in a resource-rich environment.

Much more work is needed to test the various suggestions presented here. Some estimates of water-use efficiency for poikilohydric terrestrial plants comes from carbon isotope ratio measurements (see Raven, 1981b) but more direct measurements of these and other resource-use efficiencies of terrestrial rhizophytes are needed (cf. Schulze, 1982). An example is the effect of nutrient deficiency on water-use efficiency (Woodward, 1699; Tanner & Sinclair, 1983; Power, 1985; Walker & Richards, 1985), where there are theoretical reasons for expecting a lower water-use efficiency in plants whose growth rate is nutrient-limited (Handley & Raven, in preparation).

Solute transport within terrestrial rhizophytes

The homoiohydric vascular plants transport soil-derived solutes to transpirational termini via the transpiration stream. The composition of the transpiration stream differs from that of the soil solution in a number of ways (cf. Raven, 1983). Some nutrient elements (e.g. N, P, K) are more concentrated in the transpiration stream than in the soil solution, while others (e.g. Cl^-, Na^+) are less concentrated. Some elements are chemically changed *en route* to the xylem. NH_4^+ and N_2 are invariably converted to organic N before transfer to the xylem, while NO_3^- may be converted to organic form (root reduction) or transported as such (shoot reduction). These constraints on the composition of xylem sap are related to shoot demand for resources, and the constraints on correcting any imbalances in the xylem-supplied solutes by means of phloem retranslocation, excretion through glands, loss via organ abscission etc. (see Raven, 1983, 1985a, 1986b; Raven & Smith, 1976).

Potentially ecologically important aspects of xylem transport relate to water-use efficiency, N sources and the location of NO_3 reductase.

The very high water-use efficiency of CAM plants [1 g of dry matter produced per (about) 50 g water lost in transpiration, compared with about 300 for C_4 and about 500 for C_3] means that, with similar elemental composition on a dry weight basis, the concentration of nutrients in the xylem sap must be substantially higher in the CAM plants. This may lead to problems of solubility of the required mixture of calcium, phosphate and organic acid anions; the relatively low pH of xylem sap may help (see Raven, 1986c). The data of Clark, Holland & Smith (1986) on the composition of the C_3 vine, *Actinidia chinensis*, shows that, at least under 'root pressure' conditions (with solutes more concentrated than in the transpiration stream), the activities of dissolved calcium, phosphate, organic anions and protons mean that the sap is supersaturated with respect to calcium and phosphate ions, and that some inhibitor(s) of crystallization is present which prevents precipitation of apatite. This may be the case for transpiratory movement of xylem sap in CAM plants.

Raven (1985a) discusses variations in light- and water-use efficiency attendant on different combinations of biochemical and biophysical pH regulation mechanisms of plants growing with NO_3^- as N source when assimilation of NO_3^- is mainly in roots or mainly in shoots. The regulation of loading of N-containing solutes into the root xylem is clearly very important in controlling these processes and, hence, in possible optimization of the use of a limiting resource (light in shaded but damp environments; water in a sunny, arid environment).

The importance of specific xylem loading for the nutrition of shoots of endohydric plants should not blind us to the many unknowns related to the mechanisms and regulation of xylem loading. It is not clear, for example, how essential a morphologically recognizable endodermis is for this process; an endodermis is not readily identified in many early fossil vascular plants or in ectohydric bryophytes (see Raven, 1984c).

The xylem stream delivers solutes to transpirational termini which, in many plants, are essentially mature, non-growing photosynthetic organs. Passage of xylem-borne solutes to weakly transpiring growing parts of the shoot involves both retranslocation in the phloem and xylem-to-xylem transfer such that xylem fluid passing to weakly transpiring, growing regions is enriched in solutes relative to streams destined for non-growing regions. Both xylem-to-phloem and xylem-to-xylem transfer are available for phloem-mobile elements such as N, P and K: the relatively phloem-immobile Ca^{2+} can, presumably, only use the xylem-to-xylem option. B presents even more difficulties, boric acid having properties which predispose it to accumulate at transpirational termini (Raven, 1980). Whether these redistributional problems are especially severe in plants of a given life form or habitat is not clear.

The ectohydric rhizophytes present a very different picture with respect to solute transport from rhizoids to shoots. Here, the shoots presumably receive solutes in the *external* transpiration stream at concentrations very similar to those found in the soil solution. How plant (shoot) parts relatively remote from the soil respond to this imposition is not clear.

The other main long-distance transport pathway in terrestrial rhizophytes, the phloem (leptome of mosses), is analogous in structure and, probably, in mechanism to the phloem-like tissue of the larger phaeophyte algae. This pathway is symplastic and thus has constraints on the composition of the moving solution which differ from those on the (apoplastic) xylem (Raven 1977b, 1985a). We have already seen that the low symplastic total Ca^{2+} concentration limits the utility of the phloem in redirecting Ca^{2+} dumped at transpirational termini. Furthermore, it is constrained in its direction of movement, which is invariably from sources to sinks for reduced carbon. How far these constraints have differential effects on plants from different habitats, or with different life forms, is unclear.

CONCLUSIONS

The diverse organisms discussed in this article show that the impact of water relations and of nutrient solute transport on plant behaviour varies greatly between life forms and habitats. It could be argued that it is problems of extraplant solute fluxes which largely determine that planktophytes shall be small ($< 50\,\mu$m diameter), though not, on the basis of present knowledge, *how* small. The capacity (which may be selectively advantageous) to 'shade out' other plants by large differentiated attached parts (despite the 'shading out' of their juveniles!) implies the occurrence of specialized mass flow mechanisms of long-distance intraplant solute transport in aquatic haptophytes. The need for resource (e.g. N, P) transport from rhizoids or roots to shoots, as well as of photosynthate in the opposite direction, adds to the intraplant solute transport requirements of differentiated aquatic rhizophytes. Water relations are also significant discriminants in aquatic environments. Energy requirements for osmotic (volume or turgor) regulation vary between habitats (freshwater, marine or estuarine) and life forms (wall-less or walled cells), while dehydration during emersion for intertidal plants adds new challenges and constraints on growth and survival.

The problems of water loss during CO_2 fixation at the expense of atmospheric CO_2 are exacerbated in fully terrestrial plants, where resupply of water via regular shoot submersion does not occur. Many of the ecophysiological constraints on the behaviour of terrestrial rhizophytes can be related to the acquisition and transport of water and the regulation of water loss *vis-à-vis* CO_2 fixation. The constraints limit the resource (water, light, nitrogen) use efficiency of growth, as well as the specific growth rate under resource-saturated *and* resource-limited conditions. The constraints on growth rate and efficiency are greatest for larger, homoiohydric plants, which are desiccation-intolerant in the vegetative phase but are ameliorated (or even reversed) by the greater resource availability to a tall (unshaded) plant with deep-penetrating roots. These plants dominate most terrestrial habitats and are the only plants more than 1 to 2 m high. Other life forms (homoiohydric, desiccation-tolerant plants; ecto- and endohydric poikilohydric plants) are significant and, not uncommonly, dominant members of the flora where maximum plant height is less than 1 to 2 m.

Much of the foregoing deals with the differences *between* life forms; much remains to be done here in terms of quantifying the differences in potential for growth rate and for resource-use efficiency, which were qualitatively, or semi-quantitatively, discussed. Even more remains to be quantified with respect to differences in the water and transport relationships of plants within the broad categories delineated here. It is likely that many of the categories will contain representatives of the 'R', 'C', and 'S' categories of Grime (1979) which may accordingly have different selective priorities with respect to the rate of resource acquisition relative to the efficiency of use of already acquired resources. The different Grimean categories may also have different selective priorities with respect to resource acquisition relative to resource retention, with corresponding diversion of already acquired resources to making defence compounds as opposed to producing resource-acquiring machinery or a morphology better able to resist grazing or mechanical damage than to maximize the rate of resource acquisition (cf. Littler, Littler & Taylor, 1983). It is clear that natural selection's effects on different plants may represent different 'perceptions' of the parable of the talents.

Finally, it is important to remember that demonstration of quantitative differences in water relations and of transport between plants does not, of itself, indicate significance for the differences in terms of natural selection (Osmond *et al.*, 1980).

ACKNOWLEDGEMENTS

The authors wish to acknowledge receipt of research support for the investigation of plant transport processes and water relations from A.F.R.C., N.E.R.C. and S.E.R.C. (UK) and N.S.F. (USA) under grant numbers INT-8501918 and CEE-83-15819.

REFERENCES

BAUMAN, JR, R. W. & JONES, B. R. (1986). Electrophysiological investigations of the red alga *Griffithsia pacifica*. *Journal of Phycology*, **22**, 49–56.
BRADSHAW, A. D. (1987). Comparisons – its scope and limits. In: *Frontiers of Comparative Plant Ecology*. (Ed. by I. H. Rorison, J. P. Grime, R. Hunt, G. A. F. Hendry & D. H. Lewis), *New Phytologist*, **106** (Suppl.), 3–21. Academic Press, London.
CLARK, C. J., HOLLAND, P. T. & SMITH, G. S. (1986). Chemical composition of bleeding xylem sap from kiwifruit vines. *Annals of Botany*, **58**, 353–362.
COWAN, I. R. (1986). Economics of carbon fixation in higher plants. In: *On the Economy of Plant Form and Function* (Ed. by T. J. Givnish), pp. 133–170. Cambridge University Press, Cambridge.
EDWARDS, D. S. (1986). *Aglaophyton major*, a non-vascular land plant from the Devonian Rhynie Chert. *Botanical Journal of the Linnean Society*, **93**, 173–204.
EDWARDS, D., FANNING, U. & RICHARDSON, J. B. (1986). Stomata and sterome in early land plants. *Nature*, **323**, 438–440.
EVANS, G. C. (1972). *The Quantitative Analysis of Plant Growth*. Blackwell Scientific Publications, Oxford.
GAFF, G. D. (1981). The biology of resurrection plants. In: *The Biology of Australian Plants* (Ed. by J. S. Pate & A. J. McComb), pp. 114–146. University of West Australia Press, Nedlands.
GIVNISH, T. J. (1984). Leaf and canopy adaptations in tropical forests. In: *Physiological Ecology of Plants of the Wet Tropics* (Ed. by E. Medina, H. Mooney & C. Vasquez-Yanes), pp. 51–84. W. Junk, The Hague.
GIVNISH, T. J. (1986). Biomechanical constraints on self-thinning of plant populations. *Journal of Theoretical Biology*, **119**, 139–146.
GIVNISH, T. J. & VERMEIJ, G. J. (1976). Sizes and shapes of liane leaves. *American Naturalist*, **110**, 743–778.
GRIER, C. C. & LOGAN, R. S. (1977). Old-growth *Pseudotsuga menziesii* communities of a Western Ontario watershed: biomass distribution and production budgets. *Ecological Monographs*, **47**, 373–400.
GRIER, C. C. & WARING, R. H. (1976). Conifer foliage mass related to sapwood area. *Forest Science*, **20**, 205–206.
GRIME, J. P. (1979). *Plant Strategies and Vegetation Processes*. John Wiley, Chichester.
HARRIS, G. P. (1986). *Phytoplankton Ecology: Structure, Function and Fluctuations*. Chapman & Hall, London.
HEINE, R. W. (1971). Hydraulic conductivity in trees. *Journal of Experimental Botany*, **22**, 503–511.
HUBER, B. (1956). Die Gefässleitung. In: *Encyclopedia of Plant Physiology, Volume III* (Ed. by M. Rüfelt), pp. 541–582. Springer-Verlag, Berlin.
KEELEY, J. E., OSMOND, C. B. & RAVEN, J. A. (1984). *Stylites*, a vascular land plant without stomata absorbs CO_2 via its roots. *Nature*, **310**, 694–695.
KIRK, J. T. O. (1983). *Light and Photosynthesis in Aquatic Ecosystems*. Cambridge University Press, Cambridge.
LITTLER, M. M., LITTLER, D. S. & TAYLOR, P. R. (1983). Evolutionary strategies in a tropical barrier reef system: functional-form groups of marine macroalgae. *Journal of Phycology*, **19**, 229–237.
LITTLER, M. M., LITTLER, D. S., BLAIR, S. M. & NORRIS, J. N. (1985). Deepest known plant life discovered on an uncharted seamount. *Science*, **227**, 57–59.
LITTLER, M. M., LITTLER, D. S., BLAIR, S. M. & NORRIS, J. N. (1986). Deep-water plant communities from an uncharted seamount off San Salvador Island, Bahamas: distribution, abundance and primary productivity. *Deep-Sea Research*, **33**, 881–892.
LUTHER, H. (1949). Vorschlag zu einer ökologischen Grundeinteilung der Hydrophyten. *Acta Botanica Fennica*, **44**, 1–15.
MACFARLANE, J. J. & RAVEN, J. A. (1985). External and internal CO_2 transport in *Lemanea*: interactions with the kinetics of ribulose bisphosphate carboxylase. *Journal of Experimental Botany*, **36**, 610–622.
MILBURN, J. A. (1979). *Water Flow in Plants*. Longman, London.
NIKLAS, K. J. (1978). Morphometric relationships and rates of evolution among palaeozoic vascular plants. *Evolutionary Biology*, **11**, 509–543.

NIKLAS, K. J. & O'ROURKE, T. D. (1982). Growth patterns of plants that maximize vertical growth and minimize internal stresses. *American Journal of Botany*, **69**, 1367–1374.

OATES, B. R. (1985). Photosynthesis and amelioration of desiccation in the intertidal saccate alga *Colpomenia peregrina*. *Marine Biology*, **89**, 109–119.

OATES, B. R. (1986). Components of photosynthesis in the intertidal saccate alga *Halosaccion americanum* (Rhodophyta, Palmariales). *Journal of Phycology*, **22**, 217–223.

ODUM, H. T. & PINKERTON, R. C. (1955). Time's speed regulater: the optimum efficiency for maximum power output in physical and biological systems. *American Scientist*, **43**, 331–343.

OSMOND, C. B., BJÖRKMAN, O. & ANDERSON, D. J. (1980). *Physiological Processes in Plant Ecology. Toward a Synthesis with* Atriplex. Springer-Verlag, Berlin.

POWER, J. F. (1985). Nitrogen- and water-use efficiency of several cool-season grasses receiving ammonium nitrate for 9 years. *Agronomy Journal*, **77**, 189–192.

RAVEN, J. A. (1977a). The evolution of vascular land plants in relation to supracellular transport processes. *Advances in Botanical Research*, **5**, 153–219.

RAVEN, J. A. (1977b). H^+ and Ca^{2+} in phloem and symplast: relation to relative immobility of the ions to the cytoplasmic nature of the transport pathways. *New Phytologist*, **79**, 465–480.

RAVEN, J. A. (1980). Short and long-distance transport of boric acid in plants. *New Phytologist*, **84**, 231–249.

RAVEN, J. A. (1981a). Nutritional strategies of submerged benthic plants: the acquisition of C, N and P by rhizophytes and haptophytes. *New Phytologist*, **88**, 1–30.

RAVEN, J. A. (1981b). Ribulose bisphosphate carboxylase activity in terrestrial vascular plants: significance of O_2 and CO_2 diffusion. In: *Commentaries in Plant Sciences*, vol. 2 (Ed. by H. Smith), pp. 109–125. Pergamon Press, Oxford.

RAVEN, J. A. (1982). The energetics of freshwater algae: energy requirements for biosynthesis and osmoregulation. *New Phytologist*, **92**, 1–20.

RAVEN, J. A. (1983). Phytophages of xylem and phloem: a comparison of animal and plant sap-feeders. *Advances in Ecological Research*, **13**, 135–234.

RAVEN, J. A. (1984a). *Energetics and Transport in Aquatic Plants*. A. R. Liss, New York.

RAVEN, J. A. (1984b). A cost–benefit analysis of photon absorption by photosynthetic unicells. *New Phytologist*, **98**, 593–625.

RAVEN, J. A. (1984c). Physiological correlates of the morphology of early vascular plants. *Botanical Journal of the Linnean Society*, **88**, 105–126.

RAVEN, J. A. (1985a). Regulation of pH and generation of osmolarity in vascular land plants: costs and benefits in relation to efficiency of use of water, energy and nitrogen. *New Phytologist*, **101**, 25–77.

RAVEN, J. A. (1985b). Physiology and biochemistry of pteridophytes. *Proceedings of the Royal Society of Edinburgh, B*, **86**, 37–44.

RAVEN, J. A. (1986a). Evolution of plant life forms. In: *On the Economy of Plant Form and Function* (Ed. by T. J. Givnish), pp. 421–476. Cambridge University Press, Cambridge.

RAVEN, J. A. (1986b). Biochemical disposal of excess H^+ in plants? *New Phytologist*, **104**, 175–206.

RAVEN, J. A. (1986c). Long distance transport of calcium. In: *The Molecular and Cellular Aspects of Calcium in Plant Development* (Ed. by A. Trewavas), pp. 241–250. Plenum Press, New York.

RAVEN, J. A. (1987a). Physiological consequences of extremely small size for autotrophic organisms in the sea. In: *Physiological Ecology of Photosynthetic Picoplankton in the Sea* (Ed. by T. R. Platt). *Canadian Bulletin of Fisheries and Aquatic Science*. (In press.)

RAVEN, J. A. (1987b). Algae. In: *Solute Transport in Plant Cells and Tissues* (Ed. by D. A. Baker & J. L. Hall). Pitman, London. (In press.)

RAVEN, J. A. (1987c). Limits to growth. In: *Microalgal Biotechnology* (Ed. by M. A. Borowitzka & L. J. Borowitzka). University Press, Cambridge. (In press.)

RAVEN, J. A. & RICHARDSON, K. (1984). Dinophyte flagella: a cost–benefit analysis. *New Phytologist*, **98**, 259–276.

RAVEN, J. A. & RUBERY, P. H. (1982). Co-ordination of development: hormone receptors, hormone action and hormone transport. In: *The Molecular Biology of Plant Development* (Ed. by H. Smith & D. Grierson), pp. 28–48. Blackwell Scientific Publications, Oxford.

RAVEN, J. A. & SMITH, F. A. (1976). Nitrogen assimilation and transport in vascular land plants in relation to intracellular pH regulation. *New Phytologist*, **76**, 415–431.

RAVEN, J. A., GRIFFITHS, H. & BEARDALL, J. (1982). Inorganic C-sources for *Lamenea, Cladophora* and *Ranunculus* in a fast-flowing stream: measurements of gas exchange and of carbon isotope ratio and their ecological implications. *Oecologia*, **53**, 68–78.

RAVEN, J. A., JOHNSTON, A. M., MACFARLANE, J. J., BIN SURIF, M. & McINROY, S. (1987a). Diffusion and active transport of inorganic carbon species in freshwater and marine macroalgae. In: *Progress in Photosynthesis Research*, vol. 4 (Ed. by J. Biggins), pp. 333–340. Nijhoff/Junk, The Hague.

RAVEN, J. A., MACFARLANE, J. J., JOHNSTON, A. M., BIN SURIF, M. & McINROY, S. (1987b). Inorganic carbon transport in relation to habitat, and resource use efficiency, in the macroalgae *Lemanea mamillosa*

(Rhodophyta) and *Ascophyllum nodosum* (Phaeophyta). In: *Proceedings of the Seventh International Workshop on Plant Membrane Transport* (Ed. by N. A. Walker). Australian Academy of Science, Canberra. (In press.)

REYNOLDS, C. S. (1984). *The Ecology of Freshwater Phytoplankton.* University Press, Cambridge.

SAUTER, J. J. (1984). Detection of embolization of vessels by a double staining technique. *Journal of Plant Physiology*, **116**, 331–342.

SCHULZE, E.-D. (1982). Plant life forms and their carbon, water and nutrient relations. In: *Encyclopedia of Plant Physiology (New Series) Volume 12B* (Ed. by O. L. Lange, P. S. Nobel, C. B. Osmond & H. Ziegler) pp. 615–676. Springer-Verlag, Berlin.

SOURNIA, A. (1982). Form and function in marine phyoplankton. *Biological Reviews*, **57**, 347–394.

STERNBERG, L. DA S. L., DENIRO, M. J., MCJUNKIN, D., BERGER, R. & KEELEY, J. E. (1985). Carbon, oxygen and hydrogen isotope abundances in *Stylites* reflect its unique physiology. *Oecologia*, **67**, 598–600.

TAJIRI, S. & ARUGA, Y. (1984). Effect of emersion on the growth and photosynthesis of the *Porphyra yezoensis* thallus. *Japanese Journal of Phycology*, **32**, 134–156.

TANNER, C. B. & SINCLAIR, T. R. (1983). Efficient water use in crop production: research or re-search. In: *Limitations to Efficient Water Use in Crop Production* (Ed. by H. M. Taylor, W. R. Jordan & T. R. Sinclair), pp. 1–27. Publication of the American Society of Agronomy, American Soil Science Society and the American Crop Science Society.

TYREE, M. T. & DIXON, M. A. (1986). Water stress induced cavitation and embolism in some woody plants. *Physiologia Plantarum*, **66**, 397–405.

VALENTINE, H. T. (1985). Tree-growth models: derivations employing the pipe-model theory. *Journal of Theoretical Biology*, **117**, 579–585.

WALKER, G. K. & RICHARDS, J. E. (1985). Transpiration efficiency in relation to nutrient status. *Agronomy Journal*, **77**, 263–269.

WARING, R. H., GHOLZ, M. L., GRIER, C. C. & PLUMMER, M. L. (1977). Evaluating stem conducting tissue as an estimator of leaf area in four woody angiosperms. *Canadian Journal of Botany*, **55**, 1474–1477.

WARING, R. H., SCHROEDER, P. E. & OREN, R. (1982). Application of the pipe model theory to predict canopy leaf area. *Canadian Journal of Forest Research*, **12**, 556–560.

WATT, W. B. (1986). Power and efficiency as indexes of fitness in metabolic organisation. *American Naturalist*, **127**, 629–653.

WOODWARD, J. (1699). Some thoughts and experiments concerning vegetation. *Philosophical Transactions of the Royal Society of London*, **21**, 193–227.

WOODHOUSE, R. M. & NOBEL, P. S. (1982). Stipe anatomy, water potentials, and xylem conductances in seven species of ferns (Filicopsida). *American Journal of Botany*, **69**, 135–140.

WOOLHOUSE, H. W. (1981). Aspects of the carbon and energy requirements of photosynthesis considered in relation to environmental constraints. In: *Physiological Ecology. An Evolutionary Approach to Resource Use* (Ed. by C. R. Townsend & P. Calow), pp. 51–85. Blackwell Scientific Publications, Oxford.

ZIMMERMANN, M. H. & MILBURN, J. A. (1982). Transport and storage of water. In: *Encyclopedia of Plant Physiology (New Series), Volume 12B* (Ed. by O. L. Lange, P. S. Nobel, C. B. Osmond & H. Ziegler), pp. 135–151. Springer-Verlag, Berlin.

APPENDIX

A glossary of terms used in the article.

Apoplast. That space outside the plasmalemma but inside the plant. Defined in this way it includes the aqueous apoplast (cell walls, functional xylem cell lumina) and the gaseous apoplast (intercellular gas spaces, 'cavitated' xylem cell lumina).

Diel. Alternating by day and by night [Synonymous with 'diurnal' as often used but contrasting with 'diurnal' (= of, in, done by, active in, the day) as opposed to 'nocturnal' (= of, etc. the night)].

Desiccation-intolerant. Unable to survive (at a specified stage in the life-cycle) equilibration with an atmospheric relative humidity of less than $\sim 5\%$.

Desiccation-tolerant. Able to survive (at a specified stage in the life-cycle) equilibration with an atmospheric relative humidity of less than $\sim 5\%$.

Ectohydric. Able to transport a significant fraction of the water transpired by the plant from the substratum *externally*.

Endohydric. Equipped with a specialized internal, apoplastic water mass flow pathway and with a limited capacity for external transport of liquid water.

Haptophyte. A benthic plant which is small relative to the particle size of the substratum to which it is attached.

Homoiohydric. An emersed plant with substantial control over its degree of hydration despite varying soil water supply and evaporative demand by the atmosphere.

Planktophyte. A microscopic plant which occurs free-floating in a water body.

Poikilohydric. An emersed plant with little control over its degree of hydration with varying soil water supply and evaporative demand by the atmosphere.

Rhizophyte. A benthic plant which is large relative to the particle size of the substratum in which it is partially embedded.

Symplast. In the restricted sense, the cytosol, including the cytosol phase of intercellular connections (plasmodesmata; sieve pores in phloem).

New Phytol. (1987) **106** (Suppl.), 235–249   235

GROWTH AND PARTITIONING

By RODERICK HUNT AND PHILIP S. LLOYD*

*Unit of Comparative Plant Ecology (NERC), Department of Botany,
The University, Sheffield S10 2TN, UK*

SUMMARY

Innate growth potential and innate patterns of allocation to different plant parts are both subject
to physiological limitations. Yet the ways in which these attributes not only vary between species
but also exhibit plasticity in response to environmental change are important determinants of
ecological behaviour and contribute strands to Plant Strategy Theory. In this paper, we review
the evidence for innate variation in growth potential and partitioning and consider the
implications of this variation for the development of high-level ecological syntheses.

Key words: Relative growth rate, growth potential, partitioning, innate variation, Strategy
Theory.

INTRODUCTION

As we near the end of this symposium, we move a little towards synthesis, for it
is time to consider some of the net effects of phenomena already described. Growth
and partitioning are relatively high-level physiological processes which contribute
important strands to Plant Strategy Theory. We need an appropriate form of access
to them. Within the context of comparative plant ecology, we also need to
remember that the species is the vehicle of the means for evolutionary action – and
the subject of ecological theorization in its own right. To provide a suitable focus
for attention, we thus consider the attributes and actions of the individual plant
which become evident when it is grown in isolation.

This detached viewpoint necessarily neglects the behaviour of individuals in the
sometimes bewildering richness of natural communities (see Grime, 1984) or in
the serried and often equally crowded ranks of the artificial populations known as
crop stands (see Benjamin & Hardwick, 1986). But apart from convenience, the
justification for concentrating upon the individual plant as an entity in itself lies
in the fact that, under natural conditions, it clearly enjoys a degree of self-
determination which is remarkably above that possessed by the individual animal.
Because of this we see a weakening – perhaps sometimes to vanishing point
– of the truly emergent properties which would otherwise distinguish the plant
population as a level of organization (Feibleman, 1955; Allen & Starr, 1982) in
its own right.

The chief conveyance for our tour around the comparative ecology of growth
and partitioning within the individual plant is the most resilient of overall indices
of plant growth. Schoolchildren now learn (JMB, 1985) that V. H. Blackman's
hopes of plants possessing an 'efficiency index' which obeyed Lord Kelvin's
'compound interest law' were quickly dashed (Blackman, 1919; Briggs, Kidd &

---

* Unit of Comparative Plant Ecology 1964–1973, deceased 1975.

0028-646X/87/05S235 + 15 $03.00/0

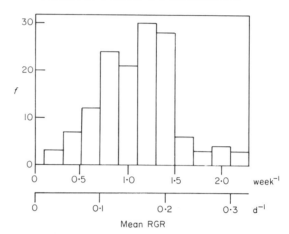

Fig. 1. The frequency distribution of mean relative growth rate among 143 British herbaceous and woody species grown under productive, controlled conditions.† Measurements were made in the seedling phase over the period 2 to 5 weeks following germination. Species are representative of all of the dry terrestrial habitats of the Sheffield region. [data from Grime & Hunt (1975) with augmentation.]

West, 1920). However, what became known as 'relative growth rate'* (West, Briggs & Kidd, 1920) still provided a convenient and equitable comparator of performance for widely differing types of plant. It was then, as now, a particularly appropriate vehicle for the broader kind of physiological survey, which of course is especially useful to ecology. Indeed, the very *in*constancy of RGR (Evans, 1972) provides fertile ground for the development of studies aimed at disentangling both the separate and the joint influences of nature and nurture on the growth of the individual plant.

Innate Variations in Growth Potential

Overall variation

Variations in growth potential between plants grown as spaced individuals are considerable. In the very short term, instantaneous RGR in seedlings of ruderal vascular species can exceed 0.5 d^{-1} under productive, controlled conditions† (Hunt, unpublished data), implying a doubling time for dry matter of not more than 1.4 d. Such a performance approaches that of the eutrophic alga, *Anabaena cylindrica* (Fogg, 1949). On the other hand, the growth potential of young rosettes of the dune wintergreen, *Pyrola rotundifolia* ssp. *maritima*‡ under a similar regime is only *c.* 0.007 d^{-1} (R. Hunt & J. F. Hope Simpson, unpublished data), with a doubling over 99 d. Between these specimen extremes falls a rich variety of intermediate examples, which may be reviewed along the following lines.

Variations between species

The largest study of innate variations between species is that of Grime & Hunt

* The rate of dry matter production per unit of dry matter.
† The phrase 'productive, controlled conditions' is used throughout to denote a uniform regime supplying 35 W m^{-2} PAR over a 16 or 18 h day at 20/15 °C day/night, with minerals supplied by the full 'Long Ashton' nutrient solution (Hewitt, 1966).
‡ Nomenclature for British vascular plants follows that of Clapham, Tutin & Warburg (1981).

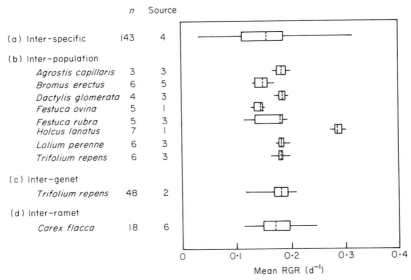

Fig. 2. Variation in mean relative growth rate in the vegetative phase within four levels of organization: (a) between-species variation within the large sample illustrated in Figure 1; (b) between-population variation within collections of natural or commercial populations of eight different species; (c) between-genet variation within a collection of individuals drawn from a single natural population; (d) between-ramet variation within a collection of similar individuals cloned from a single genet. Parts (b) to (d) are scaled to correspond with (a), allowing for small differences in conditions of measurement and stages of growth. Results are given in the form of 'box and whisker' diagrams, which identify the ranges and quartiles of each distribution (Tukey, 1977; p. 39). Sources are: (1) A. J. M. Baker (unpublished data); (2) Burdon & Harper (1980); (3) Elias & Chadwick (1979); (4) Grime & Hunt (1975), with augmentation; (5) Law (1972); (6) Hunt (1984). (Synthesis from Hunt, 1984.)

(1975), which has subsequently been augmented by further unpublished observations. The frequency distribution of mean RGR over the period 2 to 5 weeks after germination is statistically near-normal (Fig. 1). But within the whole sample, species of annual life history have a predominantly high bias in mean RGR and species of woody habit a low bias. Grasses and forbs both show a wide range of growth potential.

Variations within other levels of organization

Very approximately, the 10-fold range of variation in mean RGR between species (Fig. 1) is complemented by a two-fold range at each of the three levels, the population, the genet and the ramet (Fig. 2). Mostly, these distributions are also near-normal, so the probability is small that unusually fast- or slow-growing individuals exist to disturb the broad picture presented by Figure 1. Such occurrences would, of course, have a considerable evolutionary potential but little current ecological relevance.

Variations with developmental state

General. All the material discussed above was grown in a fully established but juvenile state. Variations between species clearly exist in other states, both earlier and later. As between levels of organization, consistent trends may occur.

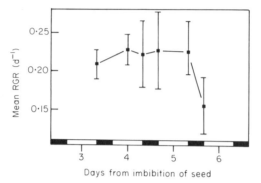

Fig. 3. The growth of *Holcus lanatus* in relation to the timing of seedling emergence. All germinated individuals, if any, were removed at 8 h intervals from imbibition and transferred as cohorts to a productive, controlled environment. The mean relative growth rate of each cohort over the subsequent 8 d (measured exactly from the time of emergence) is shown in relation to that time, with 95 % limits. (Previously unpublished data.)

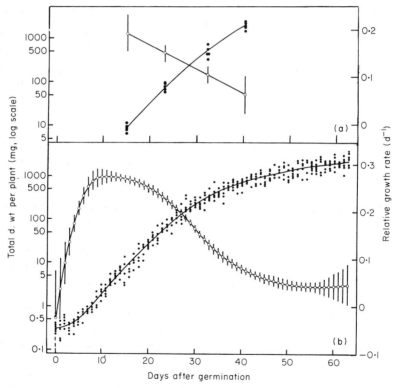

Fig. 4. Curves fitted to progressions of total dry weight per plant (●, left-hand axes), with the derived instantaneous relative growth rates (○, with 95 % limits, right-hand axes) for *Holcus lanatus* grown in a productive, controlled environment. (a) contains data from the screening programme of Grime & Hunt (1975) (see Fig. 1); (b) contains data obtained by harvesting daily from germination onwards. Curve-fitting in (a) is by the simple polynomial methods of Hunt & Parsons (1974) and, in (b), by the splined polynomials of Parsons & Hunt (1981). (Previously unpublished data.)

The pre-emergent state. Crop species are known to contain intrinsic variations in pre-emergent (embryo) growth rate. But here, and even more so in native species, the quantitative information available is very sparse (Benjamin & Hardwick, 1986). Both the intrinsic 'vigour' of the embryo (Wurr & Fellows, 1984)

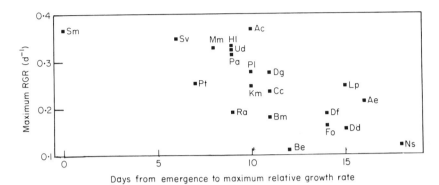

Fig. 5. The inverse relation between the value of instantaneous maximum relative growth rate and the age of seedling at which this is achieved. Values were obtained by fitting splined polynomials to data obtained from daily harvesting [see the example given in Fig. 4(b)]. The correlation is -0.741, significant at $P < 0.001$. The 21 species are: (Ac) *Agrostis capillaris*, (Ae) *Arrhenatherum elatius*, (Bm) *Briza media*, (Be) *Bromus erectus*, (Cc) *Cynosurus cristatus*, (Dg) *Dactylis glomerata*, (Dd) *Danthonia decumbens*, (Df) *Deschampsia flexuosa*, (Fo) *Festica ovina*, (Hl) *Holcus lanatus*, (Km) *Koeleria macrantha*, (Lp) *Lolium perenne*, (Mm) *Matricaria matricarioides*, (Ns) *Nardus stricta*, (Pl) *Plantago lanceolata*, (Pa) *Poa annua*, (Pt) *Poa trivialis*, (Ra) *Rumex acetosa*, (Sv) *Senecio vulgaris*, (Sm) *Stellaria media* and (Ud) *Urtica dioica*. (Previously unpublished data.)

and the activity of the microflora (Maude, 1973) seem implicated. It is likely that such factors will also contribute to intrinsic variations between native species — perhaps especially so given the wider anatomical and physiological ranges which they display.

The newly emerged state. Native species have not necessarily been selected for uniformity of germination rate and relatively wide variations in performance are known (Grime *et al.*, 1981). For example, germination in a batch of *Holcus lanatus* commonly begins at three days after imbibition and then spreads over the succeeding two days. Subsequently, the late-germinating individuals have a significantly lower RGR ($P < 0.05$) than the population mean (Fig. 3) and there is a suggestion that early germinating individuals are also of lesser growth potential. Similar effects have been observed in other species such as *Deschampsia flexuosa*, where the time to first germination in a collection, and the subsequent duration of germination, are both about five times greater than in *H. lanatus*.

The early juvenile state. The screening experiments of Grime & Hunt (1975) examined growth over approximately the period 15 to 40 d from germination [Fig. 4(a)]. Though useful for broad comparisons, this sampling policy was often not fine enough to uncover the true daily maximum in seedling RGR. Daily sampling [Fig. 4(b)] reveals that, in fast-growing species, this maximum can occur relatively early in the life-cycle. A value of 0.33 d^{-1} is obtainable for *H. lanatus* at only 9 d from germination. Further measurements, involving many species, have shown that the pattern of change in RGR illustrated in Figure 4(b) is almost general; a period of hyperexponential growth is followed by a short-lived plateau in RGR and then by a long decline towards the plant's maturity. In large-seeded species, or in those which are very fast-growing, the hyperexponential phase is reduced.

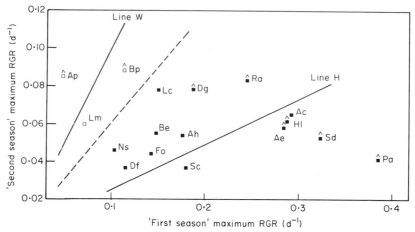

Fig. 6. Instantaneous maximum relative growth rate (RGR) in a 'second season' (P. S. Lloyd & S. R. Band, unpublished data) as a function of that observed in a 'first season', or early seedling phase (Grime & Hunt, 1975). The second season's data were obtained from young plants overwintered in an experimental garden before being transferred to a glasshouse and harvested sequentially during the following spring; the first season's data are those obtained for seedlings grown under productive, controlled conditions by Grime & Hunt (1975). Young seedlings of *Arrhenatherum elatius* grown (without overwintering) under both regimes provide a method of correcting for the difference between the two environments: the line of equivalence (– – – –) is of slope 0·60. The 17 species are: (Ac) *Agrostis capillaris*, (As) *Arrhenatherum elatius* (overwintered), (Ah) *Arabis hirsuta*, (Ap) *Acer pseudoplatanaus*, (Be) *Bromus erectus*, (Bp) *Betula pendula*, (Df) *Deschampsia flexuosa*, (Dg) *Dactylis glomerata*, (Fo) *Festuca ovina*, (Hl) *Holcus lanatus*, (Lc) *Lotus corniculatus*, (Lm) *Lathyrus montanus*, (Na) *Nardus stricta*, (Pa) *Poa annua*, (Ra) *Rumex acetosa*, (Sc) *Scabiosa columbaria*, (Sd) *Silene dioica*. The line (W) fitted to data for the three woody species (Ap, Bp, Lm) is of slope 0·91; the line (H) fitted to data for the 14 herbaceous species is of slope 0·24. The symbol (^) indicates that no true maximum RGR was recorded in the second season, the value plotted being the final point on an otherwise rising trend.

Most significantly, fast-growing species also reach their maxima earlier. Figure 5 contrasts the ruderal *Stellaria media*, *Senecio vulgaris*, and *Matricaria matricarioides*, small-seeded (0·08 to 0·35 mg) with high maxima in RGR (0·33 to 0·37 d⁻¹) attained at, or very soon following, germination (0 to 8 d), with the stress-tolerant *Nardus stricta*, *Danthonia decumbens*, and *Festuca rubra*, larger-seeded (0·38 to 0·87 mg) with a lower maximum RGR (0·12 to 0·16 d⁻¹) attained later (at 14 to 18 d from germination).

Beyond the juvenile state. To see whether or not the juvenile growth rates bear any relation to rates of growth later in the life-cycle, experiments have been performed at Sheffield in which species of known early growth potential were established in summer, overwintered in an experimental garden, and then harvested sequentially from under glass during the following spring (Fig. 6). *Arrhenatherum elatius* was included in this study and plants of this species were also grown and harvested *as seedlings* in the spring, revealing that the glasshouse environment was only 0·6 as productive as that of the growth-room (Fig. 6, solid line). Even with this correction in mind, the 14 herbaceous species in the main study showed consistently slower recovery growth after overwintering, with the 'second season' rates averaging only 40 % of the first. This was to be expected for purely

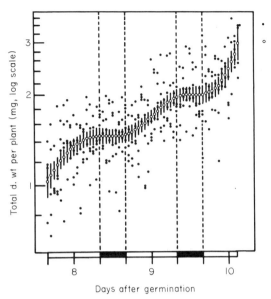

Fig. 7. The diurnal progression of total dry weight per plant in *Holcus lanatus* over the interval 8 to 10 d after germination [i.e. when daily relative growth rate is maximal, see Fig. 4(b)]. Plants were grown under productive, controlled conditions in individual 20 cm³ containers, with hourly harvesting. The fitted splined polynomial is shown, with 95 % limits. (From Hunt, 1980.)

developmental reasons [see Fig. 4(b)]. However, the full spring growth potential of the faster-growing species such as *Poa annua*, *Silene diocia* and *Agrostis capillaris* is unlikely to have been attained completely (see the legend to Fig. 6) and the true ranking of species would be similar in both seasons. Three woody species, on the other hand, grew 50 % faster than expected in their 'second season', a result clearly influenced by their large seed reserves. However, the ranking against one another was again similar.

Diurnal variations. Preliminary work involving hourly harvesting of *H. lanatus*, both day and night, has revealed that instantaneous RGR during the light phase of productive, controlled conditions can reach 0·54 d⁻¹ against the background of a 24 h mean value of 0·30 d⁻¹ (Fig. 7; from Hunt, 1980). Such rates indicate the degree to which species of high growth potential and competitive strategy may exploit periods of transient richness of resources in the field, even if such conditions ultimately demand respiratory repayments and would induce 'fatigue' if sustained experimentally for any great length of time.

Environmentally induced variations
    The lower extremes. A full discussion of this topic lies beyond our scope. But we remember that little or nothing of a species' potential in the laboratory may

necessarily be realized in the field. Though the *Pyrola* mentioned in the Introduction may seldom be induced to grow at rates beyond *c*. 0·007 d⁻¹, even this performance is high in relation to its attainment in the field. Figure 8 records that one clonal colony of this plant increased in size from eight rosettes to 326 over

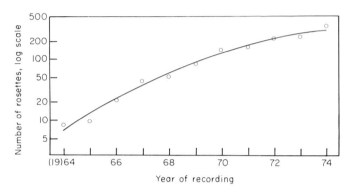

Fig. 8. The growth curve for the number of rosettes of the dune wintergreen, *Pyrola rotundifolia*, present in midsummer at a clonal colony on Braunton Burrows, North Devon. Each rosette weighs *c*. 150 mg and the mean relative growth rate in dry weight of the colony over the 10 years studied was *c*. 0·001 d⁻¹, averaged over all seasons. (From Hunt, Hope-Simpson & Snape, 1985.)

a period of 10 years, a mean RGR for dry weight of *c*. 0·001 d⁻¹. Had its full potential of 0·007 d⁻¹ been sustained over all of this period, the colony of rosettes at this site alone would ultimately have numbered 150 billion, neglecting mortality. Yet growing at only 0·001 d⁻¹ (net), and restricted to highly specialized sites, this species is increasing its geographical range spectacularly (Kay, Roberts & Vaughan, 1974).

   The upper extremes. How critical the environmental conditions are for the attainment of maximum physiological growth potential may only be settled exactly by elaborate experimentation. The standard 'productive, controlled conditions' are by no means absolutely optimal. In particular, an enhanced radiation input (50 W m⁻²) produces still higher and earlier maximum RGRs in *H. lanatus*, although it exerts scarcely any change in *D. flexuosa* [Fig. 9(a)]. Averaged over many species, the standard regime produces only 81 % of the growth seen under the enhanced regime [Fig. 9(b)] but the effect of the enhanced regime is also to preserve rankings between species and to magnify species differences towards the upper end of the range.

   Enhanced radiation and enhanced temperature (25 °C) together affect the instantaneous maximum RGR and its timing relatively little in species such as *H. lanatus* (Fig. 10). Work in a temperature gradient tunnel at Sheffield (Fig. 11) reveals that the temperature-response curves of both fast- and slow-growing C₃ species are relatively flat-topped. However, for some C₄ species, there is an exponential rise in RGR with increasing temperature over this range (J. P. Grime & J. M. L. Mackey, unpublished data).

INNATE VARIATIONS IN PARTITIONING

Introduction
   It has been known since Roman times that the relative sizes of the above- and

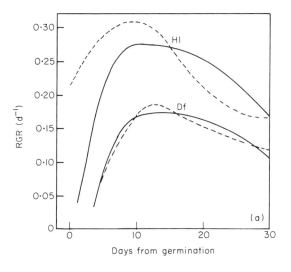

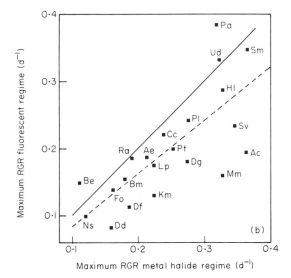

Fig. 9. (a) Progress curves for instantaneous relative growth rate (RGR) in *Deschampsia flexuosa* (Df) and *Holcus lanatus* (Hl) obtained from daily harvesting under a fluorescent and tungsten lighting regime (35 W m⁻² PAR, ———) or under a metal halide and tungsten regime (50 W m⁻², ————), both at 20/15 °C. LSD = 0·02 d⁻¹. See also Figure 4(b). (Previously unpublished data.) (b) The maximum RGR observed by Grime & Hunt (1975) under a fluorescent and tungsten lighting regime (35 W m⁻² PAR) as a function of the instantaneous maximum RGR [see (a)] observed under a metal halide and tungsten regime (50 W m⁻²) (previously unpublished data). The correlation coefficient is 0·703 ($P < 001$) but the mean trend (————) is of slope 0·812, significantly below the line of equivalence (———). The 21 species are: (Ac) *Agrostis capillaris*, (Ae) *Arrhenatherum elatius*, (Bm) *Briza media*, (Be) *Bromus erectus*, (Cc) *Cynosurus cristatus*, (Dg) *Dactylis glomerata*, (Dd) *Danthonia decumbens*, (Df) *Deschampsia flexuosa*, (Fo) *Festuca ovina*, (Hl) *Holcus lanatus*, (Km) *Koeleria macrantha*, (Lp) *Lolium perenne*, (Mm) *Matricaria matricarioides*, (Ns) *Nardus stricta*, (Pl) *Plantago lanceolata*, (Pa) *Poa annua*, (Pt) *Poa trivialis*, (Ra) *Rumex acetosa*, (Sv) *Senecio vulgaris*, (Sm) *Steelaria media* and (Ud) *Urtica dioica*.

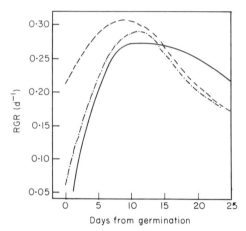

Fig. 10. Progress curves for instantaneous relative growth rate in *Holcus lanatus* [see Fig. 4(b)] obtained from daily harvesting under: (———) a fluorescent and tungsten lighting regime (35 W m^{-2}) at 20 °C, (————) a metal halide and tungsten regime (50 W m^{-2}) at 20 °C and (—·—·—·—) a metal halide and tungsten regime (50 W m^{-2}) at 25 °C. LSD = 0·02 d^{-1}. (Previously unpublished data.)

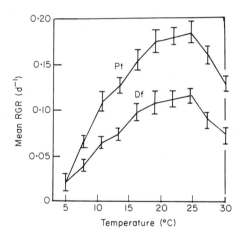

Fig. 11. Mean relative growth rate in *Deschampsia flexuosa* (Df) and *Poa trivialis* (Pt) as a function of mean daily temperature. Measurements spanned 4 weeks of the seedling phase; 95 % limits are added. (Data of G. Mason, unpublished data, from experiments performed in a temperature-gradient tunnel described by Mason, Grime & Lumb, 1976.)

below-ground parts of plants are strongly influenced by external environmental conditions (Crist & Stout, 1929). They also depend upon innate properties of the organism. Ecologically meaningful variation in partitioning has been observed when sampling natural vegetation (Bray, 1963; Monk, 1966). Experimental evidence clearly confirms that changes in many individual external conditions, or in resources, have a consistent and predictable influence on the allocation of dry weight into above-ground parts ('shoot') and below-ground parts ('root'). Ledig & Perry (1965) and Hunt & Burnett (1973) listed many references and reviews in this field.

The general picture which has emerged is that any growth-limiting condition

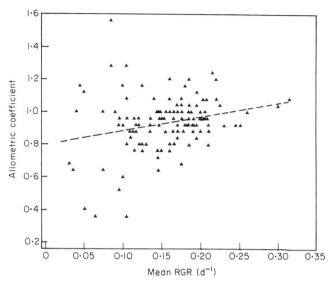

Fig. 12. The relation between the allometric coefficient [mean relative growth rate (RGR) for roots/mean RGR for shoots, K] and the mean RGR for the whole plant (R). The sample consists of 132 British woody and herbaceous species grown under productive, controlled conditions. The fitted line, which is significant at $P < 0.01$, is $K = 0.81 + 0.84R$. (Unpublished data from the screening programme of Grime & Hunt, 1975.)

or resource (in the classical sense of being one which will bring about a sustained increase in plant performance if it alone is alleviated artificially) will also induce a change in the resource partitioning of the plant. This will result in proportionally increased allocation of dry weight in favour of that part of the plant which draws most upon the growth-limiting part of the environment. As plain examples: nutrient-limited plants commonly become more rooty, shaded plants more shooty.

Similar pressures, arising from spatially patchy environments, also induce partitioning *within* root and shoot systems (see Grime, Crick & Rincon, 1986). The present chapter, however, deals with partitioning phenomena only at the level of whole root and shoot systems.

Root–shoot partitioning

Because the instantaneous ratio between root and shoot dry weight in the individual plant is simultaneously subject to genetic, ontogenetic and environmental control, in many ecological studies the most useful broad measure of partitioning is not this instantaneous ratio but the ratio of mean relative growth rates over an extended period of time, the allometric coefficient, K. If R_W and S_W are root and shoot dry weights respectively, the allometric coefficient may be evaluated as the linear regression of $\log_e R_W$ on $\log_e S_W$ over the whole period under study (Pearsall, 1927; Troughton, 1955). Thus constituted, K is relatively stable, often sufficiently so to represent the whole of the vegetative phase of growth without serious loss of information – yet it remains responsive to genetic and environmental influences.

Innate variations between species

In the juvenile state. In Grime & Hunt's (1975) sample of species, the relation between allometric coefficient and mean RGR is not clear cut (Fig. 12). There is a significant trend towards 'rootiness' in fast-growing species ($P < 0.05$) but the scatter diagram is basically triangular. Fast-growing species are (just) innately 'rooty' rather than 'shooty' but slow-growing ones may be either.

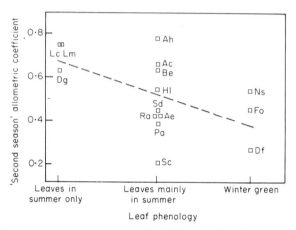

Fig. 13. Partitioning in a 'second season' (P. S. Lloyd & S. R. Band, unpublished data) as a function of leaf phenology. The allometric coefficient is the ratio of root and shoot mean relative growth rates. The fitted regression line is significant at $P < 0.05$ but depends upon the assumption of equal spacing between the three states on an otherwise non-parametric abscissa. The 15 species are: (Ac) *Agrostis capillaris*, (Ae) *Arrhenatherum elatius*, (Ah) *Arabis hirsuta*, (Be) *Bromus erectus*, (Df) *Deschampsia flexuosa*, (Dg) *Dactylis glomerata*, (Fo) *Festuca ovina*, (Hl) *Holcus lanatus*, (Lc) *Lotus corniculatus*, (Lm) *Lathyrus montanus*, (Ns) *Nardus stricta*, (Pa) *Poa annua*, (Ra) *Rumex acetosa*, (Sc) *Scabiosa columbaria* and (Sd) *Silene dioica*.

Beyond the juvenile state. After overwintering in an experimental garden (see Fig. 6), the allometric coefficient for re-growth under glass shows a marked reduction (with a mean value for 15 non-tree species of 0.52 in the 'second season' as against 0.89 for the same species in their juvenile state). Re-growth is thus relatively more shoot-dominated, as might be expected. Surprisingly, however, it is in the winter green species that this shift is most pronounced (Fig. 13).

Environmentally induced variations

The general relation between innate patterns of partitioning and growth potential (Fig. 12) becomes clearer when a smaller sample of species is considered (R. Hunt, A. O. Nicholls & F. A. Fathy, unpublished). Data from 18 British grasses grown in the juvenile state in a cumulative series of similar glasshouse experiments at Sheffield (Fig. 14) not only confirm clearly that innately fast-growing species are innately more root-orientated in their resource partitioning ($P < 0.01$) but they also establish that this innate trend is consistently respected during environmentally induced variations. In a series of treatments involving 'no stress' (with full light and nutrients), 'root stress' (with the concentration of potassium reduced to one-hundredth), 'shoot stress' (with irradiance reduced to one-fifth) and 'root and shoot stress' (with a combination of the two), the order of magnitude of the allometric coefficient was always root stress > control > root and shoot stress > shoot stress. The echelons of four points per species in Figure 14 reveal this clearly.

MODELLING GROWTH AND PARTITIONING

In seeking a high-level synthetic model to account for the coarse control of both growth and partitioning by the plant's external environmental conditions, Hunt & Nicholls (1986) suggested a model based upon simple *a priori* principles. This approach complemented that of the existing physiologically based models (for

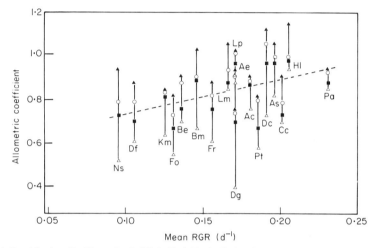

Fig. 14. Partitioning (R. Hunt, A. O. Nicholls & F. A. Fathy, in preparation) as a function of mean relative growth rate under productive, controlled conditions (Grime & Hunt, 1975) in 18 British grasses. The four allometric coefficients shown for each species are (from top to bottom) for treatments imposing 'root stress', 'no stress', 'root and shoot stress', and 'shoot stress' (see text). Plants were grown in a glasshouse over a period of 5 weeks. The fitted line has a slope of 1·524 units and is significant at $P < 0.01$; the overall mean value of K is 0·825. The 18 species are: (Ac) *Agrostis capillaris*, (As) *Agrostis stolonifera*, (Ae) *Arrhenatherum elatius*, (Be) *Bromus erectus*, (Bm) *Briza media*, (Cc) *Cynosurus cristatus*, (Dc) *Deschampsia cespitosa*, (Df) *Deschampsia flexuosa*, (Dg) *Dactylis glomerata*, (Fo) *Festuca ovina*, (Fr) *Festuca rubra*, (Hl) *Holcus lanatus*, (Km) *Koeleria macrantha*, (Lm) *Lolium perenne* × *multiflorum*, (Lp) *Lolium perenne*, (Ns) *Nardus stricta*, (Pa) *Poa annua* and (Pt) *Poa trivialis*.

references, see Robinson, 1986) and was designed to be particular relevance to comparative plant ecology. The model formally invoked the concept of environmental *stress*, defined as external constraints on dry weight production, and referred to *disturbance*, the external destruction of dry weight already formed (see Grime, 1984 for further discussion of these concepts).

Hunt & Nicholls held that the partitioning of environmental resources into below- and above-ground plant parts was jointly controlled by the absolute amounts of below- and above-ground environmental stress, by the below-/above-ground environmental stress ratio, and by the growth potential of the species itself. Under any particular set of conditions, any shortfall from the species' innate growth potential was explained in terms of the total of, and ratio between, below- and above-ground stress. Hyperbolic functions linked the variables under consideration. Although arbitrary in degree, these functions were held to be correct in kind, providing biologically plausible linkages between cause and effect.

The 'hyperbolic' model was simplified by Hunt, Nicholls & Fathy (in preparation) in order to reveal more clearly the role played by growth potential itself and by environmental stress and disturbance. With inputs consisting of observed mean relative growth rates and observed allometric coefficients, the model predicts for each species and treatment the levels of stress which exist below and above ground and which are 'perceived' by the plant. Although in the modified form of the model, innate growth potential plays no part in the prediction of stress 'perceived', its influence upon the 'perception' is considerable. For example, the 'stress tolerant' *Nardus stricta* and *Festuca ovina* respectively 'perceived' 0·159 and 0·164 artificial units of stress in a control treatment, where the relatively more 'competitive' *Deschampsia cespitosa* and *H. lanatus* respectively 'perceived' 0·609 and 0·506 units. The model clearly reflects the important principle that plants of unequal growth potential find equal environments unequally stressful.

Conclusions

Much of the reasoning behind the quantitative work of the unit of Comparative Plant Ecology at the 'strategic' and evolutionary levels of plant ecology has depended upon comparisons of attributes made across many species. If, in respect of the important attributes investigated, variation within species were to approach or exceed variation between species, then the quality of the 'broad picture' being built up at the higher levels of organization would be seriously affected. Naturally, no complete validation of this assumption can be made without obtaining full information on all possible aspects of the problem, in which case the 'broad-picture' approach would itself be redundant. However, this exploration of variation in relative growth rate and its derivates, involving investigations con-ducted in moderate detail under different development states and conditions, and at different levels of organization, has provided at least some insight into the security of the approach which has been undertaken. Coherent patterns of species-specific performances are indeed to be found in most instances and, together with the knowledge that normal or log-normal frequency distributions appear to be the rule, they give increased confidence to programme of research which draw their material randomly from natural situations in order to attempt a conceptual synthesis from representative data.

Acknowledgements

We thank many collaborators who have contributed to this work over the years, particularly Professor J. P. Grime and Drs J. G. Hodgson and J. F. Hope-Simpson, and I. T. Parsons. Indispensable support has been given by Mrs A. M. N. Ruttle, A. M. Neal, S. R. Band, Miss C. R. V. Rathey and Mrs H. J. Hucklesby. Professor J. P. Grime commented valuably upon the manuscript.

References

Allen, T. F. H. & Starr, R. B. (1982). *Hierarchy: Perspectives for Ecological Complexity*. University of Chicago Press, Chicago & London.
Benjamin, L. R. & Hardwick, R. C. (1986). Sources of variation and measures of variability in even-aged stands of plants. *Annals of Botany*, **58**, 757–778.
Blackman, V. H. (1919). The compound interest law and plant growth. *Annals of Botany*, **33**, 353–360.
Bray, J. R. (1963). Root production and the estimation of net productivity. *Canadian Journal of Botany*, **41**, 65–72.
Briggs, G. E., Kidd, F. & West, C. (1920). A quantitative analysis of plant growth. I. *Annals of Applied Biology*, **7**, 103–123.

BURDON, J. J. & HARPER, J. L. (1980). Relative growth rates of individual members of a plant population. *Journal of Ecology*, **68**, 953–957.

CLAPHAM, A. R., TUTIN, T. G. & WARBURG, E. F. (1981). *Excursion Flora of the British Isles*, 3rd Edn. Cambridge University Press, London.

CRIST, J. W. & STOUT, G. J. (1929). Relation between top and root size in herbaceous plants. *Plant Physiology*, **4**, 63–85.

ELIAS, C. O. & CHADWICK, M. J. (1979). Growth characteristics of grass and legume cultivars and their potential for land reclamation. *Journal of Applied Ecology*, **16**, 537–544.

EVANS, G. C. (1972). *The Quantitative Analysis of Plant Growth*. Blackwell Scientific Publications, Oxford.

FEIBLEMAN, J. K. (1955). Theory of integrative levels. *British Journal for the Philosophy of Science*, **5**, 59–66.

FOGG, G. E. (1949). Growth and heterocyst production in *Anabaena cylindrica* Lemm. II. In relation to carbon and nitrogen metabolism. *Annals of Botany*, **13**, 241–159.

GRIME, J. P. (1984). The ecology of species, families and communities of the contemporary British flora. *New Phytologist*, **98**, 15–33.

GRIME, J. P. & HUNT, R. (1975). Relative growth-rate: its range and adaptive significance in a local flora. *Journal of Ecology* **63**, 393–422.

GRIME, J. P., MASON, G., CURTIS, A. V., RODMAN, J., BAND, S. R., MOWFORTH, M. A. G., NEAL, A. M. & SHAW, S. (1981). A comparative study of germination characteristics in a local flora. *Journal of Ecology*, **69**, 1017–1059.

GRIME, J. P., CRICK, J. G. & RINCON, J. E. (1986). The ecological significance of plasticity. In: *Plasticity in Plants* (Ed. by D. H. Jennings & A. J. Trewavas), pp. 5–29. Cambridge University Press, Cambridge.

HEWITT, E. J. (1966). *Sand and Water Culture Methods Used in the Study of Plant Nutrition*. Commonwealth Agricultural Bureaux, Farnham Royal.

HUNT, R. (1980). Diurnal patterns of dry weight increment and short-term plant growth studies. *Plant, Cell and Environment*, **3**, 475–478.

HUNT, R. (1984). Relative growth rates of ramets cloned from a single genet. *Journal of Ecology*, **72**, 299–306.

HUNT, R. & BURNETT, J. A. (1973). The effects of light intensity and external potassium level on root/shoot ratio and rates of potassium uptake in perennial ryegrass (*Lolium perenne* L.). *Annals of Botany*, **37**, 519–537.

HUNT, R. & NICHOLLS, A. O. (1986). Stress and the coarse control of root–shoot partitioning in herbaceous plants. *Oikos*, **47**, 149–158.

HUNT, R. & PARSONS, I. T. (1974). A computer program for deriving growth functions in plant growth analysis. *Journal of Applied Ecology*, **11**, 297–307.

HUNT, R., HOPE-SIMPSON, J. F. & SNAPE, J. (1985). Growth of the dune wintergreen (*Pyrola rotundifolia* ssp. *maritima*) at Braunton Burrows in relation to weather factors. *International Journal of Bio-meteorology*, **29**, 323–334.

JMB (1985). *GCE Revised Syllabuses 1988: Biology (Advanced)*. Joint Matriculation Board, Manchester.

KAY, Q. O. N., ROBERTS, R. H. & VAUGHAN, I. M. (1974). The spread of *Pyrola rotundifolia* L. subsp. *maritima* (Kenyon) E. F. Warb. in Wales. *Watsonia*, **10**, 61–67.

LAW, R. (1972). *Features of the biology and ecology of* Bromus erectus *and* Brachypodium sylvaticum *in the Sheffield region*. Ph.D. thesis, University of Sheffield, U.K.

LEDIG, F. T. & PERRY, T. O. (1965). Physiological genetics of the shoot–root ratio. *Proceedings of the Society of American Foresters, 1965*, pp. 39–43.

MASON, G., GRIME, J. P. & LUMB, A. H. (1976). The temperature-gradient tunnel: a versatile controlled environment. *Annals of Botany*, **40**, 137–142.

MAUDE, R. B. (1973). Seed-borne diseases and their control. In: *Seed Ecology* (Ed. by W. Heydecker), pp. 325–335. Butterworth, London.

MONK, C. (1966). Ecological importance of root/shoot ratios. *Bulletin of the Torrey Botanical Club*, **93**, 402–406.

PARSON, I. T. & HUNT, R. (1981). Plant growth analysis: a curve-fitting program using the method of B-splines. *Annals of Botany*, **48**, 341–352.

PEARSALL, W. H. (1927). Growth studies. VI. On the relative sizes of growing organs. *Annals of Botany*, **41**, 549–556.

ROBINSON, D. (1986). Compensatory changes in the partitioning of dry matter in relation to nitrogen uptake and optimal variations in growth. *Annals of Botany*, **58**, 841–848.

TROUGHTON, A. (1955). The application of the allometric formula to the study of the relationship between the roots and shoots of young grass plants. *Agricultural Progress*, **30**, 1–7.

TUKEY, J. W. (1977). *Exploratory Data Analysis*. Addison-Wesley, Reading, Massachusetts.

WEST, C., BRIGGS, G. E. & KIDD, F. (1920). Methods and significant relations in the quantitative analysis of plant growth. *New Phytologist*, **19**, 200–207.

WURR, D. C. & FELLOWS, J. R. (1984). Further studies on the slant test as a measure of the vigour of crisp lettuce seedlots. *Annals of Applied Biology*, **105**, 575–580.

New Phytol. (1987) **106** (Suppl.), 251–263

INTERSPECIFIC VARIATION IN PLANT ANTI-HERBIVORE PROPERTIES: THE ROLE OF HABITAT QUALITY AND RATE OF DISTURBANCE

By PHYLLIS D. COLEY

Department of Biology, University of Utah, Salt Lake City, Utah 84112, USA

SUMMARY

General patterns of herbivory and plant defence are summarized for a range of tree species studied in a lowland rain forest of Panama. Species growing in comparable micro-habitats differed by more than three orders of magnitude in the rates of herbivorous damage to mature leaves. Over 70% of this interspecific variation was statistically accounted for by differences in leaf characteristics. Species having tough, fibrous leaves with low nutritional value suffered little herbivory. Concentrations of immobile defences such as tannins and fibre were higher in species with long-lived leaves. Among species, there was also a significant negative correlation between growth rate and defence, and a positive correlation between growth and herbivory. These results suggest that differences in defence among species are due to interspecific differences in intrinsic growth rates and not to differences in apparency.

Theories of plant defence based on interspecific differences in growth rate and apparency are combined in a single general model of plant defence. This model follows Grime's triangular classification of plant strategies and assumes that quality of habitat and rate of disturbance are both important determinants of plant defence. Predictions are made as to the types of defensive properties expected in stress-tolerant, competitive and ruderal species.

Key words: Herbivory, plant defence, habitat quality, disturbance, plant growth, apparency.

INTRODUCTION

It is now accepted that plants have evolved an enormous variety of physical and chemical properties which are effective deterrents against herbivores. Every plant species has a suite of secondary metabolites whose primary function is defence. Although chemical defences have received most attention, physical and pheno-logical characteristics are also varied and are important in reducing herbivory. Therefore, if defensive options are both diverse and ubiquitous, why are some species of plants better defended than others? Are there generalities in the interspecific patterns of herbivory and plant defence, and can we understand the proximate and evolutionary mechanisms responsible for these patterns?

The first part of this paper will describe patterns of herbivory and defence for tree species in a tropical rain forest and then discuss why these patterns might have evolved. The second part will examine alternative models for the evolution of defences and suggest that many of these views are actually quite similar or, in some cases, complementary. An attempt will be made to combine these views into general predictions of interspecific variation in defence.

PATTERNS OF HERBIVORY IN TROPICAL TREES

This work was carried out in a lowland tropical forest on Barro Colorado Island in Panama. [For a description of the site, see Leigh, Rand & Windsor, (1982).] To document general community-wide patterns of herbivory and defensive

0028-646X/87/05S251 + 13 $03.00/0

properties, I measured 47 of the most common species of canopy tree (Coley, 1983). These exhibited a range of shade tolerance, from species which could persist in the shaded forest understorey to those which required light gaps for establishment and continued growth. All measurements were made on saplings growing in light gaps to limit micro-environmental differences and variation in the availability of herbivores.

Patterns of herbivory were quantified as the rate of damage to mature leaves. These rates (percentage leaf area lost per unit time) were determined by repeatedly measuring marked leaves at the beginning and end of a 6-week period in each of the early wet, late wet and dry seasons. Estimates of rates are essential for comparing damage between species, because leaf longevity can vary by an order of magnitude (Coley, 1981, 1987a). A single measure of damage is useless if one cannot determine how long the leaf has been available to herbivores.

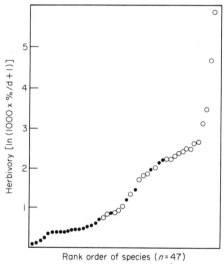

Fig. 1. Mean annual rates of herbivory (percentage loss of leaf area per day—specially scaled) on mature leaves of 47 lowland rain forest tree species on Barro Colorado Island, Panama. Species are divided into shade-tolerant species (●) and light-demanding species (○).

There was enormous variation among the 47 Panamanian species in their palatability to herbivores (Fig. 1). Although species were growing as neighbours in the same light gaps, potentially susceptible to the same herbivores, there were differences of more than three orders of magnitude in rates of herbivory on mature leaves.

PATTERNS OF DEFENCE IN TROPICAL TREES

Why do herbivores show such strong preference for some species over others? To determine whether this was due to differences between species in characteristics of leaves, I measured several defensive and nutritional features which could potentially influence palatability (Table 1; Coley, 1983). These included the concentrations of two types of condensed tannins, since these polyphenolic compounds are ubiquitous, often in large concentrations (Swain, 1979). Although there is considerable controversy surrounding the mode of action and effectiveness

Table 1. *Herbivory, growth and defensive characteristics of mature leaves for 47 lowland rain forest trees*

Leaf characteristic†	Correlation with herbivory	Light-gap specialists		Shade-tolerant	
		Mean	(Range)	Mean	(Range)
Herbivory (% d⁻¹)	—	0·24	(0·02–0·9)	0·04	(0·0003–0·17)*
Growth rate (cm year⁻¹)	+0·520*	96	(23–475)	37	(1–86)*
Tannins-VAN (% d. wt)	−0·112	0·8	(0·03–3·4)	2·7	(0·04–7·8)*
Tannins-PRO (% d. wt)	−0·128	1·7	(0–7·5)	4·8	(0·07–13·2)*
Tannin-VAN/nitrogen	−0·161	0·6	(0·01–4·7)	1·2	(0·01–4·2)*
Tannin-PRO/nitrogen	−0·184	0·9	(0–5·4)	2·1	(0·01–7·1)*
Fibre-NDF (% d. wt)	−0·278*	43·6	(27·2–64·2)	51·3	(32·8–67·5)*
Fibre-ADF (% d. wt)	−0·424*	29·5	(19·7–47·7)	37·4	(15·6–49·9)*
Lignin (% d. wt)	−0·223	10·5	(3·3–18·5)	12·5	(3·3–20·8)
Cellulose (% d. wt)	−0·473*	17·6	(10·2–30·4)	23·4	(12·2–30·3)*
Toughness (g)	−0·515*	392	(251–723)	622	(278–1179)*
Pubescence (No. hairs mm⁻²)	+0·635*	5·1	(0–18·0)	0·5	(0–3·4)*
Water (%)	+0·507*	70	(57–82)	62	(49–77)*
Nitrogen (% d. wt)	+0·287*	2·5	(1·7–3·1)	2·2	(1·2–3·1)
r² – multiple regression	+0·70*	0·86*		0·74*	

Correlation coefficients are given for each characteristic and herbivory. Mean values and ranges are given for 23 gap specialists and 24 shade-tolerant species. The multiple r^2 coefficients are given for all species, for light-gap and for shade-tolerant species, respectively. * $P < 0.05$.

† For abbreviations, see text.

of tannins in insect guts (Bernays, 1981; Martin & Martin, 1982, 1984; Zucker, 1983), it is likely that they serve some defensive function (Feeney, 1968, 1976; Rhoades & Cates, 1976; Coley, 1986). Tannins were measured using the vanillin (VAN) and proanthocyanidin butanol (PRO) assays (Coley, 1983). I also measured four components of fibre: cellulose, acid-detergent fibre, neutral-detergent fibre and lignin. Fibre not only provides structural support for the leaf but it also reduces its digestibility for herbivores (Van Soest, 1978; Milton, 1979; Swain, 1979). A related measure, leaf toughness, was quantified as the weight needed to punch a 3 mm rod through the leaf. Pubescence was measured as the density of hairs on the lower surface of the leaf. Water and nitrogen contents give an indication of the nutritional quality of leaves.

Table 1 records the correlation between each mature leaf characteristic with herbivory. The magnitude of the correlation coefficient indicates the statistical importance or relative effectiveness of each character in reducing herbivory. High toughness and fibre content and low nutritional value are the characteristics most negatively correlated with herbivory.

I suggest that, although toughness provides structural support for the leaf, a major selective pressure for tougher leaves is herbivory. Toughness provides an effective defence against a broad range of invertebrate and vertebrate herbivores as well as pathogens (Tanton, 1962; Feeny, 1970; Grubb, 1977; Milton, 1979; Oates, Waterman & Choo, 1980; Mansfield, 1982; Coley, 1983; Raupp, 1985). We should therefore see tough leaves in species where strong herbivore defence is important. The observed occurrence of tough leaves in dry or nutrient-poor sites (Loveless, 1961; Grubb, 1977; Medina, 1983; Specht & Moll, 1983) may therefore not be an adaptation to abiotic stress, as has often been assumed, but it may instead be due to the increased importance of protection against herbivores.

None of the estimates of tannin nor the tannin : protein ratios are significantly correlated with herbivory (Table 1). This is perplexing, since their concentrations in mature leaves of some species are as high as 13·2% of dry weight (Table 1; Coley, 1983). The lack of correlation could be due to the difficulty of measuring tannins, to variation in effectiveness of different types of tannins or to the fact that tannins may be more effective against pathogens than insects (Mandels & Reese, 1965; Kuč, 1972; Bernays, 1981; Mansfield, 1982; Martin & Martin, 1982, 1984; Callow, 1983; Zucker, 1983; Waterman, Ross & McKey, 1984).

Pubescence is the only characteristic to show a strong positive relation with herbivory (Table 1). This is not because hairs are good to eat but because pubescence is negatively correlated with other defences (Coley, 1983). The presence of hairs may therefore be a simple visual way to identify poorly defended species.

Although the characteristics measured (Table 1) are only a subset of the defences of each species, they appear to be the most important ones for reducing herbivory. More than 70% of the variation in herbivory among species is explained statistically by differences in these characteristics (see r^2 values, Table 1). In other words, the amount of investment in defence determines the rates at which species are eaten. Species with tough, fibrous leaves of low nutritional values have the lowest rates of herbivory. Species poorly defended in this way do not escape damage by other means (Feeny, 1976; Rhoades & Cates, 1976) but suffer much herbivory.

SELECTION FOR PLANT DEFENCE: AMOUNT

For the Panamanian trees, most of the variation between species in herbivore preference can be explained by characteristics of leaves recorded in Table 1. The amount of damage is directly related to the degree of investment in defences. Therefore, if defences reduce herbivory, why are all species not well defended? The following section presents possible adaptive explanations for why some species have evolved more defence than others.

Apparency theory

The major theory for the last 10 years to explain defensive differences among species was presented by Feeny (1976) and Rhoades & Cates (1976). This 'apparency theory' suggests that some species are poorly defended because they rely on escaping discovery by herbivores. They predict that only species which are easily found by herbivores need to invest in defences. The theory implies that species should have similar rates of damage, with some species (unapparent) minimizing damage by escaping and others (apparent) by chemical defences.

In the case presented above for tropical trees, apparency theory does not seem to explain the observed differences in defence. Firstly, there were no differences among species in the distribution of damage within and among plants, suggesting that species did not differ in apparency and were equally discovered by herbivores (Coley, 1983). Secondly, all species were not eaten at the same rates (Fig. 1) and, finally, the species with little defence did not escape discovery by herbivores as they suffered high rates of damage (Table 1; Coley, 1983).

Growth rate hypothesis

As an alternative to apparency theory, it has been proposed that plant species differ in their defences because they differ in their intrinsic rate of growth (Coley, Bryant & Chapin, 1985). This is primarily because of two reasons.

Firstly, the immediate costs of defence are actually smaller for slow-growing species (Coley *et al.*, 1985; Gulmon & Mooney, 1986). This is illustrated in Figure 2(a) where a fast and a slow-growing species each have two morphs with either 0 or 10% investment in defence. Concentrations of defensive compounds in tissues are therefore similar for the defended morphs of both species. However, because investments in defence reduce growth by a particular percentage, this translates to a much larger difference in absolute size between a defended and undefended fast grower than for a defended and undefended slow grower [Fig. 2(a)]. This relatively low opportunity cost of defence for slow-growing species suggests that they should exhibit higher investments in defence.

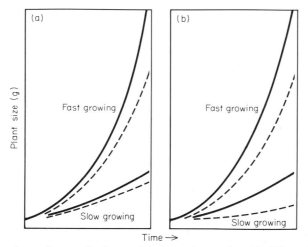

Fig. 2. Simulated growth curves for fast- and slow-growing species with different levels of defence and herbivory. Growth equations are based on Coley *et al.* (1985). (a) Growth rates of defended morphs are reduced by 10% owing to costs of defence. Defended morphs of both fast- and slow-growing species have similar concentrations of defensive compounds in their tissues. (b) Herbivory is simulated by removing identical amounts of leaf tissue from the fast- and slow-growing species. (a) ——, Undefended; ----, defended. (b) ——, No herbivory; ----, herbivory.

Secondly, because herbivores eat amounts and not percentages, their removal of a certain number of grammes per day will represent a larger fraction of the net production of a slow-growing species than of a fast-growing one [Fig. 2(b)]. Therefore, because the relative impact of herbivory is higher in intrinsically slow-growing species, we again expect to see larger concentrations of defensive compounds.

Evidence from trees in Panama (Coley, 1983, 1987a) indicates an inverse relationship between intrinsic growth rate and defence (Fig. 3). There is a strong negative correlation between the average annual growth rate and the investment in defences ($r = 0.69$, $P < 0.001$, $n = 40$). Investment in defence is estimated by a linear combination of defences measured for each species (Table 1, excluding nitrogen and water). These data indicate that more slowly growing species have more chemical and structural defences. Additional evidence is seen in the positive correlation between rates of growth and herbivory ($r = 0.52$, $P < 0.001$, $n = 40$), again suggesting that more slowly growing species are better defended. In Figure 3, growth rate is expressed as the annual increase in height. The relationships are the same using other measures of growth such as annual leaf production or the maximum growth rate seen for an individual of each species. Therefore, despite the overwhelming variation found in natural systems, growth rate statistically explains a substantial amount of the interspecific differences in herbivory and defence for trees on Barro Colorado Island.

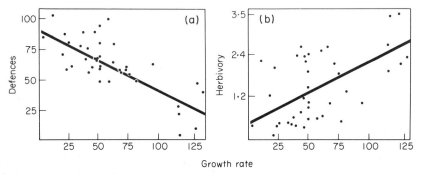

Fig. 3. Defences and herbivory as a function of growth rate for Panamanian trees. Growth is the mean annual increase in height (cm); defence is a linear combination in arbitrary units of chemical and physical defences in Table 1 (not including nitrogen and water) (see Coley, 1983); herbivory is the log of the percentage of leaf area eaten per day (as in Fig. 1). (a) $r = -0.69$, $P < 0.001$. (b) $r = 0.52$, $P < 0.001$.

Resource availability and growth rate

Species differ in their intrinsic rate of growth because they are adapted to habitats with different amounts of resources. Intrinsic growth rates have evolved in response to the degree of limitation by resources (Grime, 1977, 1979; Chapin, 1980). Species adapted to resource-rich environments or micro-habitats have fast intrinsic growth rates, whereas species adapted to resource-limited environments grow slowly even if they are moved to good conditions (Grime & Hunt, 1975). The availability of resources in a particular habitat therefore selects for appropriate growth rates and these in turn determine appropriate levels of defence (Coley et al., 1985). A prediction from this hypothesis is that species adapted to resource-rich environments should have little defence.

This relationship can be tested with the data on Panamanian trees (Coley, 1983). On Barro Colorado Island, light is the major resource which limits growth as irradiance on the forest floor is less than 1 % full sunlight (Chazdon & Fetcher, 1984; T. A. Kursar & P. D. Coley, unpublished data). This contrasts with gaps made by fallen trees where irradiance is much greater.

The 47 species studied exhibit a range of shade tolerance but can be loosely grouped into two general classes: gap specialists and shade-tolerant species (Croat, 1978; Foster & Brokaw, 1982; Coley, 1983; Brokaw, 1985). Since gap specialists are adapted to micro-habitats rich in resources, the growth rate hypothesis (Coley et al., 1985) predicts that they should have higher rates of growth and herbivory than shade-tolerant species. On average, the former have twice the average growth rate of the latter and are eaten six times more rapidly (Fig. 1, Table 1). Leaves of gap specialists also have significantly lower concentrations of condensed tannins, are less fibrous and only half as tough (Table 1). In addition to being less well-defended chemically, light-gap species have a higher nutritional value for herbivores as they contain significantly more water and protein. All these characteristics are consistent with the observations that light-gap species suffer higher rates of herbivory.

Another test of the negative relationship between investment in defence and availability of resources is seen by comparing communities which differ in soil nutrients (Coley, 1987b). I examined seven lowland tropical forests with similar rainfall regimes but which differed in the nutrient availability of the soils. Leaf toughness was used as a measure of investment in defence because it is highly positively correlated with other defences and is easy to quantify in the field. Leaf toughness was measured for a random sample of 200 mature leaves per site. Data on soil quality at these sites were not taken but sites can be ranked fairly confidently. At one extreme are forests which stand on alluvial or relatively young soils eroded from recent mountain-building events. At the other extreme are sites on leached sands from the ancient Guyana shield. Leaves in the severely nutrient-limited forests are twice as tough as leaves from richer sites. McKey's work in two African forests shows that another defence, condensed tannin, is better developed in the nutrient-poor forest (McKey *et al.*, 1978). Other studies in tropical, temperate and arctic areas have also found more defence for species adapted to resource-limited conditions (Grime, MacPherson-Stewart & Dearman, 1968; Janzen, 1974; Bryant & Kuropat, 1980; McKey & Gartlan, 1981; Bryant, Chapin & Klein, 1983; Waterman, 1983; Coley *et al.*, 1985; Owen-Smith & Cooper, 1987; Bryant, 1987).

Resource availability and intraspecific defence variation

The above arguments for the influence of resources on interspecific variation in defence do not apply to variation within a species. Although it is tempting to assume that it would be adaptive for individuals growing under resource-limited conditions to have more defence, this is not always the case. Instead, it appears that intraspecific variation in defence is determined by the balance of resources, not their absolute level. In their carbon/nutrient hypothesis, Bryant *et al.* (1983) suggest that resources present in excess of growth demands are put into defence. For example, in sunny conditions with limiting nutrients, carbon will be relatively in excess, and carbon-based defences such as tannins and terpenoids will increase. Conversely, in shaded conditions, carbon-based defences decrease. Similar patterns are seen for nitrogen-based defence and nitrogen availability. These trends are often exactly the reverse of those seen among species. Intraspecific variation can therefore be explained as a response to imbalances in source/sink relations rather than as an adaptive solution to herbivory and availability of resources.

SELECTION FOR PLANT DEFENCE: TYPE

Evolutionarily, plants have made two types of response, the quantitative (how much to commit to defence) and the qualitative (what type of defence to construct). In the previous section, I presented evidence in support of the theory that, in many cases, intrinsic growth rates determine the optimal amount of resources that a plant should commit to defence. The following section will discuss factors which influence the adaptive value of different types of defence. Are there patterns of investment within the enormous diversity of secondary metabolites? For example, can we make predictions as to why some species have alkaloids and some have tannins?

The relative costs of different types of compounds should be an important consideration (Mooney & Gulmon, 1982). Traditionally, secondary compounds have been considered expensive if they are present in large amounts (Feeny, 1976; Rhoades & Cates, 1976). This ignores the potentially high costs of turnover associated with many small molecular weight compounds such as alkaloids, cardiac glycosides and monoterpenes (Coley et al., 1985). The biological half-life of these 'mobile' defences can be in the order of hours or days (Robinson, 1974; Waller & Nowacki, 1978; Croteau & Johnson, 1984). They must be continually synthesized to maintain the same concentration in the leaves. The total costs of defending a leaf with mobile defences therefore depends on the leaf's lifetime. This contrasts with defence associated with immobile compounds such as tannins or lignins which have extremely low turnover rates (Walker, 1975; E. Haslam, pers. comm.). Since these immobile defences are often present in large concentrations, the initial costs of construction can be substantial. Furthermore, they cannot be reclaimed upon leaf senescence (McKey, 1979). The total cost of defending a leaf with immobile defences is therefore independent of its lifetime.

Considering costs of both construction and maintenance, mobile defences should be more cost-effective in species with short-lived leaves and immobile defences favoured in species with long-lived leaves. For leaves of the trees studied in Panama, there is a significant increase in immobile defences with increasing lifetime (Fig. 4; Coley, 1987a). Both condensed tannin and fibre content are positively correlated with leaf life. Similar conclusions were reached by McKey (1984) studying ant plants in Africa.

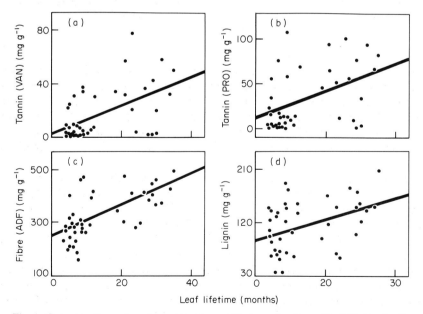

Fig. 4. Concentrations of condensed tannin and fibre as a function of leaf lifetime for Panamanian tree species. (a) $r = 0.51$, $P < 0.001$; (b) $r = 0.41$, $P < 0.01$; (c) $r = 0.64$, $P < 0.001$; (d) $r = 0.41$, $P < 0.01$.

RECONCILING PLANT DEFENCE THEORIES

In the examples I have given and in other comparative studies, species of plant show large differences in defence but not in apparency. In these cases at least, growth rate rather than apparency seems to be causally related to the level and type of defence. However, are there other conditions under which plants do escape from herbivores? Could these lead to selection for lowered defence? Can these two views of apparency and growth rate be reconciled into a single theory of defence?

Before attempting to answer these questions, I would like to examine components of the habitat that are important in the evolution of characteristics other than defences in plants. Since it is often difficult to test the adaptive significance of evolutionary patterns experimentally, support must come from the convergence of various lines of thinking. Many workers, interested in classifying life history strategies of plants have converged on essentially two important properties of the habitat: its quality and rate of disturbance (Fig. 5). The recurring appearance of these two concepts in a variety of theories suggest that they have some universal importance.

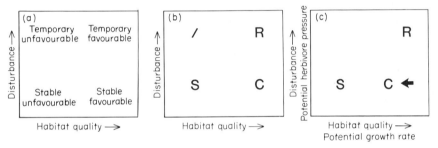

Fig. 5. Properties of habitat (a) and species [(b), (c)] as a function of quality of habitat and rates of disturbance. C, competitive; R, ruderal; S, stress-tolerant. [(a) and (b) based on Grime (1977, 1979).]

Disturbance can be defined as the frequency at which a habitat is destroyed and is the inverse of stability. The importance of disturbance was early recognized in the $r-K$ continuum (MacArthur & Wilson, 1967). Disturbance has since been referred to as 'durational stability' by Southwood (1977), and as the 'intensity of disturbance' by Grime (1977). Predictability or emphemerality of habitats were also central to the theories of herbivore defence of Feeny (1976) and Rhoades & Cates (1976).

Habitat quality, the second important property, refers to the degree of resource limitation or the favourableness for growth. This concept has appeared in Grime's work as the 'intensity of stress' (Grime, 1977, 1979), in Chapin's as 'resource availability' (Chapin, 1980), and in Southwood's as 'habitat adversity' (Southwood, 1977). Janzen (1974), McKey (1979) and Coley *et al.* (1985) have all considered habitat quality to be important in determining plant defence.

A combination of habitat quality and the rates of disturbance, if simplified, defines four types of habitat: (1) stable and of low quality, (2) stable and of high quality, (3) temporary and of high quality, and (4) temporary and of low quality [Grime, 1977; Fig. 5(a)]. Selection would lead to different life histories in each of these four general habitats. Grime (1977, 1979) suggests that temporary, low quality habitats are unsuitable for life, that stable, low quality habitats should select for stress-tolerant species (S), stable, high quality habitats for competitive species (C), and temporary, high quality habitats for ruderal species (R) [Fig. 5(b)].

In his triangular scheme of vegetation classification, Grime (1977, 1979) extends these ideas to emphasize three types of selection pressures: stress, disturbance and competition. Although conceptually extremely similar, I would like to retain the original axes of disturbance and habitat quality, since these are the two major organizing principles behind many defence theories. The effects of competition do, however, enter indirectly. Because of competition among individuals in the stable, high quality habitats, the actual quality seen by each individual will decrease (Grime, 1977, 1979). Competition therefore has the effect of moving the position of competitive species (C) left along the habitat quality gradient [Fig. 5(c)].

The properties of habitat and species outlined in Figure 5(c) are an attempt to reconcile the various defence theories and to relate them to existing vegetation strategies. This formulation suggests that both habitat quality and disturbance rates have important influences on the evolution of plant defences. Disturbance will influence the degree to which species are apparent to herbivores and hence the necessity for physical or chemical defences (Feeny, 1976; Rhoades & Cates, 1976). Habitat quality will influence the intrinsic growth rates of species which in turn influence the costs and benefits of defence (Janzen, 1974; McKey, 1979; Coley et al., 1985; Gulmon & Mooney, 1986). The axes of Figure 5(c) have therefore been relabelled to include the biotic components of herbivore pressure and intrinsic growth rate. A very similar conclusion was reached by Southwood, Brown & Reader (1986).

This formulation (Fig. 5) does suggest that most interspecific differences in defence are due to differences in intrinsic growth rates and that escape is only an important component for ruderal species. Disagreements over the relative importance of growth rate vs escape probably arose because studies compared different habitat types. For example, the work of Coley (1983) with tropical trees compared species along the S–C continuum where the important factor affecting defence investment was intrinsic growth rate as determined by habitat quality. The classic comparison of oaks and crucifers (Feeny, 1976) was between S and R species which therefore encompassed differences in both escape and growth rate.

GRIME'S VEGETATION STRATEGIES AND PLANT DEFENCE

Following the division by Grime (1977) of plants into stress-tolerant, competitive and ruderal groupings, I have outlined the types of defensive properties which might be associated with each of the three strategies (Table 2). Stress-tolerant species have slow intrinsic growth rates and long-lived leaves, so they should be well defended by immobile defences such as tannins and fibre. Despite high herbivore pressure, they will therefore receive little tissue damage. I would expect to see little quantitative variation in defence within individuals because the potentially negative impact of herbivory is always high, because immobile defences are not reclaimable (McKey, 1979) and because slow growth prohibits fast metabolic adjustments.

Competitive species have substantially faster growth rates and shorter leaf lifetimes than stress-tolerant species (Table 2). Fast growth rates would select for low to intermediate levels of defence and short leaf lifetimes would favour mobile defences (Coley et al., 1985). Low defence levels coupled with high potential herbivore pressure would lead to high rates of herbivory. Although allocation of

Table 2. *Predicted defensive characteristics of stress-tolerant, competitive and ruderal species*

Characteristics of growth and defence	Stress-tolerant	Competitive	Ruderal
Intrinsic growth rate	Slow	Medium-high	High
Tissue lifetimes	Long	Intermediate	Short
Apparency to herbivores	High	High	Low
Potential rates of herbivory	High	High	Low
Actual rates of herbivory	Low	High	Low
Amount of defence	High	Low–medium	Low
Type of defence	Immobile	Mobile	Mobile
Defence: turnover rates	Low	High	High
Ability to reclaim defences	Low	High	High
Defence: plasticity	Low	High	Low
Defence: induction	Low	High	Low

resources to defence would in general be low, I would expect to see large variation within individuals over time. Firstly, because of strong competition, there should be selection to maximize growth at the expense of defence. The costs of defence should only be incurred when necessary. We might therefore expect competitive species to show an enhanced ability to induce production of defences in response to herbivore attack. Secondly, strong competition would favour individuals that could effectively absorb nutrients when they were available and store them until needed. Mobile defences can be used as temporary storage for nutrients available in excess of requirements for growth (Mooney, Gulman & Johnson, 1983). Allocation to defence would therefore rise and fall according to storage requirements.

In ruderal species, rates of growth and leaf turnover are extremely high. Potential herbivore pressure should be less than in the more stable habitats. Overall allocation to defence as well as actual herbivory should therefore be low. I would expect little plasticity in defence since levels are already low and since the combination of fast growth and low herbivore pressure would rarely give rise to a situation where defence would be cost-effective. The only occasion under which we might expect to see an increase in defence might be for temporary storage of nutrients early in the growing season.

ACKNOWLEDGEMENTS

I am particularly indebted to J. P. Grime, P. P. Feeny and T. R. E. Southwood for provocative and influential discussions. I take full responsibility for any misconceptions presented here. I also thank T. A. Kursar.

REFERENCES

BERNAYS, E. A. (1981). Plant tannins and insect herbivores: an appraisal. *Ecological Entomology*, **6**, 353–360.
BROKAW, N. V. L. (1985). Gap-phase regeneration in a tropical forest. *Ecology*, **66**, 682–687.
BRYANT, J. P. (1987). Resource availability: implications for constitutive and long-term inducible defences in foliage of woody plants. In: *Chemical Mediation of Coevolution* (Ed. by K. Spencer), AIBS. (In press.)
BRYANT, J. P. & KUROPAT, P. L. (1980). Selection of winter forage by subarctic browsing vertebrates: the role of plant chemistry. *Annual Review of Ecology and Systematics*, **11**, 261–285.
BRYANT, J. P., CHAPIN III, F. S. & KLEIN, D. R. (1983). Carbon/nutrient balance of boreal plants in relation to vertebrate herbivory. *Oikos*, **40**, 358–368.

CALLOW, J. A. (Ed.) (1983). *Biochemical Plant Pathology*. John Wiley, New York.

CHAPIN III, F. S. (1980). The mineral nutrition of wild plants. *Annual Review of Ecology and Systematics*, **11**, 261–285.

CHAZDON, R. L. & FETCHER, N. (1984). Light environments of tropical forests. In: *Physiological Ecology of Plants of the Wet Tropics* (Ed. by E. Medina, H. A. Mooney & C. Vazquez-Yanes), pp. 27–36. Junk, Boston.

COLEY, P. D. (1981). *Ecological and evolutionary responses of tropical trees to herbivory : A quantitative analysis of grazing damage, plant defences and growth rates*. Ph.D. dissertation, University of Chicago, Illinois.

COLEY, P. D. (1983). Herbivory and defensive characteristics of tree species in a lowland tropical forest. *Ecological Monographs*, **53**, 209–233.

COLEY, P. D. (1986). Costs and benefits of defense by tannins in a neotropical tree. *Oecologia*, **70**, 238–241.

COLEY, P. D. (1987a). Effects of plant growth rate and leaf lifetime on the amount and type of anti-herbivore defence. *Oecologia*, **71**. (In press.)

COLEY, P. D. (1987b). Patrones en las defensas de las plantas: por que los herbivoros prefieren ciertas especies? *Revista de Biologia Tropical*. (In press.)

COLEY, P. D., BRYANT, J. P., CHAPIN III, F. S. (1985). Resource availability and plant anti-herbivore defence. *Science*, **230**, 895–899.

CROAT, T. D. (1978). *Flora of Barro Colorado Island*. Stanford University Press, Stanford, California.

CROTEAU, R. & JOHNSON, M. A. (1984). Biosynthesis of terpenoids in glandular trichomes. In: *Biology and Chemistry of Plant Trichomes* (Ed. by E. Rodriguez, P. L. Healey & I. Menta), pp. 133–185. Plenum Press, New York.

FEENY, P. P. (1968). Effects of oak leafs tannins on larval growth of the winter moth *Operophtera brumata*. *Journal of Insect Physiology*, **14**, 805–807.

FEENY, P. P. (1970). Seasonal changes in oak leaf tannins and nutrients as a cause of spring feeding by winter moth caterpillars. *Ecology*, **51**, 565–581.

FEENY, P. P. (1976). Plant apparency and chemical defence. In: *Biochemical Interactions Between Plants and Insects* (Ed. by J. Wallace & R. L. Mansell). *Recent Advances in Phytochemistry*, **10**, pp. 1–40. Plenum Press, New York.

FOSTER, R. B. & BROKAW, N. V. L. (1982). Structure and history of vegetation of Barro Colorado Island. In: *The Ecology of a Tropical Forest : Seasonal Rhythms and Long-term Changes* (Ed. by E. G. Leigh, A. S. Rand & D. M. Windsor), pp. 67–81. Smithsonian Institution Press, Washington, D.C.

GRIME, J. P. & HUNT, R. (1975). Relative growth-rate: its range and adaptive significance in a local flora. ecological and evolutionary theory. *American Naturalist*, **111**, 1169–1194.

GRIME, J. P. (1979). *Plant Strategies and Vegetation Processes*. J. Wiley & Sons, Chichester.

GRIME, J. P. & HUNT, R. (1975). Relative growth-rate: its range and adaptive significance in a local flora. *Journal of Ecology*, **63**, 393–422.

GRIME, J. P., MACPHERSON-STEWART, S. F. & DEARMAN, R. S. (1968). An investigation of leaf palatability using the snail *Cepaea nemoralis*. *Journal of Ecology*, **56**, 405–420.

GRUBB, P. J. (1977). Control of forest growth and distribution on wet tropical mountains: with special reference to mineral nutrition. *Annual Review of Ecology and Systematics*, **8**, 83–107.

GULMON, S. L. & MOONEY, H. A. (1986). Costs of defence on plant productivity. In: *On the Economy of Plant Form and Function* (Ed. by T. J. Givnish), pp. 681–689. Cambridge University Press, Cambridge.

JANZEN, D. H. (1974). Tropical blackwater rivers, animals and mast fruiting by the Dipterocarpaceae. *Biotropica*, **6**, 69–103.

KUČ, J. (1972). Phytoalexins. *Annual Review of Plant Pathology*, **10**, 207–232.

LEIGH, E. G., RAND, A. S. & WINDSOR, D. M. (Eds) (1982). *Ecology of a Tropical Forest : Seasonal Rhythms and Long-term Changes*. Smithsonian Institution Press, Washington, D.C.

LOVELESS, A. R. (1961). A nutritional interpretation of sclerophylly based on differences in the chemical composition of sclerophyllous and mesophytic leaves. *Annals of Botany*, **25**, 168–184.

MACARTHUR, R. H. & WILSON, E. O. (1967). *The Theory of Island Biogeography*. Princeton University Press, Princeton, New Jersey.

MANDELS, M. & REESE, E. T. (1965). Inhibition of cellulases. *Annual Review of Phytopathology*, **3**, 85–102.

MANSFIELD, J. W. (1982). The role of phytoalexins in disease resistance. In: *Phytoalexins* (Ed. by J. A. Bailey & J. W. Mansfield), pp. 253–288. John Wiley, New York.

MARTIN, J. S. & MARTIN, M. M. (1982). Tannin assays in ecological studies: lack of correlation between phenolics, proanthocyanidins and protein-precipitating constituents in mature foliage of six oak species. *Oecologia*, **54**, 203–211.

MARTIN, M. M. & MARTIN, J. S. (1984). Surfactants: their role in preventing the precipitation of proteins by tannins in insect guts. *Oecologia*, **61**, 342–345.

McKEY, D. G. (1979). The distribution of secondary compounds within plants. In: *Herbivores : Their Interactions with Secondary Plant Metabolites* (Ed. by G. A. Rosenthal & D. H. Janzen), pp. 55–133. Academic Press, New York.

McKEY, D. B. (1984). Interaction of the ant-plant *Leonardoxa africana* (Caesalpiniaceae) with its obligate inhabitants in a rainforest in Cameroon. *Biotropica*, **16**, 81–99.

McKey, D. B. & Gartlan, J. S. (1981). Food selection by black colobus monkeys (*Colobus satanus*) in relation to plant chemistry. *Biological Journal of the Linnean Society*, **16**, 115–146.

McKey, D. B., Waterman P. G., Mbi, C. N., Gartlan, J. S. & Strusaker, T. T. (1978). Phenolic content of vegetation in two African rain-forests: ecological implications. *Science*, **202**, 61–64.

Medina, E. (1983). Adaptations of tropical trees to moisture stress. In: *Tropical Rain Forest Ecosystems: A Structure and Function* (Ed. by F. B. Golley), pp. 225–237. Elsevier, Amsterdam.

Milton, K. (1979). Factors influencing leaf choice by howler monkeys: a test of some hypotheses of food selection by generalist herbivores. *American Naturalist*, **114**, 362–378.

Mooney, H. A. & Gulmon, S. L. (1982). Constraints on leaf structure and function in reference to herbivory. *BioScience*, **32**, 198–206.

Mooney, H. A., Gulmon, S. L. & Johnson, N. D. (1983). Physiological contraints on plant chemical defences. In: *Plant Resistance to Insects* (Ed. by P. A. Hedin), pp. 21–36. American Chemical Society Symposium, Series 208, Washington D.C.

Oates, J. F., Waterman, P. G. & Choo, G. M. (1980). Food selection by the south Indian leaf-monkey, *Presbytis johnii*, in relation to leaf chemistry. *Oecologia*, **45**, 45–56.

Owen-Smith, N. & Cooper, S. M. (1987). Classifying African savanna trees and shrubs in terms of their palatability for browsing ungulates. In: *Plant/Herbivore Interactions* (Ed. by J. Flinders & F. D. Provenza), pp. 43–47. Proceedings of the Fourth Annual Wildland Shrub Symposium.

Raupp, M. J. (1985). Effects of leaf toughness on mandibular wear of the leaf beetle, *Plagiodera versicolora*. *Ecological Entomology*, **10**, 73–79.

Rhoades, D. F. & Cates, R. G. (1976). Towards a general theory of plant anti-herbivore chemistry. In: *Biochemical Interactions Between Plants and Insects* (Ed. by J. Wallace & R. L. Mansell). *Recent Advances in Phytochemistry, Volume 10*, pp. 168–213. Plenum Press, New York.

Robinson, T. (1974). Metabolism and function of alkaloids in plants. *Science*, **184**, 430–435.

Southwood, T. R. E. (1977). Habitat, the templet for ecological strategies? *Journal of Animal Ecology*, **46**, 337–365.

Southwood, T. R. E., Brown, V. K. & Reader, P. M. (1986). Leaf palatability, life expectancy and herbivore damage. *Oecologia*, **70**, 544–548.

Specht, R. L. & Moll, E. J. (1983). Mediterranean-type heathlands and sclerophyllous shrublands of the world: an overview. In: *Mediterranean-type Ecosystems* (Ed. by F. J. Kruger, D. T. Mitchell & J. U. M. Jarvis), pp. 41–65. Springer-Verlag, Berlin.

Swain, T. (1979). Tannins and lignins. In: *Herbivores: Their Interaction with Secondary Plant Metabolites* (Ed. by G. A. Rosenthal & D. H. Janzen), pp. 657–682. Academic Press, New York.

Tanton, M. Y. (1962). The effect of leaf 'toughness' on the feeding of the larvae of the mustard beetle, *Phaedon cochleariae* Fabricus. *Entomologia Experimentalis et Applicata*, **5**, 74–78.

Van Soest, P. J. (1978). Dietary fibers: their definition and nutritive properties. *American Journal of Clinical Nutrition*, **31**, S12–S20.

Walker, J. R. L. (1975). *The Biology of Plant Phenolics*. Crane, Russak & Company, New York.

Waller, G. R. & Nowacki, E. K. (1978). *Alkaloids: Biology and Metabolism*. Plenum Press, New York.

Waterman, P. G. (1983). Distribution of secondary metabolites in rain forest plants: toward an understanding of cause and effect. In: *Tropical Rain Forest: Ecology and Management* (Ed. by S. L. Sutton, T. C. Whitmore & C. Chadwick), pp. 167–179. Blackwell Scientific Publications, Oxford.

Waterman, P. G., Ross, J. A. M. & McKey, D. B. (1984). Factors affecting levels of some phenolic compounds, digestibility, and nitrogen content of the mature leaves of *Barteria fistulosa* (Passifloraceae). *Journal of Chemical Ecology*, **10**, 387–401.

Zucker, W. V. (1983). Tannins: does structure determine function? An ecological perspective. *American Naturalist*, **121**, 335–365.

New Phytol. (1987) **106** (Suppl.), 265–281

THE COMPARATIVE ECOLOGY OF FLOWERING

By Q. O. N. KAY

Ecology and Evolution Research Group, School of Biological Sciences, University College of Swansea, Singleton Park, Swansea SA2 8PP, UK

Summary

The major roles of flowering are in sexual reproduction and in fruit and seed production. Flowering competes with vegetative growth and vegetative reproduction for resources, and involves heavy resource costs which are often much greater than those required for seed production alone. Interactions between plants dependent upon flowering may be competitive or mutualistic. Theoretical models suggest that these interactions may often be of primary importance in determining whether different species or genotypes coexist, or displace or mutually exclude one another, and may thus be responsible for the heavy investment in flowering, but very few cases have been analyzed satisfactorily. The existence of simultaneously or sequentially flowering guilds of species with similar floral characteristics may be a consequence of mutualistic interactions. Flowering strategies and the range of possible interactions dependent upon flowering are potentially very diverse but are constrained by the physical and biotic environment, sometimes very severely.

Key words: Flowering, resource allocation, pollination, competition, mutualism.

GENERAL INTRODUCTION

Flowering plays a central role in plant reproduction and the establishment, survival and spread of plant species and communities often depend upon its success. The ecological interactions that are related to flowering are correspondingly important and are also unusually diverse. Until recently, great interest in relationships between zoophilous plants and their pollinators has diverted attention from the interactions within and between plant species that are involved in flowering. The former interactions may take place at several levels; for example, at one level variation in floral structure, sex expression, compatibility relationships or timing of flowering may affect gene flow and recombination between individuals or subpopulations while, at another level, variation between individuals in frequency or intensity of flowering may affect the balance between vegetative and sexual reproductive success in the same plants. The latter interactions are also diverse and range from mutualism (facilitation; Rathcke, 1983), for example attraction or support of pollinating insects or birds by simultaneously or sequentially flowering species, to various types of competitive relationships which may lead either to character displacement or to exclusion or extinction of a species. These interspecific interactions may be of major importance both in interspecific competition and in the control of community structure, overriding the vegetative interactions that have more commonly been studied. It has recently been recognized by field biologists that interspecific interactions during flowering offer exceptional opportunities for testing many current hypotheses about general ecological and evolutionary processes (Real, 1983). Work in this field is now very active.

0028-646X/87/05S265 + 17 $03.00/0

THE COST OF FLOWERING

Introduction

Plant demographers have often regarded flowering simply as a necessary step towards seed production. Harper & White (1974) wrote 'The ideal speedy life-cycle might include a seed that germinates to expose a green flower that immediately proceeds to leave several seeds that germinate without delay. The flower would need to photosynthesize sufficiently to stock the new seeds with reserves and to support a root for the necessary mineral uptake.' Although a few species capable of rapid seed production during brief periods favourable for growth may exceptionally approach this ideal in suitable circumstances, for example *Chenopodium rubrum** growing in 8 h days (Harper, 1977), even desert ephemerals commonly possess flowers that have little or no effective photosynthetic capacity and show many features that go well beyond the minimum requirements for seed production. Indeed, it might equally be argued that the ideal speedy life-cycle for a plant would exclude the costly process of flowering and would simply involve rapid vegetative multiplication of a plant consisting of a single basic photosynthetic unit – a frond, thallus or cell. Life-cycles largely or entirely of this type are common in algae. Among flowering plants, they are approached only in a few floating aquatic species in *Lemna* and related genera, in which flowering is rare or absent. However, predominantly or entirely vegetative reproductive strategies are fairly common in herbaceous perennial angiosperms, especially among grasses, sedges and aquatics (Richards, 1986). In habitats where flowering is unusually disadvantageous or unsuccessful, some species or genotypes may flower little if at all, for example *Cymodocea serrulata* (R. Br.) Aschers. & Magnus and some other marine angiosperms (den Hartog, 1970; Kay, 1971c), or may have the flowers partly or completely replaced by bulbils or plantlets, for example *Polygonum viviparum* (Law, Cook & Manlove, 1983) and the grasses *Festuca vivipara* and *Deschampsia alpina*.

Flowering and any subsequent seed production require the diversion of resources from vegetative growth (Evenson, 1983) and may be expensive not only in terms of energy requirements but also in other respects. Genotypes or species in which flowering or seeding is reduced or absent may thus, at least for a period, have an advantage in vegetative competition with species which flower or fruit more freely. In some cases where comparison is possible, the cost of flower production alone appears to be much less than the additional cost of fruiting. Fruit trees, for example, show reduced growth and depressed vigour after a year of heavy fruit yield but not after years in which such yield was prevented by frosts or other factors following equally heavy flowering. In several dioecious species, the cost of flowering is much less than half the total cost of fruiting if fruit set is high but may be greater if fruit set is reduced. In these and other allogamous species, the absolute costs of flowering appear to be higher than in autogamous species. The cost of flowering relative to that of fruiting will increase if fruiting is reduced by external or internal factors. In any case, it is clear that the direct and indirect costs of flowering may represent a heavy burden during the peak flowering period, especially in allogamous species.

---

* Except where authorities for binomials are given, nomenclature follows Clapham, Tutin & Warburg (1981).

Allocation of resources in dioecious species

Dioecious species, in which males usually flower more profusely and for a longer period than females but do not fruit, often show sexual niche specialization and other adaptations which have been interpreted as reducing the advantage of the non-fruiting males in intraspecific competition, although other interpretations are possible (Kay & Stevens, 1986). Measurements of relative growth and resource allocation support the assumption that fruiting females allocate a much greater proportion of resources to reproduction than do males. This is despite intrasex competition among males resulting in heavy and prolonged flowering and, when allocation to flowers has been measured separately, females allocating less to flowers than males. In the subdioecious tree *Fraxinus excelsior*, Rohmeder (1967) observed that females showed substantially smaller growth increments than males, possibly as a consequence of the added burden of resource allocation to seed production. In the dioecious shrub *Simmondsia chinensis* (Link) Schneider, a greater proportion of resources was allocated to reproduction to females than in males when seed set was greater than 30% (Wallace & Rundel, 1979) and, in the dioecious herb *Silene alba*, a similar effect occurred when seed set was greater than 20% (Gross & Soule, 1981). In the stoloniferous dioecious herb *Fragaria chiloensis* (L.) Duchn., more resources are devoted to sexual reproduction in females (8·8% dry weight) than in males (1·5%), although female inflorescences before fruiting were only about half as costly as male inflorescences. Higher allocations to male functions than to female functions at anthesis have also been found in many non-dioecious species (e.g. Lovett Doust & Harper, 1980; Devlin & Stephenson, 1985). Outbreeders also show higher relative allocations than inbreeders (Lovett Doust & Cavers, 1982). The cost of the male function may be even greater than that indicated by comparison of biomass because of the high metabolic cost of pollen production, which is possibly greater than that of producing female parts or seeds (Goldman & Willson, 1986).

Allocation of costs to structures other than fruits and seeds

High reproductive investments in structures other than fruits or seeds occur in some species with hermaphrodite flowers. In the annuals *Senecio vulgaris* (Harper & Ogden, 1970) and *Chrysanthemum segetum* (Howarth & Williams, 1972) flowers (inflorescences) comprise a surprisingly high proportion of the dry weight of the plant during the peak flowering period. In the autogamous *S. vulgaris*, which had inconspicuous flowers, the peak percentage allocation of dry weight to flowers was about half the final allocation to seeds but, in the allogamous, showy flowered *C. segetum*, it was slightly greater than the final percentage allocation to seeds. Partly analogous comparisons between autogamous and allogamous flowers were made by Schemske (1978) and Waller (1979, 1980) who compared allocations of energy to cleistogamous and chasmogamous flowers in *Impatiens capensis* Meerb. Chasmogamous seeds were two to three times as costly as cleistogamous seeds. Pitelka (1977) noted high relative costs of flowering in studies of the caloric content and dry weight of different organs in three *Lupinus* spp. In the annual *L. nanus* Dougl., the total allocation to reproductive effort was 61%, with 29% in seeds but, in the perennial *L. variicolor* Steud., there was 18% reproductive effort with only 5% in seeds and, in the shrub *L. arboreus* Sims, 20% reproductive effort with only 6% in seeds. It should be remembered that measurements of allocations of dry weight may substantially underestimate the metabolic costs of flowering.

Secretion of nectar, in particular, can represent a large, undetected investment. In *Asclepias syriaca* L., for example, up to 37% of daily photosynthate may be secreted as nectar sugar, with a total of 2·8% of photosynthetically assimilated carbon being routed through nectaries during the lifetime of the plant. In *Medicago sativa*, the allocation to nectar may be about 20% of the final dry weight energy content of the plant and thus almost twice that allocated to the total seed crop (Southwick, 1984).

High relative costs also occur in sexually reproducing *Ranunculus* spp. (buttercups). In *R. bulbosus*, the peak allocation of dry weight to flowers and flowering stems before achene set was 42%. Later in the flowering period the allocation was 34 to 35%, with only a further 14 to 17% in achenes. In *R. acris*, allocation to flowers and flowering stems was 43 to 46%, to achenes only 10 to 14% (Sarukhán, 1976). At the other extreme, in the palm *Astrocaryum mexicanum* Liebm., *c.* 35% of total dry matter was allocated to reproduction, of which 30% was in fruit and only 5% in other structures involved in flowering (Sarukhán, 1980).

Differential allocation of resources to sexual and vegetative reproduction

The resources allocated to sexual reproduction have been compared with those allocated to vegetative (asexual) reproduction in many herbaceous perennial species, (Evenson, 1983) both from different habitats and between related groups of species. In general, allocation of resources to vegetative reproduction is less variable than that to sexual reproduction within a species. Differences between species in total reproductive effort, which in some cases are very great, have often been interpreted in terms of *r*- and *K*-selection theory, sometimes with difficulty. However, higher allocations to total reproductive effort have usually been found, as expected, in disturbed or seral habitats, and higher allocations to vegetative reproduction in closed communities than in open sites. Although differences between species in breeding system, in relative allocation to flowering and fruiting, in growth patterns and in means of vegetative reproduction sometimes make comparisons difficult, these studies have provided much of value in assessing the absolute and relative costs of flowering in relation to vegetative growth and vegetative reproduction. For example, the allocation of dry weight to vegetative reproduction in *Circaea quadrisulcata* (Maxim.) Franch. & Sav. was constant at about 12% (Struik, 1965). In a habitat in which it flowered and fruited abundantly, the allocation to sexual reproduction was about half as great. In a mesic habitat in which flowering and fruiting were reduced, it was only a quarter as great. In contrast, Sarukhán (1976) found allocations to total reproduction (vegetative and sexual) of about 48 to 60% in each of three species of buttercup, two of which (*R. bulbosus* and *R. acris*) only reproduced sexually, while the third, *R. repens*, reproduced mainly vegetatively (its allocation to sexual reproduction was 0 to 9%). Hancock & Bringhurst (1980) recorded an even higher allocation to vegetative reproduction, about 86% in both sexes, in the dioecious strawberry, *F. chiloensis*, which has a stoloniferous pattern of growth similar to that of *R. repens*. Allocations to sexual reproduction in *F. chiloensis* were only 8·8% in females and 1·5% in males. However, the very high allocations to vegetative reproduction found here should be interpreted with caution. It is debatable whether stoloniferous growth in species like *R. repens* and *F. chiloensis*, extreme examples of mobile perennials in which short-lived modular ramets are produced in sequence as a genet moves through an open habitat, can be equated with vegetative spread by a long-lived fixed perennial species.

The comparative study of the reproductive strategies of five perennial composites made by Bostock & Benton (1979) helped to clarify relative allocations to vegetative and sexual reproduction in species with a range of patterns of growth. They investigated *Achillea millefolium*, *Artemisia vulgaris*, *Cirsium arvense*, *Taraxacum officinale* and *Tussilago farfara*. Total allocations of dry weight to reproduction ranged from 11 % in *A. vulgaris* to 46 % in *T. farfara*., with the allocations to sexual reproduction ranging from only 2 to 3 % in *A. millefolium* and in the wind-pollinated *A. vulgaris* to 7 % in the dioecious *C. arvense* and 24 to 26 % in *T. farfara* and *T. officinale*. (Although *T. officinale* is agamospermous, flower and seed production was regarded as sexual reproduction for the purposes of this comparison.) The dry weight allocations to vegetative reproduction ranged from 9 % in *A. vulgaris* to 13 % in *T. officinale*, 15 % in *C. vulgare*, 20 % in *T. farfara* and 26 % in *A. millefolium*. As in several other cases, allocations to vegetative reproduction were relatively constant while allocations to sexual reproduction varied.

The study of the balance between vegetative and sexual reproduction in five populations of *Mimulus primuloides* Benth. at different altitudes made by Douglas (1981) showed intraspecific variation in reproductive strategy and resource allocation. The vegetative reproductive biomass was greatest (56 %) at an intermediate altitude; it was 41 % at the lowest site and 46 % at the highest. Allocation to sexual reproduction was small at the four lower sites, ranging from 2 to 4 %, but a high rate of flower production appeared to be genetically fixed in the population growing at the highest altitude, where allocation to sexual reproduction was 14 %, although establishment of seedlings was apparently rare.

Indirect costs of flowering

In addition to the direct resource costs of flowering, flowering may also involve additional indirect costs, especially in species with showy zoophilous inflorescences. The production of inflorescences usually disrupts or terminates the normal pattern of vegetative growth of a shoot. Flowers may reduce the photosynthetic efficiency of the plant by shading the leaves and reducing the water supply to photosynthetic tissues. The weight and wind resistance of flowers, especially when wet, may break stems or cause wind-throw. Fallen petals or dead flowers may accumulate on the plant and act as foci for infection. Conspicuous flowers may attract the attention of phytophagous insects or vertebrate grazers. Heavy or clumsy pollinators, especially birds, may damage the plant and pollinators may sometimes act as vectors for pathogens or pests. Clearly, flowering, considered as part of an overall strategy of growth and reproduction, involves very substantial costs which must be balanced by corresponding selective advantages.

THE ROLES OF FLOWERING

Flowering is generally agreed to have two major roles. The first is in sexual reproduction; meiosis, formation of male and female gametophytes, and fertilization take place in the flower, resulting in genetic recombination. Pollen grains containing the male gametophytes may be dispersed from flower to flower by various means, thus enabling gene flow between individuals. Viral infections may also be eliminated. No satisfactory alternative to sexual reproduction exists for the generation and recombination of variation in flowering plants, although some genetic material may occasionally be transmitted by viral vectors or other means and, in some agamospermous (seed apomict) groups, high rates of somatic mutation may generate variation (e.g. Mogie, 1985). However, flowering plants show a wide range of breeding systems, in some of which sexual reproduction is reduced, modified or even completely eliminated (Richards, 1986).

The second major role of flowering is in fruit and seed production. Seeds and the fruits in which they may be contained have important (although not necessarily essential) roles in dispersal and establishment, in gene flow within and between populations, in perennation and in survival through exceptionally unfavourable conditions. They may also produce large numbers of genetically different individuals from which successful genotypes may be chosen by natural selection. Seed production is often retained when sexual reproduction has been modified or abandoned, for example in autogamous species or cleistogamous flowers and in subsexual groups (e.g. *Rosa canina sensu lato*) and agamospermous groups (e.g. *T. officinale*). Various forms of vegetative reproduction exist as alternatives to seed production and may partly or completely replace it in iteroparous perennial plants, especially in habitats or at stages of habitat or population development in which flowering is disadvantageous or ineffective, but seed production is an essential stage in the life-cycles of annuals and semelparous perennials (Harper, 1977).

Flowering can also play a number of supplementary roles that are normally, but not necessarily, combined with its major roles of sexual reproduction and seed production. These mainly relate to intra- or interspecific competition in zoophilous species. For example, flower production may reduce the reproductive output of competing species, or of competing individuals within the same species, by attracting so many of the available pollinators that seed set is reduced in competitors (exploitative competition) or flowers may provide habitats for flower or seed predators or pathogens that have a more detrimental effect on competitors. Genetically incompatible pollen may be transferred to the stigmas of competitors, thus reducing the success of compatible pollen (improper pollen transfer; Rathcke, 1983). Cross-pollination of competitors may produce inviable or sterile hybrid zygotes. It has occasionally been suggested that the flowers of some agamospermous species which retain their zoophilous adaptations (e.g. *T. officinale*) may still perform some of these roles. There is increasing evidence that some or all of these mechanisms may be involved in competitive interactions and act as selective forces causing character displacement. On the other hand, similar interactions often appear to be positive rather than negative, leading to the production of guilds of species with similar flowering characteristics (see below).

Flower production can also be advantageous in some minor roles that have no direct connection with reproduction. Many occasionally or habitually nectar-feeding birds are also insectivorous and may remove harmful insects after initial attraction by nectar-producing flowers, for example warblers and blue tits feeding on *Salix* spp. (Kay, 1985). Similar effects may occur in nectarless spring-flowering anemophilous trees in which the male flowers may be the food item initially attracting partly insectivorous birds. On balance this may be advantageous. The remains of inflorescences may provide protection against herbivores or wind, as for example in *Euphorbia acanthothamnos* Heldr. & Sart. and other shrubs of arid habitats and in many grasses, or when shed may contribute to allelopathic interactions (Harborne, 1982). The flowers of distasteful species may sometimes act as aposematic warning signals to herbivores; for example, it is possible that yellow-flowered legumes, which have flowers closely resembling one another in the spectrum visible to vertebrates but which differ in UV patterns visible only to their insect pollinators, may form an aposematic guild of distasteful Müller mimics recognized partly by their flowers and associated with some palatable Batesian mimics (Kay, 1987). Minor roles like these are perhaps relatively rare but, when they occur, they may be of considerable ecological importance, especially when other factors are evenly balanced.

Strategies and Interactions

Wind pollination

Biotic pollination (zoophily) requires the attraction, usually by rewards, of suitable animal pollinators drawn from a range of potential pollinators which, although they have widely different characteristics and requirements, actively transport pollen in a wide variety of ways. In contrast, wind pollination (anemophily) is a relatively passive and uniform process controlled primarily by microclimatic factors. The adaptations of wind-pollinated flowers which tend to be correspondingly uniform most commonly include the production of large numbers of pollen grains with appropriate aerodynamic characters by inflorescences which are well exposed to wind. A corresponding exposure of stigmatic surfaces traps airborne pollen with maximum efficiency (Niklas, 1985; Niklas & Buchmann, 1985). Dioecy, self-incompatibility or the separation of male and female functions in time, and the release of pollen at a time of day and season when dispersal to conspecific plants are also optimal. For effective pollination, populations of wind-pollinated species must be relatively closely spaced and must, at least at the time of flowering, grow in a community with a relatively open structure that minimizes loss of pollen on non-stigmatic surfaces and is accessible to wind. There must be sufficiently long periods of relatively low humidity and suitable winds, preferably indicated by clear environmental cues that co-ordinate flowering (Whitehead, 1983). These requirements impose severe constraints on the range of strategies of flowering, on population structure and on the suitability of habitats for wind-pollinated species. Further constraints are imposed by the comparative inefficiency of wind as a pollen vector (most wind-pollinated species produce few seeds per flower, often only one relatively large seed) and its unidirectional nature (upwind plants or populations may receive no pollen from elsewhere and gene flow along ecological gradients may be impaired).

Within a species, the seasonal timing of release of anemophilous pollen may be governed by community structure. Species of deciduous woodland normally flower prevernally. Species of grassland and other open habitats have a wider range of flowering times and the flowering period of a species of this type is often determined by the annual phasing of growth in that species, with a peak in early summer in Britain. Low-growing species, however, typically flower early in the year, before taller herbs appear, for example the diminutive grass, *Mibora minima*, which flowers in February, and the small sedge, *Carex caryophyllea*, which flowers in April in Britain. In a uniform habitat, the effects of selection for simultaneity of flowering will be reduced both by the uncertainties of the weather (in some years individuals flowering earlier or later than the norm may be at an advantage) and by sequential maturation of male and female flowers or organs, especially in species lacking self-incompatibility. In a non-uniform habitat, selection for the reduction of gene flow between genotypically differentiated habitat forms may lead to divergence in the periods of flowering of the different genotypes (McNeilly & Antonovics, 1968), or to autogamy in local genotypically distinct populations. If genotypic variation is clinal, flowering periods may also be clinal in time, either as a consequence of selection or as a more or less automatic consequence of climatic gradients, for example along an altitudinal range.

Few interactions between species mediated by flowering can take place in wind-pollinated species. Although it is theoretically possible for inappropriate pollination to reduce the success of compatible pollen in wind-pollinated species, the small numbers of pollen grains that are normally intercepted by the stigma are present at such low density on the stigmatic surface that interaction of this type is unlikely to have a significant effect. The chief potential form of interaction is probably hybridization between related species. Very low levels of hybridization will not reduce reproductive output perceptibly and may help to generate variation by enabling interspecific gene flow to take place. This is thought to be the main source of variation in the mainly zoophilous or autogamous New Zealand species of *Epilobium* (Raven & Raven, 1976). But if significant numbers of inviable embryos or hybrid seeds are produced, effective reproductive output will be reduced. If hybrid zygotes are formed equally freely by two species hybridizing in this way, the effects will be more severe in the species that contributes fewer pollen grains at a given site, and this minority species may eventually be eliminated. If hybrid zygotes are formed less freely by one species, the point of balance will be shifted in its favour. Hybridization of this type is particularly likely to take place in mixed stands of wind-pollinated species because the wind disperses pollen grains impartially, and because physical and seasonal factors constrain many species to flower simultaneously. If it is not reduced or prevented by divergence in flowering time (if this is possible), either during the day [as described by Philipson (1937) for *Agrostis* spp.] or during the year, or prevented by interspecific incompatibility barriers (Levin, 1971), the species may have mutually exclusive distributions as a consequence of the instability of mixed stands.

Biotic pollination

General introduction. Biotically pollinated flowers show an astonishingly wide range of adaptations for pollination, and a correspondingly wide range of pollin- ation strategies. In zoophilous species with appropriate adaptations, large numbers of pollen grains can be transferred between scattered individual plants reliably and rapidly. However, in order to attract and reward (or sometimes trap or deceive) animal pollinators high investments are often required, especially in obligate outbreeders. Although potential animal pollinators are usually available in terre- strial habitats suitable for the growth of vascular plants, the diversity and abundance of these pollinators varies from site to site and season to season, and several specialized groups of pollinators have limited geographic or ecological distributions. The range of potential pollinators is particularly small on isolated oceanic islands and in some exposed or marginal habitats, and this may impose severe constraints on the breeding systems of species in these areas.

Even in favourable, biotically diverse habitats, for example temperate or tropical forest or submontane grassland, varying degrees of interaction for the services of pollinators usually exist. These interactions may be positive (facilitation or mutualism) or negative (competition), or may combine positive and negative features. For example, plants acting together may both support and attract pollinators in local areas (e.g. Thomson, 1981, 1982). Yet individuals may compete with one another for the services of these pollinators (Rathcke, 1983). Competition will tend to increase the resource allocations that are required for the attraction and reward of pollinators. The heterogeneity of the environment, both in space (microtopographic and biotic diversity) and in time (variation in weather conditions, seasonal changes and biotic succession) leads to variation and uncertainty in the availability of pollinators. This may also increase the resource allocations that are required for successful flowering, both by reducing the proportion of flowers that are adequately pollinated, and possibly by selection for more costly multiple floral adaptations that attract a wider range of pollinators. However, one of the most striking features of the interactions among plant species that share pollinators is the high frequency of positive interactions, and this may also be in part a consequence of environmental heterogeneity.

Competitive (unilaterally or bilaterally negative), mutualistic (unilaterally or bilaterally positive) or parasitic (positive on one side, negative on the other) interactions between zoophilous plants during pollination can take place in a number of ways, and a variety of terms have been adopted to describe some of the possible interactions of this type (Rathcke, 1983; Waser, 1983b). These mechanisms can operate at one or more of several levels; at the levels of pollinator attraction, modification of pollinator behaviour, pollinator support, nectar or pollen removal, floral damage or modification, pollen transfer, stigma function and pollen growth, zygote formation, or at successive post-zygotic stages.

Competitive interactions. The simplest type of potentially competitive interaction occurs when a pollinator visits two or more species indiscriminately, thus reducing the amount of pollen transferred to conspecifics. In an interaction that may be of this type, the presence of *Claytonia virginica* L. reduces outcrossing and pollen flow in *Stellaria pubera* Michx. (Campbell & Motten, 1981; Campbell, 1985). However, depletion of nectar in an interaction of this type may have the opposite effect, leading to increased pollen flow because pollinators have to forage more extensively to secure the same reward. If a pollinator finds one species or genotype within a species more attractive than another, and so visits it more, it has often been presumed that the less attractive species will consequently suffer reduced relative reproductive success (Waser, 1983b). In practice, the consequent discrimination means that this may not be the case. The dispersal of pollen in, for example, an attractive and abundant species visited preferentially by *Apis* may be relatively ineffective in ensuring reproductive success for the species if, as a result, it becomes the only one visited by the primary pollinator. Meanwhile, casual or secondary pollinators which may be more effective in interplant pollen dispersal (for example *Bombus* spp. or Lepidoptera) will continue to visit the less attractive species. The reproductive success of the species that is less attractive to the primary pollinator may then actually increase, either because of decreased pollen depletion by the primary pollinator with a constant level of visitation by the secondary pollinators, or because the 'less attractive' species, no longer depleted of resources by the primary pollinator, becomes more attractive to the secondary pollinators. Complex, dynamic and often apparently paradoxical interactions of this type are probably common at this level, although few cases have been analyzed.

Competitive interactions between, or sometimes within, species which are mediated by pollen transfer are also potentially important and their possible consequences are more predictable. In the simplest case, pollinators may visit two or more species without discrimination, transferring pollen between plants and depleting pollen reserves equally. If the transfer of equal quantities of compatible pollen is required for full seed set by each plant species in the mixture, the component that contributes the smallest proportion of pollen (normally the minority component) will be the first to have its seed set reduced if the level of pollinator visitation to the mixture is insufficient. Mixtures of this type may be unstable, with the minority component being eliminated in a few generations (Levin & Anderson, 1970). Waser (1978b, 1983b) considers, however, that this type of interaction is unlikely to lead to actual extinction except when plant species share edaphic requirements as well as pollinators. Another possible negative interaction dependent upon pollen transfer occurs when the presence of foreign pollen on the stigma reduces the success of compatible pollen, either through reduction of the receptive surface of the stigma or as a consequence of physiological interference with the growth of compatible pollen. This may reduce reproductive success even when pollinator visitation is not limiting. Interactions between related species or cytotypes in which cross-pollination results in the formation of inviable hybrid zygotes or hybrid seed may have a similar effect. Again, minority forms will be at a greater disadvantage unless crossability differs (Lewis, 1961; Levin & Schaal, 1970), although hybridization may, of course, be an important source of genetic variation (e.g. Raven & Raven, 1976).

Competitive reproductive interactions of these and other types will tend either to lead to evolutionary divergence that reduces competition for pollinators [for example, by divergence in flowering time or other floral characteristics (Robertson, 1924; Waddington, 1979; Waser, 1983b)] or to differential extinction (MacArthur & Levins, 1967) and mutually exclusive patterns of distribution. Such patterns appear to exist in some groups of outbreeding annual weeds with similar flowers, for example among mayweeds, *Anthemis* and *Matricaria* spp. and related genera, which have closely similar daisy-like flowers visited mainly by Diptera (Kay, 1969, 1971a, b), and among annual *Brassica* and *Sinapis* spp. and related genera, which have closely similar yellow cruciferous flowers visited by bees, syrphids and Lepidoptera (Ford, 1986).

Mutualistic interactions. Positive, mutualistic interactions (and also, for the parasites, parasitic interactions) may have the opposite of the above effects, promoting sympatry and increasing the degree of similarity of the flowers of different species, although in these cases various adaptations may reduce inter-specific pollen transfer or prevent hybrid zygote formation. Despite the diversity of floral adaptations and pollination syndromes that exist among zoophilous flowers, a striking feature of many species-rich, natural plant communities is that many zoophilous species, often the majority, fall into a limited number of groups, each of which shares a common basic type of flower (Kugler, 1970). In some cases many or most of the members of such a group (guild) belong to the same family, for example yellow Fabaceae or white-rayed, daisy-like Asteraceae. However, it is common for a group to include species which are unrelated to one another yet have flowers which are similar, perhaps closely similar, in appearance and reward although they have completely different anatomical structures, for example the yellow bowl-flowers of *Ranunculus* spp., *Potentilla* spp., *Blackstonia perfoliata* and *Crepis capillaris*, or the purple or red zygomorphic flowers of many members of the Lamiaceae and Scrophulariaceae, or the yellow or red ornithophilous brush-flowers of members of the Myrtaceae, Mimosaceae, Proteaceae and other families, or, at a different level, the unusual yet extraordinarily similar flowers of the simultaneously flowering and sympatric endemic Cretan chasmophytes *Procopiania cretica* (Willd.) Guşuleac (Boraginaceae) and *Petromarula pinnata* (L.) A. DC.

(Campanulaceae). This pattern of variation has often been interpreted in terms of pollination syndromes representing adaptations to particular types of pollinator (Proctor & Yeo, 1973; Faegri & van der Pijl, 1979). This interpretation provides a convenient conceptual and nomenclatural framework for the classification and discussion of floral adaptations but it is to a large extent based on assumptions rather than observations. While the concept of pollination syndromes can often be useful, it should not be interpreted too literally. The floral guilds that can be observed in the field, for example the various types of yellow and white bowl-flower, are more numerous than the limited range of assumed pollination syndromes and may combine their characters in a variety of ways. As Wyatt (1983) has pointed out, recent observations show that even many apparently highly specialized flowers receive visits from a diversity of pollinators. Less specialized flowers are normally visited by a range of pollinators which may vary in time and with site and are sometimes completely inappropriate to their presumed pollination syndrome (e.g. Waser, 1978a, 1979).

An alternative interpretation of the existence of florally similar guilds of species that do not coincide with specialized pollination syndromes is that they are, at least in part, a consequence of mutualistic (facilitative) interactions between species that share the same, perhaps rather diverse, range of pollinators, a range that may vary at different times and in different places. If these pollinators show some degree of flower constancy (Waser, 1986), mutualistic interactions may lead to selection for similarity, but not for specialization. Eventually, a series of guilds, distinct in appearance but each with generalized adaptations for pollination, could be produced. Proctor & Yeo (1973) cite the resemblances between yellow-flowered buttercups (*Ranunculus* spp.), cinquefoils (*Potentilla* spp.) and *Helianthemum nummularium*, and between white-flowered *Ranunculus alpestris*, *Dryas octopetala* and *Leucanthemopsis alpina* (L.) Heywood as possible examples of resemblances due to this type of interaction. The number of species or individuals in a guild arising in this way may be limited because mutualistic interactions of this type are often rather delicately balanced and may become competitive if floral density increases or if pollinator numbers fall.

In the simplest model of a mutualistic interaction of this type, florally similar members of a guild, flowering simultaneously and providing similar resources, jointly attract, condition or support pollinators that are shared between the members of the guild. If it is assumed that pollinators are foraging optimally, this model can provide a stable, mutualistic interaction only when resources are so limited that pollinators must move between members of the guild and can be replaced by competitive interactions as density of resources increases (Rathcke, 1983). This condition also applies to intraspecific interactions. Observations of foraging behaviour, pollinator energetics and the availability of floral rewards in natural communities have led to the proposal (Heinrich & Raven, 1972; Heinrich, 1975a, b), which is now widely accepted (Waser, 1983a), that floral rewards within species are indeed limited, being great enough to maintain the desired pollinator populations but not so great that they are satiated by one or a few plants without moving between conspecifics. If this proposal is correct, and it is supported by many field observations of rapid depletion of nectar and pollen even in abundant species (e.g. Montgomerie & Gass, 1981; Corbet *et al.*, 1984; Kay *et al.*, 1984), mutualistic interactions of this type under resource-limited conditions may be common. However, it is difficult to demonstrate the increased fitness that should result, and very few examples of floral mutualism have been fully analyzed and documented. Little (1983) cites only three, although Dafni (1984) refers to about 20 possible cases, some specific and others more general (e.g. Proctor & Yeo, 1973).

Mutualistic interactions will be strengthened if the resource provision by the species or genotypes that are involved is sequential in time, either during the day, or during the season (e.g. Waser & Real, 1979), or even from one year to another (Rathcke, 1983). Close floral similarity will, however, be favoured only if the same individual pollinators are involved and they are able to retain the search images that they have formed for earlier species in the flowering sequence. A different type of mutualistic interaction may take place between species or genotypes that supply complementary resources, particularly nectar and pollen (e.g. Thomson, 1978), although other resources, for example basking places, may sometimes be involved. This type of interaction may be important for the pollination of some zoophilous dioecious species in which male flowers have pollen as the chief or only attractant, e.g. *Viscum album* L. (Kay, 1986). However, it is probably of only minor significance in relationships between species and has received little attention (Kevan & Baker, 1983).

The floral similarities between groups of plants that gain mutual advantage by jointly attracting or supporting pollinators are analogous to Müllerian mimicry. Although this term has sometimes been used to describe them it is rather inappropriate (Little, 1983). Unilateral relationships in which a species that may provide no reward obtains pollinator service by mimicking more rewarding species, analogous to Batesian mimicry, are called floral food deception mimicries by Little (1983) and resource parasitism by Rathcke (1983). Mimicry of rewarding species by orchids that provide no reward have been described for *Oncidium* spp. by Nierenberg (1972), for *Epidendrum* spp. by Boyden (1980) and Dodson (unpublished data, cited by Little, 1983) and for *Orchis* by Dafni & Ivri (1981); several more possible cases are cited by Little (1983) and Dafni (1984). Mimicry of a more rewarding species by a less rewarding species, which may be of more general occurrence and intergrades with mutualism, has been suggested for *Euphrasia micrantha* and *Calluna vulgaris* by Yeo (1968) and for two desert annual species, *Mohavea confertiflora* Jepson and *Mentzelia involucrata* S. Wats. by Little (1983). Mimicry of this type may also exist within species, for example, mimicry of rewarding male flowers by unrewarding female flowers in the dioecious *Carica papaya* L. (Baker, 1976) and of nectar-bearing by nectarless flowers in *Lobelia cardinalis* L. (Brown & Kodric-Brown, 1979).

ECOLOGICAL CORRELATIONS AND CONSTRAINTS

Broad ecological correlations between the flowering adaptations and the habitat of a species or group of species are often apparent, as are the constraints that may preclude or reduce the effectiveness of particular flowering adaptations in unsuitable habitats (Regal, 1982). Wind pollination is most common among the dominant species of temperate forests, especially at high elevations and in cooler areas, and among the dominant species of grassland or grassland-like (rush and sedge) communities. Here, the disadvantages of the inefficiency of wind as a pollen transfer mechanism are minimized because these communities typically have few dominant species, with individuals of the same species fairly closely spaced, producing few, relatively large seeds, and with growth forms that enable them to expose flowers to the wind. Its advantages are correspondingly maximized, especially its ability to transfer pollen between large numbers of individuals at a time of year or in sites where potential animal pollinators may be absent, or present in insufficient numbers for the requirements of a dense population of flowering plants. Less abundant species in the same communities and understorey plants in temperate forest tend to be biotically pollinated, especially those in which individuals are widely spaced and those which produce large numbers of seeds per flower. If pollinators are characteristically absent or scarce, autogamy, agamo-spermy and vegetative reproduction are likely to be more frequent among the less abundant species.

Biotic pollination is predominant in tropical forest, in many temperate shrub communities of mesic and semi-arid habitats (e.g. species-rich heathland and Mediterranean maquis or garrigue) and to a lesser extent in communities of open or disturbed habitats (e.g. steep slopes, cliffs and river banks, and also ruderal sites) and in aquatic or marshland communities. Here, greater abundance and diversity of both plant and animal species, wider spacing of individuals and requirements for high reproductive capacity are obvious factors favouring biotic pollination.

Similar correlations exist between environmental factors and the periods of flowering of species and communities (Rathcke & Lacey, 1985). In temperate, deciduous forests, the majority of species, both among the wind-pollinated dominant trees and the biotically pollinated field-layer species, flower during the short pre-vernal period before the leaf canopy opens. For the field-layer species, this coincides with their period of maximum photosynthesis and growth. Competition for the services of the pollinators that are available is potentially intense and the opportunities for sequentially mutualistic relationships among field-layer species are reduced by the shortness of the flowering period. Relationships of this type may, however, exist between biotically pollinated minor tree and liane species in temperate forests which flower over a much longer period, often in sequence, and may or may not show close floral resemblance. An example of a potentially mutualistic sequence of this kind, from early spring to autumn among native woodland species in southern England, is the sequence formed by *Prunus spinosa*, *P. avium*, *Crataegus monogyna*, *Sorbus* spp., *Rosa* spp., *Euonymus europaeus*, *Rhamnus catharticus*, *Acer compestre*, *Ilex aquifolium*, *Tila platyphyllos*, *T. cordata*, *Rubus* spp., *Clematis vitalba* and *Hedera helix*. Early-flowering field-layer herbs, e.g. *Anemone nemorosa*, may form part of this sequence, which is ultimately limited by frost in spring and autumn. Similar sequences may occur among the biotically pollinated herbs and shrubs of temperate mesic grassland and heathland communities, and in arctic and alpine habitats in which flowering is often dependent on the seasonal phasing of growth of each species in relation to temperature, water stress or edaphic factors related to waterlogging.

Climatic and biotic constraints, sometimes of rather different kinds, are also important in seasonally arid Mediterranean habitats and in tropical forests. In both habitats biotic diversity is very great, there is a wide range of both plants and pollinators, and high degrees of specialization to a particular pollinator are rather frequent. Plants dependent upon a particular species of pollinator, for example orchids in the genus *Ophrys*, may thus be limited by the ecological and geographic range of that pollinator. In Mediterranean habitats, summer aridity constrains many species to flower during the autumn, winter and spring. Both the floral adaptations and the geographic and altitudinal ranges of winter-flowering species may be limited by the availability of pollinators. Fruiting in the winter-flowering dioecious Mediterranean subshrub *Ruscus aculeatus*, for example, is sparse as a consequence of inadequate pollination at the northern limit of its range in Britain

(Kay & Page, 1985). A conspicuous feature of many Mediterranean habitats is the number of species that flower simultaneously or nearly so, especially during the spring when many of the species involved are annuals and many interactions mediated by flowering are possible.

In tropical forests, the structure of the community imposes unusual constraints on pollination systems so that those of tropical trees are of unusual interest. Simultaneous mass flowering of the individuals of a scattered species is a frequent system. Highly opportunistic bird and bat pollinators may disperse locally to areas of abundant floral resources (Rathcke & Lacey, 1985). Pollinator populations may show rapid responses to flowering. In Malaysia, where groups of dipterocarp tree species flower together every seven to 10 years, the thrips that are their major pollinators may increase in abundance more than 1000-fold as the first species flowers and then be available for subsequently flowering species (Chan & Appanah, 1980). Mutualistic interactions between tree species may occur in both cases.

Constraints on flowering and its success in the extremely hot and arid climate of the southern fringe of the Saharan desert and in the extremely exposed and rainy climate of the Faeroe Islands have been discussed by Hagerup (1932, 1951). In the Faeroes, the usual pollinators of many biotically pollinated species are scarce or absent, and Hagerup reported that many normally outbreeding species, for example *Lotus corniculatus*, *Trifolium repens* and *Ranunculus acris*, were selfed, in *R. acris* by a rain pollination mechanism, while obligatorily insect-pollinated species, for example the dioecious *Silene dioica*, were limited to areas close to the bird colonies or human settlements where their dipteran pollinators bred. Hagerup's conclusions, which have frequently been cited, though often with some caution (e.g. Proctor & Yeo, 1973), are only partly supported by the evidence he provided but the ecological conditions that he describes in the Faeroes are widespread on the north-western Atlantic fringes of Europe and the reproductive ecology of many of the species that grow in these extreme conditions requires re-investigation. Although species and genotypes that show vivipary and other forms of asexual reproduction are more frequent in these conditions, the factors that control the balance between sexual and asexual reproduction are far from clear (Law *et al.*, 1983).

ACKNOWLEDGEMENTS

I am grateful to Martin Ford and David Stevens for helpful comments and discussion.

REFERENCES

BAKER, H. G. (1976). 'Mistake' pollination as a reproductive system with special reference to the Caricaceae. In: *Tropical Trees: Variation, Breeding and Conservation* (Ed. by J. Burley & B. T. Styles), pp. 161–169. Academic Press, New York.
BOSTOCK, S. J. & BENTON, R. A. (1979). The reproductive strategies of five perennial Compositae. *Journal of Ecology*, **67**, 91–107.
BOYDEN, T. C. (1980). Floral mimicry by *Epidendrum ibaguense* (Orchidaceae) in Panama. *Evolution*, **34**, 135–136.
BROWN, J. H. & KODRIC-BROWN, A. (1979). Convergence, competition and mimicry in a temperate community of hummingbird-pollinated flowers. *Ecology*, **60**, 1022–1035.
CAMPBELL, D. R. (1985). Pollen and gene dispersal. The influences of competition for pollination. *Evolution*, **39**, 418–431.
CAMPBELL, D. R. & MOTTEN, A. F. (1981). Competition for pollination between two spring wildflowers. *Bulletin of the Ecological Society of America*, **62**, 99.

CHAN, H. T. & APPANAH, S. (1980). Reproductive biology of some Malaysian dipterocarps. I. Flowering biology. *Malaysian Forester*, **43**, 132–143.

CLAPHAM, A. R., TUTIN, T. & WARBURG, E. F. (1981). *Excursion Flora of the British Isles*, 3rd Edn. Cambridge University Press, Cambridge.

CORBET, S. A., KERSLAKE, C. J. C., BROWN, D. & MORLAND, N. E. (1984). Can bees select nectar-rich flowers in a patch? *Journal of Apicultural Research*, **23**, 234–242.

DAFNI, A. (1984). Mimicry and deception in pollination. *Annual Review of Ecology and Bystematics*, **15**, 259–278.

DAFNI, A. & IVRI, Y. (1981) Floral mimicry between *Orchis israelitica* Baumann and Dafni (Orchidaceae) and *Bellevalia flexuosa* Boiss. (Liliaceae). *Oecologia (Berlin)*, **49**, 229–232.

DEVLIN, B. & STEPHENSON, A. G. (1985). Sex differential floral longevity, nectar secretion and pollinator foraging in a protandrous species. *American Journal of Botany*, **72**, 303–310.

DOUGLAS, D. A. (1981). The balance between vegetative and sexual reproduction of *Mimulus primuloides* (Scrophulariaceae) at different altitudes in California. *Journal of Ecology*, **69**, 295–310.

EVENSON, W. E. (1983). Experimental studies of reproductive energy allocation in plants. In: *Handbook of Experimental Pollination Biology* (Ed. by C. E. Jones & R. J. Little), pp. 249–274. Van Nostrand Reinhold, New York.

FAEGRI, K. & VAN DER PIJL, L. (1979). *The Principles of Pollination Ecology*, 3rd Edn. Pergamon, Oxford.

FORD, M. A. (1986). *Pollinator-mediated interactions between outbreeding annual weeds*. Unpublished Ph.D. thesis, University of Wales.

GOLDMAN, D. A. & WILLSON, M. F. (1986). Sex allocation in functionally hermaphroditic plants: a review and critique. *Botanical Review*, **52**, 157–194.

GROSS, K. L. & SOULE, J. D. (1981). Differences in biomass allocation to reproductive and vegetative structures of male and female plants of a dioecious, perennial herb, *Silene alba* (Miller) Krause. *American Journal of Botany*, **68**, 801–807.

HAGERUP, O. (1932). On pollination in the extremely hot air at Timbuctu. *Dansk botanisk Arkiv*, **8**, 1–20.

HAGERUP, O. (1951). Pollination in the Faeroes – in spite of rain and poverty of insects. *Biologiske Meddelelser Kongelige Danske Videnskabernes Selskab*, **18**, 1–48.

HANCOCK, J. F. & BRINGHURST, R. S. (1980). Sexual dimorphism in the strawberry *Fragaria chiloensis*. *Evolution*, **34**, 762–768.

HARBORNE, J. B. (1982). *Introduction to Ecological Biochemistry*, 2nd Edn. Academic Press, London.

HARPER, J. L. (1977). *Population Biology of Plants*. Academic Press, London.

HARPER, J. L. & OGDEN, J. (1970). The reproductive strategy of higher plants. I. The concept of strategy with special reference to *Senecio vulgaris* L. *Journal of Ecology*, **58**, 681–698.

HARPER, J. L & WHITE, J. (1974). The demography of plants. *Annual Review of Ecology and Systematics*, **5**, 419–463.

HARTOG, C. DEN (1970). *The Sea-grasses of the World*. North-Holland, Amsterdam & London.

HEINRICH, B. (1975a). Bee flowers: a hypothesis on flower variety and blooming times. *Evolution*, **29**, 325–334.

HEINRICH, B. (1975b). Energetics of pollination. *Annual Review of Ecology and Systematics*, **6**, 139–170.

HEINRICH, B. & RAVEN, P. H. (1972). Energetics and pollination ecology. *Science*, **176**, 597–602.

HOWARTH, S. E. & WILLIAMS, J. T. (1972). Biological Flora of the British Isles. *Chrysanthemum segetum* L. *Journal of Ecology*, **60**, 573–584.

KAY, Q. O. N. (1969). The origin and distribution of diploid and tetraploid *Tripleurospermum inodorum* (L.) Schultz Bip. *Watsonia*, **7**, 130–141.

KAY, Q. O. N. (1971a). Biological Flora of the British Isles. *Anthemis cotula* L. *Journal of Ecology*, **59**, 623–636.

KAY, Q. O. N. (1971b). Biological Flora of the British Isles. *Anthemis arvensis* L. *Journal of Ecology*, **59**, 637–648.

KAY, Q. O. N. (1971c). Floral structure in the marine angiosperms *Cymodocea serrulata* and *Thalassodendron ciliatum* (*Cymodocea ciliata*). *Botanical Journal of the Linnean Society*, **64**, 423–429.

KAY, Q. O. N. (1985). Nectar from willow catkins as a food source for blue tits. *Bird Study*, **32**, 41–45.

KAY, Q. O. N. (1986). Dioecy and pollination in *Viscum album*. *Watsonia*, **16**, 232.

KAY, Q. O. N. (1987). Ultraviolet patterning and ultraviolet-absorbing pigments in flowers of the Legum-inosae. In: *Advance in Legume Systematics, Part 3* (Ed. by C. H. Stirton). Royal Botanic Gardens, Kew. (In press.)

KAY, Q. O. N. & PAGE, J. (1985). Dioecism and pollination in *Ruscus aculeatus*. *Watsonia*, **15**, 261–264.

KAY, Q. O. N. & STEVENS, D. P. (1986). The frequency, distribution and reproductive biology of dioecious species in the native flora of Britain and Ireland. *Botanical Journal of the Linnean Society*, **92**, 39–64.

KAY, Q. O. N., LACK, A. J., BAMBER, F. C. & DAVIES, C. R. (1984). Differences between sexes in floral morphology, nectar production and insect visits in a dioecious species, *Silene dioica*. *New Phytologist*, **98**, 515–529.

KEVAN, P. G. & BAKER, H. G. (1983). Insects as flower visitors and pollinators. *Annual Review of Entomology*, **28**, 407–453.

KUGLER, H. (1970). *Blütenökologie*. Fischer, Stuttgart.

LAW, R., COOK, R. E. D. & MANLOVE, R. J. (1983). The ecology of flower and bulbil production in *Polygonum viviparum*. *Nordic Journal of Botany*, **3**, 559–565.

LEVIN, D. A. (1971). The origin of reproductive isolating mechanisms in higher plants. *Taxon*, **20**, 91–113.

LEVIN, D. A. & ANDERSON, W. W. (1970). Competition for pollination between simultaneously flowering species. *American Naturalist*, **104**, 455–467.

LEVIN, D. A. & SCHAAL, B. A. (1970). Corolla colour as an inhibitor of interspecific hybridization in *Phlox*. *American Naturalist*, **104**, 273–283.

LEWIS, H. (1961). Experimental sympatric populations of *Clarkia*. *American Naturalist*, **95**, 155–168.

LITTLE, R. J. (1983). A review of floral food deception mimicries with comments on floral mutualism. In: *Handbook of Experimental Pollination Biology* (Ed. by C. E. Jones & R. J. Little), pp. 294–309. Van Nostrand Reinhold, New York.

LOVETT DOUST, J. & CAVERS, P. B. (1982). Biomass allocation in hermaphroditic flowers. *Canadian Journal of Botany*, **60**, 2530–2534.

LOVETT DOUST, J. & HARPER, J. L. (1980). The resource cost of gender and maternal support in an andromonoecious umbellifer, *Smyrnium olusatrum* L. *New Phytologist*, **8**, 241–264.

MACARTHUR, R. H. & LEVINS, R. (1967). The limiting similarity, convergence and divergence of coexisting species. *American Naturalist*, **101**, 377–385.

MCNEILLY, T. & ANTONOVICS, J. A. (1968). Evolution in closely adjacent plant populations. IV. Barriers to gene flow. *Heredity*, **23**, 205–218.

MOGIE, M. (1985). Morphological, developmental and electrophoretic variation within and between obligately apomictic *Taraxacum* species. *Biological Journal of the Linnean Society*, **24**, 207–216.

MONTGOMERIE, R. D. & GASS, C. L. (1981). Energy limitation of hummingbird populations in tropical and temperate communities. *Oecologia (Berlin)*, **50**, 162–165.

NIERENBERG, L. (1972). The mechanism for the maintenance of species integrity in sympatrically occurring equitant *Oncidiums* in the Caribbean. *American Orchid Society Bulletin*, **41**, 873–882.

NIKLAS, K. J. (1985). The aerodynamics of wind pollination. *Botanical Review*, **51**, 328–386.

NIKLAS, K. J. & BUCHMANN, S. L. (1985). Aerodynamics of wind pollination in *Simmondsia chinensis* (Link) Schneider. *American Journal of Botany*, **72**, 530–539.

PHILIPSON, W. R. (1937). A revision of the British species of the genus *Agrostis* Linn. *Journal of the Linnean Society (Botany)*, **52**, 73–151.

PITELKA, L. F. (1977). Energy allocation in annual and perennial lupines (*Lupinus*: Leguminosae). *Ecology*, **58**, 1055–1065.

PROCTOR, M. C. F. & YEO, P. F. (1973). *The Pollination of Flowers*. Collins, London.

RATHCKE, B. (1983). Competition and facilitation among plants for pollination. In: *Pollination Biology* (Ed. by L. Real), pp. 305–329. Academic Press, Orlando.

RATHCKE, B. & LACEY, E. P. (1985). Phenological patterns of terrestrial plants. *Annual Review of Ecology and Systematics*, **16**, 179–214.

RAVEN, P. H. & RAVEN, T. E. (1976). *The Genus Epilobium in Australasia: A Systematic and Evolutionary Study*. Bulletin 216, New Zealand Department of Scientific and Industrial Research, Wellington.

REAL, L. (1983). Introduction. In: *Pollination Biology* (Ed. by L. Real), pp. 1–5. Academic Press, Orlando.

REGAL, P. J. (1982). Pollination by wind and animals: ecology of geographic patterns. *Annual Review of Ecology and Systematics*, **13**, 497–524.

RICHARDS, A. J. (1986). *Plant Breeding Systems*. George Allen & Unwin, London.

ROBERTSON, C. (1924). Phenology of entomophilous flowers. *Ecology*, **5**, 393–407.

ROHMEDER, E. (1967). Beziehungen zwischen Frucht- bzw. Samenerzeugung und Holzerzeugung der Waldbäume. *Allgemeine Forstzeitung*, **22**, 33–39.

SARUKHÁN, J. (1976). On selective pressures and energy allocation in populations of *Ranunculus repens* L., *R. bulbosus* L. and *R. acris* L. *Annals of the Missouri Botanical Garden*, **63**, 290–308.

SARUKHÁN, J. (1980). Demographic problems in tropical systems. In: *Demography and Evolution in Plant Populations* (Ed. by O. T. Solbrig), pp. 161–188. Blackwell Scientific Publications, Oxford.

SCHEMSKE, D. W. (1978). Evolution of reproductive characteristics in *Impatiens* (Balsaminaceae): the significance of cleistogamy and chasmogamy. *Ecology*, **59**, 596–613.

SOUTHWICK, E. E. (1984). Photosynthetic allocation to floral nectar: a neglected energy investment. *Ecology*, **65**, 1775–1779.

STRUIK, G. J. (1965). Growth patterns of some native annual and perennial herbs in southern Wisconsin. *Ecology*, **46**, 401–420.

THOMSON, J. D. (1978). Effects of stand composition on insect visitation in two-species mixtures of *Hieracium*. *American Naturalist*, **100**, 431–440.

THOMSON, J. D. (1981). Spatial and temporal components of resource assessment of flower-feeding insects. *Journal of Animal Ecology*, **50**, 49–59.

THOMSON, J. D. (1982). Patterns of visitation by animal pollinators. *Oikos*, **39**, 241–250.

WADDINGTON, K. D. (1979). Divergence in inflorescence height: an evolutionary response to pollinator fidelity. *Oecologia (Berlin)*, **40**, 43–50.

WALLACE, C. S. & RUNDEL, P. W. (1979). Sexual dimorphism and resource allocation in male and female shrubs of *Simmondsia chinensis*. *Oecologia (Berlin)*, **44**, 34–39.

WALLER, D. M. (1979). The relative costs of self- and cross-fertilized seeds in *Impatiens capensis* (Balsaminaceae). *American Journal of Botany*, **66**, 313–320.

WALLER, D. M. (1980). Environmental determinants of outcrossing in *Impatiens capensis* (Balsaminaceae). *Evolution*, **34**, 747–761.

WASER, N. M. (1978a). Competition for humming bird pollination and sequential flowering in two Colorado wildflowers. *Ecology*, **59**, 934–944.

WASER, N. M. (1978b). Interspecific pollen transfer and competition between co-occurring plant species. *Oecologia (Berlin)*, **36**, 223–236.

WASER, N. M. (1979). Pollinator availability as a determinant of flowering time in ocotillo (*Fouqueria splendens*). *Oecologia (Berlin)*, **39**, 107–121.

WASER, N. M. (1983a). The adaptive nature of floral traits: ideas and evidence. In: *Pollination Biology* (Ed. by L. Real), pp. 241–285. Academic Press, Orlando.

WASER, N. M. (1983b). Competition for pollination and floral character differences among sympatric plant species: a review of evidence. In: *Handbook of Experimental Pollination Biology* (Ed. by C. E. Jones & R. J. Little), pp. 277–293. Van Nostrand Reinhold, New York.

WASER, N. M. (1986). Flower constancy: definition, cause and measurement. *American Naturalist*, **127**, 593–603.

WASER, N. M. & REAL, L. A. (1979). Effective mutualism between sequentially flowering plant species. *Nature*, **281**, 670–672.

WHITEHEAD, D. R. (1983). Wind pollination: some ecological and evolutionary perspectives. In: *Pollination Biology* (Ed. by L. Real), pp. 97–108. Academic Press, Orlando.

WYATT, R. (1983). Pollinator–plant interactions and the evolution of breeding systems. In: *Pollination Biology* (Ed. by L. Real), pp. 51–95. Academic Press, Orlando.

YEO, P. F. (1968). The evolutionary significance of the speciation of *Euphrasia* in Europe. *Evolution*, **22**, 736–747.

New Phytol. (1987) **106** (Suppl.), 283–295

BOTANICAL CONTRIBUTIONS TO CONTEMPORARY ECOLOGICAL THEORY

By J. P. GRIME and J. G. HODGSON

Unit of Comparative Plant Ecology, Department of Botany, The University, Sheffield S10 2TN, UK

SUMMARY

In the search for general principles in ecology, the role of botany has often been subordinate to those of mathematics, genetics and zoology. This paper identifies three important areas of ecological research (competition, coexistence, and relationships between anti-herbivore defence and decomposition processes) where comparative studies of autotrophic plants can play a leading part in analyses of the structure of communities and the functioning of ecosystems. With respect to a fourth research topic (ecological constraints arising from evolutionary histories), a stronger botanical response to the zoological lead is advocated.

Key words: Ecological theory, competition, species coexistence, antiherbivore defence, decomposition.

INTRODUCTION

During its early history, ecology was concerned primarily with the identification and description of plants and animals observed in their natural habitats. In taxonomically 'difficult' groups (e.g. fungi, lichens, bryophytes and numerous sections of the Invertebrata), these activities quite properly continue to receive attention. Increasingly, however, the expanding information base and the urgent need for scientific management of natural resources are prompting new and convergent lines of enquiry. Among plant and animal ecologists alike, there is growing recognition of the requirement for predictive, reductionist analyses of the properties of communities and ecosystems based upon the functional characteristics of component organisms. It can be argued, of course, that the compartment and flux models which are already available for some ecosystems adequately characterize the biomass and energetics of various trophic components, and the movement of resources between them. Real understanding of communities and ecosystems will have been achieved, however, only when their structure and characteristics can be predicted from knowledge of, firstly, the prevailing habitat conditions and, secondly, the intrinsic properties of each species and genotype, including aspects of their biochemical and anatomical organization, physiology and population dynamics.

In view of the enormous number of living organisms and the small proportion which has been the subject of ecological study, the prospects for such functional analyses of communities and ecosystems may seem remote. In this paper, we will argue to the contrary – that, because taxonomic diversity often obscures recurring ecological patterns, attainment of this objective is nearer than may at first appear. Our main purpose, however, is to suggest that comparative studies of autotrophic plants often have a leading role to play in the development of the theoretical basis necessary for the functional analysis of communities and ecosystems.

0028-646X/87/05S283+13 $03.00/0

The Interaction of Botanical and Zoological Research

Until recently, most of the impetus in the search for generalizing principles in ecology has originated from zoology. For this reason, concepts have relied upon criteria which are capable of quantitative assessment when applied to most animals. These include the breeding system, the initial size and number of offspring, the age dependency of mortalities, the size and shape of mature individuals and the timing of reproduction in relation to potential life span. Study of these variables constitutes the discipline of population biology and it has been the recent achievement of Harper (1977) and his numerous followers to extend this approach to vascular plants. There can be no doubt that this development has brought much needed precision to aspects of the study of the proximal controls operating on the distribution and abundance of plant populations. However, the application of a common methodology to studies of plants and animals has not resulted in unifying concepts immediately relevant to analyses of communities and ecosystems. It is our contention that this outcome is inevitable, since population biology or, for that matter, any specialist approach used in isolation from other disciplines cannot generate a truly comprehensive theory of functional specialization in plants or animals. In particular, the recent tendency to review botanical evidence in the light of zoological population theory is often restrictive. It has resulted in a bias towards those plants (monocarps) and those attributes (numbers and sizes of individuals) which are amenable to studies of the various kinds developed for the purposes of animal ecology. Whilst welcoming in general the benefits which have followed this cross-linkage between botanical and zoological ecology, we see lost opportunities in an emphasis which excludes aspects of study where plants exercise natural advantages over animals as subjects for ecological research. Here, we refer in particular to studies of the capture and utilization of resources, competition, temporal and spatial partitioning of resources, anti-predator defence, mutualistic interactions and regeneration. In all of these activities, considerable benefits to the research worker arise from the relative immobility of terrestrial plants, a feature which permits close comparison between the characteristics of a plant in field and laboratory, and between the nature of the environments which it exploits and from which it is excluded.

In the sections which follow, we first consider briefly three areas of current research where contributions to general ecological theory appear to be emerging from a distinctively botanical approach. The fourth section examines a topic likely to benefit from wider application of zoological principles.

Competition

The following quotations from recent publications illustrate the serious problems which have been encountered in efforts to assess the role of competition in animal populations.

Why do ecologists disagree so vehemently about the role of competition in nature?... Surely we should have reached a consensus by now, at least on whether competition is worth further empirical or theoretical study.

Welden & Slauson (1986)

As a future trend we can expect a further depreciation of competition, both intra- and inter-specific, as being a major force in ecology and evolution.

den Boer (1986)

The search for limits to similarity among co-existing competitors has been further clouded by the realization that most species (and certainly most invertebrate species) live in a world that is heterogeneous in both space and time. For species that compete for patchy and ephemeral resources, questions of relative mobility are as important as relative success when the species meet on the same patch.... Seasonality or environmental unpredictability can further complicate competitive interactions in a variety of ways.

May & Seger (1986)

It is interesting to note that the complications to which May & Seger refer also occur in analyses of the part played by competition in complex plant communities such as those occurring in ancient calcareous pastures. In view of this close parallel, it may be instructive to ask why, in general, botanists continue to express less uncertainty than zoologists about the role of competition as one of the determinants of the relative abundance of species within communities. We suggest that this is because most plant ecologists are not only familiar with vegetation in which the impact of competition is overwhelming and incontrovertible [monospecific stands of robust fast-growing perennials, e.g. *Urtica dioica** Fig. 1(a)] but they have observed also quite different situations where interventions by other factors have reduced the intensity of competition to a low value [e.g. patches of disturbed infertile soil colonized by diminutive winter annuals, such as *Erophila verna*, Fig. 1(b)]. As noted by Paine (1984), the significance of these sites is that they correspond to extremes which can provide points of reference in attempts to analyze the influence of competition in more complex types of vegetation.

The monopoly provides an idealized null state or standard against which to measure departures in the real world. In the absence of such bench-marks it is difficult to evaluate the relative importance of the many individual processes which, collectively, generate community pattern.

Paine (1984)

In comparison with the situations prevailing in many animal communities, the resources for which plants compete are relatively few and vary only quantitatively from plant to plant. This has allowed experimental evidence to play a more certain part in the elucidation of competitive mechanisms in vegetation. In consequence, there is now a growing consensus that intensities of competition and rates of competitive exclusion are higher under productive, relatively undisturbed conditions (Donald, 1958; Mahmoud & Grime, 1976; Huston, 1979; Chapin, 1980). This view has recently received strong support from field experiments in which phytometers have been used to measure the intensity of competition in vegetation established along gradients in productivity (Wilson & Keddy, 1986a, b). Theoretical analysis (Sibly & Grime, 1986), phenological studies (Al-Mufti *et al.*, 1977) and laboratory experiments (Grime, Crick & Rincon, 1986) have all contributed to a narrowing focus in the effort to identify the characteristics of plants which are responsible for high competitive ability. Importance has been attached to (1) a robust perennial life form with a strong capacity to ramify vegetatively throughout the aerial and edaphic environment, (2) the rapid commitment of captured resources to the construction of new leaves and roots, (3) high morphological plasticity during the differentiation of leaves and roots and (4) short life spans of individual leaves and roots. The combined effect of these attributes is continuous adjustment of the spatial distribution of the absorptive surfaces ('active foraging') above and below ground. This, in turn allows projection of

---

* Nomenclature follows Clapham, Tutin & Warburg (1981).

Fig. 1. (a) A monospecific stand of *Urtica dioica*. This species is also frequently associated with the pleurocarpous moss, *Brachythecium rutabulum* Hedw., which colonizes freshly deposited stem litter of *U. dioica* during cooler periods (see p. 7 of Bradshaw, 1987). Scale intervals are decimetres. (b) A fruiting individual of *Erophila verna* (together with other small winter annuals, *Arabidopsis thaliana*, extreme left and in centre background, *Cardamine hirsuta*, extreme left and *Saxifraga tridactylites*, extreme right) exploiting a shallow calcareous soil subject to summer desiccation. Scale intervals are centimetres.

leaves and roots into the resource-rich zones of the patchy and changing environment created by the activity of the plant and its competing neighbours.

The essential feature of this emerging view of competition is that it is a process which reaches its maximum intensity in environments where there is an abundance of resources. Under such conditions, a dense, rapidly expanding biomass is formed by the dominant plants. This results in the exclusion of subordinates and their confinement to zones of depleted resources created by the dominant plants and occurring simultaneously above and below ground. Competitive interactions are not, however, confined to productive, undisturbed environments and can play a critical role in communities which are of lower productivity or suffer frequent disturbance. As pointed out by Grime (1985), it is quite feasible, for example, that inconspicuous competition could determine which species among a community of slow-growing lichens on a granitic boulder becomes most abundant. This line of argument has been carried further by Welden & Slauson (1986) who propose that the intensity with which competition for resources proceeds within a plant community is unrelated to its 'importance' as a proximal control on species composition and as a factor affecting current evolution. We accept their argument that, in communities where the intensity of competition is low, differences in competitive ability may continue to contribute to differences in fitness between component species and genotypes. However, we believe that this does not justify an uncoupling of 'intensity' from 'importance', even where the latter is used in the restricted sense of its proximal control of current species composition and evolution. Even where the intensity of past competition has been so severe that it has produced a monoculture, an intense struggle between genotypes may continue [see the abundant evidence of self-thinning in stands of single species reviewed by Harper (1977)]. In such circumstances, it is competition which continues to determine the capacity of the resident dominant to monopolize capture of resources and reproduction, and it is the same process which brings about the summary failure of any invading species; it is in such monopolies that maxima in the intensity *and* importance of competition coincide.

COEXISTENCE

Under the previous heading, we have argued that, in conditions of high productivity and low disturbance of vegetation, competition between species results in the development of monocultures. This 'fight to the finish' model of competitive interactions contrasts strongly with the view, widely held among many animal ecologists and evolutionary biologists, that the effect of competition is to cause genetic divergence between competing populations which facilitates complementary exploitation of the habitat and allows co-existence of species. Botanical evidence such as that summarized by Grime (1973, 1974, 1977), Al-Mufti *et al.* (1977) and Huston (1979) has led to a quite different interpretation of the role of competition in relation to diversity. According to these authors, coexistence is possible only in conditions where the monopolistic tendencies of potential dominants are constrained by low productivity and/or physical damage. For herbaceous vegetation, this hypothesis has been formalized as the hump-backed model of floristic diversity (Grime, 1973; Connell, 1979). This model suggests that the potential for coexistence reaches a maximum in a corridor of intermediate values in standing crop plus litter (350 to 750 gm^{-2} in the British Isles). Field evidence (Al-Mufti *et al.*, 1977) confirms that above this range diversity usually falls as the vegetation becomes dominated by a small number of robust species.

Among plant and animal ecologists, debate continues concerning the mechanisms which permit high diversity in 'corridor' communities. Among botanists, the conventional 'coexistence by niche-differentiation' view has been espoused by Tilman (1982) who has proposed that specific differences in nutritional requirements and tolerances within a community may be sufficient to allow stable coexistence on a mineralogically heterogeneous soil. Four of the papers presented in this symposium (Bennett, 1987; Givnish, 1987; Hendry, 1987; Rorison, 1987) contain evidence consistent with the notion of coexistence by temporal (i.e. seasonal) niche-differentiation. Recognition of the wide range of types of seed bank and germination responses which may be represented within a single community (Thompson & Grime, 1979; Grime et al., 1981) provides circumstantial evidence in support of the hypothesis (Grubb, 1977) that differences in regeneration may contribute to the maintenance of floristic diversity. To this wealth of speculation and evidence, we must add the 'non-equilibrium' models of Whittaker & Levin (1977), Huston (1979) and Pickett (1980) which seek to explain diversity as a consequence of dynamic micro-successional mosaics generated by vegetational disturbance.

Most of these theories of coexistence are not mutually exclusive and our own observations (Grime, Hodgson & Hunt, 1987) strongly suggest that several may be operational within the same community. Many excellent opportunities now exist to test some of the hypotheses concerning diversity by experiments on natural plant communities. It is apparent, however, that some potentially important factors are not susceptible to precise measurement or manipulation in the field. Recently, this problem has been approached through an alternative strategy in which plant communities were allowed to develop from seed under controlled conditions (J. P. Grime, J. M. L. Mackey, S. H. Hillier & D. J. Read, unpublished data). Diversity indices were calculated for each of the 'communities' which developed after 12 months in low productivity microcosms providing factorial combinations of soil heterogeneity, grazing and vesicular–arbuscular (VA) mycorrhizal infection, all of which are capable in theory of promoting diversity. Both grazing and mycorrhizas increased diversity to a marked extent and ancillary measurements revealed that, in each of these treatments, the biomass of the subordinate species was raised in relation to that of the canopy dominants. The effect of grazing was due to the differential sensitivity of canopy dominants to defoliation, whereas the influence of VA mycorrhizas was shown, by studies involving $^{14}CO_2$, to be related to the export of assimilate from canopy to subordinate species through a common mycelial network. This experiment confirms the importance of damage of potential dominants as a factor maintaining diversity and suggests that the low rates of competitive exclusion occurring on infertile soils may not be simply the result of 'slow dynamics' (Huston, 1979). Transfer of resources from source (canopy dominants) to sinks (subordinates) through mycorrhizal connections [as also demonstrated in pot experiments by Francis, Finlay & Read (1986)] deserves consideration as an additional factor encouraging coexistence of species.

DEFENCE AND DECOMPOSITION

Evidence relating to a wide range of organisms and ecosystems now confirms the existence of a distinctive form of natural selection which continues to receive less attention than it deserves in the field of animal ecology. This phenomenon, described as 'adversity selection' by Whittaker (1975) and recognized under various other titles by Ramenskii (1938), Greenslade (1972a, b), Grime (1974) and Southwood (1977), is supported by a large amount of observation and experiment which has been assembled for insects (Downes, 1964, 1965; Greenslade, 1983), autotrophs (Grime, 1977, 1979) and fungi (Pugh, 1980; Cooke & Rayner, 1984). Adversity selection appears to be experienced by organisms exploiting continuously unproductive environments or niches and is conducive to extended life histories, slow growth rates, conservative mechanisms of capture and utilization of resources and low reproductive effort.

In plants, two more characteristics may be added to the list by the recognition that many species of chronically unproductive habitats are strongly defended against herbivores and are engaged in mutualistic associations (lichens and mycorrhizas). Following the arguments and evidence summarized in Grime (1979), Edwards & Wratten (1985) and earlier in this volume (Coley, 1987), it seems reasonable to suggest that both of these attributes represent additional mechanisms contributing to the efficient retention and recycling of captured mineral nutrients in organisms exploiting environments where phosphorus and/or nitrogen are severely limiting primary production.

Efficient protection of foliage (and the investment of mineral and photosynthetic capital which it represents) is not confined to the lichens, bryophytes, succulents, hair grasses, cushion plants and sclerophyllous undershrubs of deserts, tundra and mountain tops or to the species of skeletal habitats undergoing primary succession. Relatively strong anti-herbivore defences are also evident in the trees, shrubs and herbs which characterize the later stages of secondary succession in temperate and tropical forests. This supports the view that the progress of secondary succession in forests is often marked by a physiological transition (Odum, 1969; Grime, 1979, 1987) in which species exhibiting high rates of capture and loss of mineral nutrients (e.g. *Fraxinus excelsior, Sambucus nigra, Urtica dioica*) are displaced by slower-growing species with a greater capacity to resist herbivory and to retain such nutrients (e.g. *Fagus sylvatica, Quercus* spp., *Sanicula europaea*).

As Coley (1987) has shown, a major component of the anti-herbivore defences of the leaves of slow-growing species appears to be physical in nature. Particularly in colder climates, therefore, we may expect that one consequence of adversity selection will be accumulation of litter, i.e. that defences which protected the living leaf against herbivory may continue to be effective against many decomposing organisms (Grime & Anderson, 1986). Consistent with this hypothesis is the strong tendency of the litter of species, such as *F. sylvatica* and *Quercus* spp., to accumulate on forest floors in Britain and to be persistent in quantities sufficient to exert a modifying effect upon the ground flora through its physical impact (Sydes & Grime, 1981a, b). It may be no coincidence either that the majority of soils of high organic content (both acidic and calcareous) carry slow-growing vegetation of relatively low palatability to herbivores (Kubiëna, 1950).

Evolutionary History

Zoologists have long recognized the strong connections which exist between taxonomy and ecology. This is not surprising in view of the distinctive ecological specialization evident in major taxonomic groupings such as the amphibians, reptiles, birds and fish. However, despite the pioneering studies of Stebbins (1974), which indicate that higher taxa of angiosperms are to some extent ecologically specialized, few plant ecologists have allowed an evolutionary/taxo-

nomic perspective to influence their work (Hodgson, 1986c). This uncharacter-
istic failure to follow the zoological lead may be traced in part to a bias against the
use in plant taxonomy of characters of known ecological significance (Stace, 1980)
and the resulting dependence of much angiospermous taxonomy upon floral
characters, the ecological significance of which is generally obscure (Cronquist,
1968).

The reliance of taxonomy upon ecologically trivial characteristics should not
obscure the fact that, despite adaptive radiation, many families of plants show
evidence of ancestral specializations which continue to influence their current
ecology (Grime, 1984; Hodgson, 1986c). A large number of ecologically important
attributes are related to taxonomy. They include size, shape and dormancy
mechanisms of seeds (Guppy, 1912; Martin, 1946; Grime et al., 1981; Hodgson
& Mackey, 1986; Thompson, 1987), association with nitrogen-fixing micro-
organisms (as in Leguminosae), life history (Hodgson, 1986c), phenology (Givnish,
1987), wood structure and its effect on early leafing (Lechowicz, 1984) and heavy
metal tolerance (Baker, 1987).

Anatomical characteristics are under-represented in this list, perhaps because
anatomy has often been regarded by ecologists as a rather sterile, descriptive field
of study. However, when functional considerations are brought to the fore
(Givnish, 1987; Raven & Handley, 1987), it is evident that anatomical information
is essential to studies of the phylogenetic constraints which relate to ecological
specializations and currently limit the success of certain families. Particular
benefits to ecological analysis follow from the work of plant anatomists such as
Carlquist (1975) who correlated xylem structure with the water relations of the
habitat. To illustrate Carlquist's approach, Figure 2 shows that, for monocotyledonous families of the Sheffield flora, estimates of the quality of the vessel
elements for conduction of water correlate well with both phenology and the

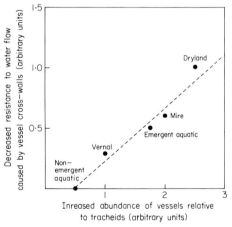

Fig. 2. Correlation between anatomical features influencing xylem resistance and the characteristic
water status of the habitat for the monocotyledonous families of the Sheffield region (Hodgson,
unpublished data). x axis (quality of xylem elements for water conduction: 0, vessels absent
(conductive tissue composed entirely of tracheids); 1, vessels in roots only; 2, intermediate between
classes 1 and 3; 3, vessels in shoots and roots. y axis (water resistance associated with vessel and
end-walls): 0, scalariform perforation plates present between vessels; 1, intermediate; 2,
perforation plates simple. Information on the structure of xylem within families was abstracted
from Wagner (1977) and Dahlgren & Clifford (1982) whose classificatory scheme was also adopted.
The values plotted refer to mean values for all families of similar ecology.

distribution of species with respect to moisture status. Families which exploit dryland habitats within the region tend to have vessel elements throughout the plant and end-walls with simple perforations. In wetland species and in vernals, whose period of aestivation coincides with potentially greater plant water deficits, efficiency of elements for conduction of water is low. In at least some parts of the plant, only tracheids are present and, where vessels do occur, they tend to have scalariform end-walls. These data suggest that features of xylem structure may have markedly constrained the adaptive radiation of many monocotyledonous families. This type of argument may be applied to other morphological character-istics of families. It has been suggested, for example, that ancestral specializations in architecture of flowers and fruits in some families (e.g. Serophulariaceae, Umbelliferae) continue to dictate numbers and sizes of seed, features which severely limit the ability of contemporary genotypes to exploit common habitats (Hodgson & Mackey, 1986).

Another important facet of evolutionary history relates to the flora of fertile habitats. From our knowledge of the structure of early angiosperms and of the morphology and ecology of their living relatives (Takhtajan, 1980; Sporne, 1980, 1982), it appears likely that flowering plants are derived from ancestors which occupied unproductive habitats (see Hodgson, 1986a, c). Hodgson (1986a, c) argues that, until recently, fertile habitats have been comparatively rare and have probably had little impact on the evolutionary development of the angiosperms. There is evidence from the Sheffield region (Hodgson, unpublished data) that the flora of fertile habitats is depauperate and that the species which exploit the common fertile habitats, created by modern land use, have arisen by recent episodes of polyploid evolution. By contrast, older polyploids and diploids tend to be restricted to less fertile habitats which are declining in abundance. In consequence, there are significant relationships between ploidy and plant strategy [*sensu* Grime (1979)] and between ploidy and abundance (Hodgson, unpublished data; Fig. 3). It is predicted that the exploitation of productive habitats by animals has been similarly constrained by evolutionary history. However, polyploidy is uncommon in the animal kingdom (Lewis, 1980) and so the genetic mechanisms of adaptation to contemporary productive habitats will be rather different from those described here for angiosperms.

Conclusions

In this short essay, we have sought quite deliberately to avoid the firm ground of established facts and specialist methodologies and we have ventured into the No Man's Land which lies between plant and animal ecology. In each of the four subject areas chosen for review, we see prospects for botanical research to play a major role in the development of fields of community and ecosystem theory which, until recently, have been populated mainly by mathematicians or zoologists. In two cases (studies of competition and of coexistence), plants provide excellent materials for experimental manipulation and hypothesis testing in field plots and laboratory microcosms. In the third subject area (defence and decomposition), a plant-centred approach again seems justified, since many of the controlling variables appear to reside in phytochemistry. The fourth topic (evolutionary history) is one in which the zoological paradigm of zones of ancestral specialization corresponding to taxa deserves to be explored with caution, following the guiding principles of Stebbins (1971, 1974, 1980, 1985). It is also a subject area where

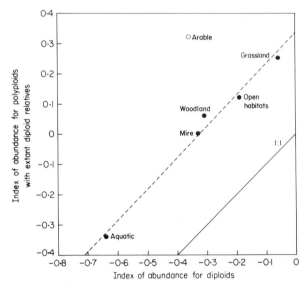

Fig. 3. The relationship for the major habitats of the Sheffield region between index of abundance
(IA) of polyploids with extant diploid relative and that of diploids (from Hodgson, unpublished
data) where

$$IA = \frac{\text{No. of common spp.} - \text{no. of (uncommon} + \text{rare} + \text{extinct) spp.}}{\text{Total no. of spp.}}$$

IA is not a totally appropriate measure for arable habitats (see Hodgson, 1986b) and data for the
habitat are therefore indicated by an unshaded symbol. A line indicating a 1:1 ratio has been
included.

botanical studies (e.g. evolution in fertile habitats) may enrich theories of animal
evolution.

An understanding of the functioning and structure of ecosystems remains a
major goal for both plant and animal ecologists. Here, botanists have a special
responsibility, since the patterns which we investigate are more tangible than
those with which the zoologists must grapple. A continuing challenge for comparative plant ecology is to influence the development of general theories of ecology
and evolution, and to present them in a coherent form relevant to the ecological
and environmental problems of the day. The earlier contributions to this symposium convince us that the approach and its practitioners are equal to the task.

ACKNOWLEDGEMENTS

During the past 25 years, we have had a the good fortune to collaborate with a
large number of talented biologists including the past and present staff of UCPE,
members of the Department of Botany of the University of Sheffield and many
research students and visiting scientists. To all these friends we offer our thanks for
both stimulus and support. Not least, we thank the Nature Conservancy Council
and the Natural Environment Research Council for their sensitivity to the needs
of long-term basic research.

REFERENCES

AL-MUFTI, M. M., SYDES, C. L., FURNESS, S. B., GRIME, J. P. & BAND, S.R. (1977). A quantitative analysis of shoot phenology and dominance in herbaceous vegetation. *Journal of Ecology*, **65**, 759–791.

BAKER, A. J. M. (1987). Metal tolerance. In: *Frontiers of Comparative Plant Ecology* (Ed. by I. H. Rorison, J. P. Grime, R. Hunt, G. A. F. Hendry & D. H. Lewis), *New Phytologist*, **106** (Suppl.), 93–111. Academic Press, New York & London.

BENNETT, M. D. (1987). Variation in genomic form in plants and its ecological implications. In: *Frontiers of Comparative Plant Ecology* (Ed. by I. H. Rorison, J. P. Grime, R. Hunt, G. A. F. Hendry & D. H. Lewis), *New Phytologist*, **106** (Suppl.), 177–200. Academic Press, New York & London.

BRADSHAW, A. D. (1987). Comparison–its scope and limits. In: *Frontiers of Comparative Plant Ecology* (Ed. by I. H. Rorison, J. P. Grime, R. Hunt, G. A. F. Hendry & D. H. Lewis), *New Phytologist*, **106** (Suppl.), 3–21. Academic Press, New York & London.

CARLQUIST, S. (1975). *Ecological Strategies of Xylem Evolution*. University of California Press, Berkeley, California.

CHAPIN, F. S. (1980). The mineral nutrition of wild plants. *Annual Review of Ecology and Systematics*, **11**, 233–260.

CLAPHAM, A. R., TUTIN, T. G. & WARBURG, E. F. (1981). *Excursion Flora of the British Isles*, 3rd Edn. Cambridge University Press, Cambridge.

COLEY, P. D. (1987). Interspecific variation in plant anti-herbivore properties: the role of habitat quality and rate of disturbance. In: *Frontiers of Comparative Plant Ecology* (Ed. by I. H. Rorison, J. P. Grime, R. Hunt, G. A. F. Hendry & D. H. Lewis), *New Phytologist*, **106** (Suppl.), 251–263. Academic Press, New York & London.

CONNELL, J. H. (1979). Tropical rain forests and coral reefs as open non-equilibrium systems. In: *Population Dynamics* (Ed. by R. M. Anderson, B. D. Turner & L. R. Taylor), pp. 141–163. Blackwell Scientific Publications, Oxford.

COOKE, R. C. & RAYNER, A. D. M. (1984). *The Ecology of Saprotrophic Fungi*. Longman, London.

CRONQUIST, A. (1968). *The Evolution and Classification of Flowering Plants*. Nelson, London & Edinburgh.

DAHLGREN, R. M. T. & CLIFFORD, H. T. (1982). *The Monocotyledons: A Comparative Study*. Academic Press, London.

DEN BOER, P. J. (1986). The present status of the competitive exclusion principle. *Trends in Ecology and Evolution*, **1**, 25–28.

DONALD, C. M. (1958). The interaction of competition for light and for nutrients. *Australian Journal of Agricultural Research*, **9**, 421–432.

DOWNES, J. A. (1964). Arctic insects and their environment. *Canadian Entomologist*, **96**, 279–307.

DOWNES, J. A. (1965). Adaptations of insects in the Arctic. *Annual Review of Entomology*, **10**, 257–274.

EWARDS, P. J. & WRATTEN, S. D. (1985). Induced plant defences against insect grazing: fact or artefact? *Oikos*, **44**, 70–74.

FRANCIS, R., FINLAY, R. D. & READ, D. J. (1986). Vesicular–arbuscular mycorrhiza in natural vegetation systems. IV. Transfer of nutrients in inter- and intra-specific combinations of host plants. *New Phytologist*, **102**, 103–111.

GIVNISH, T. J. (1987). Comparative studies of leaf form: assessing the relative roles of selective pressures and phylogenetic constraints. In: *Frontiers of Comparative Plant Ecology* (Ed. by I. H. Rorison, J. P. Grime, R. Hunt, G. A. F. Hendry & D. H. Lewis), *New Phytologist*, **106** (Suppl.), 131–160. Academic Press, New York & London.

GREENSLADE, P. J. M. (1972a). Distribution patterns of *Priochirus* species (Coeleoptera: Staphylinidae) in the Solomon Islands. *Evolution*, **26**, 130–142.

GREENSLADE, P. J. M. (1972b). Evolution in the staphylinid genus *Priochirus* (Coeleoptera). *Evolution*, **26**, 203–220.

GREENSLADE, P. J. M. (1983). Adversity selection and the habitat templet. *American Naturalist*, **122**, 352–365.

GRIME, J. P. (1973). Competitive exclusion in herbaceous vegetation. *Nature*, **242**, 344–347.

GRIME, J. P. (1974). Vegetation classification by reference to strategies. *Nature*, **244**, 310–311.

GRIME, J. P. (1977). Evidence for the existence of three primary strategies in plants and its relevance to ecological and evolutionary theory. *American Naturalist*, **111**, 1169–1194.

GRIME, J. P. (1979). *Plant Strategies and Vegetation Processes*. Wiley, Chichester

GRIME, J. P. (1984). The ecology of species, families and communities of the contemporary British flora. *New Phytologist*, **98**, 15–33.

GRIME, J. P. (1985). Towards a functional description of vegetation, In: *The Population Structure of Vegetation* (Ed. by J. White), pp. 503–514. Junk, Dordrecht.

GRIME, J. P. (1987). Dominant and subordinate components of plant communities: Implications for succession, stability and diversity. In: *Colonisation, Succession and Stability* (Ed. by A. Gray, P. Edwards & M. Crawley), pp. 413–428. Blackwell Scientific Publications, Oxford.

GRIME, J. P. & ANDERSON, J. M. (1986). Environmental controls over organism activity. In: *Forest Ecosystems in the Alaskan Taiga. A Synthesis of Structure and Function* (Ed. by K. van Cleve, C. T. Dyrness, L. A. Viereck, Flanagan, P. W. & F. S. Chapin), pp. 89–95. Springer-Verlag, Berlin.

GRIME, J. P., MASON, G., CURTIS, A. V., RODMAN, J., BAND, S. R., MOWFORTH, M. A. G., NEAL, A. M. & SHAW, S. (1981). A comparative study of germination characteristics in a local flora. *Journal of Ecology*, **69**, 1017–1059.

GRIME, J. P., CRICK, J. C. & RINCON, J. E. (1986). The ecological significance of plasticity. In: *Plasticity in Plants* (Ed. by D. H. Jennings & A. J. Trewavas), pp. 5–29. Cambridge University Press, Cambridge.

GRIME, J. P., HODGSON, J. G. & HUNT, R. (1987). *Comparative Plant Ecology : A Functional Approach to Common British Species*. Allen & Unwin, London. (In press.)

GRUBB, P. J. (1977). The maintenance of species-richness in plant communities: the importance of the regeneration niche. *Biological Reviews*, **52**, 107–145.

GUPPY, H. B. (1912). *Studies in Seeds and Fruits*. Williams & Norgate, London.

HARPER, J. L. (1977). *The Population Biology of Plants*. Academic Press, London.

HENDRY, G. A. F. (1987). The ecological significance of fructan in the contemporary flora. In: *Frontiers of Comparative Plant Ecology* (Ed. by I. H. Rorison, J. P. Grime, R. Hunt, G. A. F. Hendry & D. H. Lewis), *New Phytologist*, **106** (Suppl.), 201–216. Academic Press, New York & London.

HODGSON, J. G. (1986a). Commonness and rarity in plants with special reference to the Sheffield flora. I. The identity, distribution and habitat characteristics of the common and rare species. *Biological Conservation*, **36**, 199–252.

HODGSON, J. G. (1986b). Commonness and rarity in plants with special reference to the Sheffield flora. II. The relative importance of climate, soils and land use. *Biological Conservation*, **36**, 253–274.

HODGSON, J. G. (1986c). Commonness and rarity in plants with special reference to the Sheffield flora. III. Taxonomic and evolutionary aspects. *Biological Conservation*, **36**, 275–296.

HODGSON, J. G. & MACKEY, J. M. L. (1986). The ecological specialization of dicotyledonous families within a local flora: some factors constraining optimization of seed size and their possible evolutionary significance. *New Phytologist*, **104**, 497–515.

HUSTON, M. (1979). A general hypothesis of species diversity. *American Naturalist*, **113**, 81–101.

KUBIËNA, W. L. (1950). *The Soils of Europe*. Thomas Murky, London.

LECHOWICZ, M. J. (1984). Why do temperate deciduous trees leaf out at different times? Adaptation and ecology of forest communities. *American Naturalist*, **124**, 821–842.

LEWIS, W. H. (Ed.) (1980). *Polyploidy : Biological Relevance*. Plenum Press, New York.

MAHMOUD, A. & GRIME, J. P. (1976). An analysis of competitive ability in three perennial grasses. *New Phytologist*, **77**, 431–435.

MARTIN, A. C. (1946). The comparative internal morphology of seeds. *American Midland Naturalist*, **36**, 513–660.

MAY, R. M. & SEGER, J. (1986). Ideas in ecology. *American Scientist*, **74**, 256–267.

ODUM, E. P. (1969). The strategy of ecosystem development. *Science*, **164**, 262–270.

PAINE, R. T. (1984). Ecological determinism in the competition for space. *Ecology*, **65**, 1339–1348.

PICKETT, S. T. A. (1980). Non-equilibrium coexistence of plants. *Bulletin of the Torrey Botanical Club*, **107**, 238–248.

PUGH, G. J. F. (1980). Strategies in fungal ecology. *Transactions of the British Mycological Society*, **75**, 1–14.

RAMSKII, L. G. (1938). *Introduction to the Geobotanical Study of Complex Vegetations*. Selkhozgiz, Moscow.

RAVEN, J. A. & HANDLEY, L. L. (1987). Transport processes and water relations. In: *Frontiers of Comparative Plant Ecology* (Ed. by I. H. Rorison, J. P. Grime, R. Hunt, G. A. F. Hendry & D. H. Lewis), *New Phytologist*, **106** (Suppl.), 217–233. Academic Press, New York & London.

RORISON, I. H. (1987). Mineral nutrition in space and time. In: *Frontiers of Comparative Plant Ecology* (Ed. by I. H. Rorison, J. P. Grime, R. Hunt, G. A. F. Hendry & D. H. Lewis), *New Phytologist*, **106** (Suppl.), 79–92. Academic Press, New York & London.

SIBLY, R. M. & GRIME, J. P. (1986). Strategies of resource capture by plants – evidence of adversity selection. *Journal of Theoretical Biology*, **118**, 247–250.

SOUTHWOOD, T. R. E. (1977). Habitat, the templet for ecological strategies? *Journal of Animal Ecology*, **46**, 337–365.

SPORNE, K. R. (1980). A re-investigation of character correlations among dicotyledons. *New Phytologist*, **85**, 419–449.

SPORNE, K. R. (1982). The advancement index vindicated. *New Phytologist*, **91**, 137–145.

STACE, C. A. (1980). *Plant Taxonomy and Biosystematics*. Edward Arnold, London.

STEBBINS, G. L. (1971). *Chromosomal Evolution in Higher Plants*. Edward Arnold, London.

STEBBINS, G. L. (1974). *Flowering Plants – Evolution above the Species Level*. Edward Arnold, London.

STEBBINS, G. L. (1980). Rarity of plant species: a synthetic viewpoint. *Rhodora*, **82**, 77–86.

STEBBINS, G. L. (1985). Polyploidy, hybridization, and the invasion of new habitats. *Annals of the Missouri Botanical Garden*, **72**, 824–832.

Sydes, C. & Grime, J. P. (1981a). Effects of tree leaf litter on herbaceous vegetation in deciduous woodland. I. Field investigations. *Journal of Ecology*, **69**, 237–248.

Sydes, C. & Grime, J. P. (1981b). Effects of tree leaf litter on herbaceous vegetation in deciduous woodland. II. An experimental investigation. *Journal of Ecology*, **69**, 249–262.

Takhtajan, A. (1980). Outline of the classification of flowering plants (Magnoliophyta). *Botanical Review*, **46**, 225–359.

Thompson, K. (1987). Seeds and seed banks. In: *Frontiers of Comparative Plant Ecology* (Ed. by I. H. Rorison, J. P. Grime, R. Hunt, G. A. F. Hendry & D. H. Lewis). *New Phytologist*, **106** (Suppl.), 23–34. Academic Press, New York & London.

Thompson, K. & Grime, J. P. (1979). Seasonal variation in the seed banks of herbaceous species in ten contrasting habitats. *Journal of Ecology*, **67**, 893–921.

Tilman, D. (1982). *Resource Competition and Community Structure*. Princeton University Press, Princeton.

Wagner, P. (1977). Vessel types of the monocotyledons: a survey. *Botaniska Notiser*, **130**, 383–402.

Weldon, C. W. & Slauson, W. L. (1986). The intensity of competition versus its importance: an overlooked distinction and some implications. *The Quarterly Review of Biology*, **61**, 23–44.

Whittaker, R. H. (1975). Design and stability of plant communities. In: *Unifying Concepts in Ecology* (Ed. by W. W. van Dobben & R. H. Lowe-McConnell), pp. 169–181. Junk, The Hague.

Whittaker, R. H. & Levin, S. A. (1977). The role of mosaic phenomena in natural communities. *Theoretical Population Biology*, **12**, 117–139.

Wilson, S. D. & Keddy, P. A. (1986a). Species competitive ability and position along a natural stress/disturbance gradient. *Ecology*, **67**, 1236–1242.

Wilson, S. D. & Keddy, P. A. (1986b). Measuring diffuse competition along an environmental gradient: results from a shoreline plant community. *American Naturalist*, **127**, 862–869.

New Phytol. (1987) **106** (Suppl.), 297–300

LIST OF POSTERS PRESENTED AT THE SYMPOSIUM

M. R. Ashmore & A. K. Tickle

Department of Pure and Applied Biology, Imperial College of Science and Technology, Silwood Park, Ascot, Berks. SL5 7PY, UK
Effects of ozone pollution on native plant species

C. P. D. Birch, D. J. Read* & I. H. Rorison

*Unit of Comparative Plant Ecology (NERC) and *Department of Botany, University of Sheffield, Sheffield S10 2TN, UK*
Vesicular–arbuscular mycorrhizal infection of four species in semi-natural grassland

K. J. Brocklebank & G. A. F. Hendry

Unit of Comparative Plant Ecology (NERC), Department of Botany, The University, Sheffield S10 2TN, UK
Storage carbohydrate profiles of four woodland species

B. D. Campbell & J. P. Grime

Unit of Comparative Plant Ecology (NERC), Department of Botany, The University, Sheffield S10 2TN, UK
An experimental test of plant strategy theory

T. de Jong & P. J. L. Klinkhamer

Zoologisch Laboratorium, Leiden, The Netherlands
Population ecology of the biennials, *Cirsium vulgare* and *Cyanoglossum officinale*

J. R. Etherington & C. E. Evans

Plant Science Department, University College, Cardiff CF1 1XL, UK
Plant roots may oxidize soil and facilitate competition

C. H. Foyer

Research Institute for Photosynthesis, The University, Sheffield S10 2TN, UK
The basis for source–sink interaction in leaves: an hypothesis

R. T. Furbank, P. Horton* & D. A. Walker

*Research Institute for Photosynthesis and *Department of Biochemistry, The University, Sheffield S10 2TN, UK*
Regulation of photosynthesis after a transition in light intensity

M. J. Glendining & I. Rhodes

Welsh Plant Breeding Station, Aberystwyth, Dyfed SY23 3EB, UK
Genetic variation in compatibility in grass–clover mixtures

0028-646X/87/05S297+04 $03.00/0

J. P. Grime & R. Hunt

*Unit of Comparative Plant Ecology (NERC), Department of Botany,
The University, Sheffield, S10 2TN, UK*

Plant strategies—a model for management

J. P. Grime, J. C. Crick & J. E. Rincon

*Unit of Comparative Plant Ecology (NERC), Department of Botany,
The University, Sheffield S10 2TN, UK*

The ecological significance of plasticity

J. P. Grime, J. G. Hodgson & R. Hunt

*Unit of Comparative Plant Ecology (NERC), Department of Botany,
The University, Sheffield S10 2TN, UK*

Comparative plant ecology

J. P. Grime, J. M. L. Mackey & S. R. Band

*Unit of Comparative Plant Ecology (NERC), Department of Botany,
The University, Sheffield S10 2TN, UK*

Nuclear DNA contents, shoot phenology and species coexistence in a limestone grassland community

J. P. Grime, J. M. L. Mackey, J. M. Fletcher, S. R. Band &
N. Ruttle

*Unit of Comparative Plant Ecology (NERC), Department of Botany,
The University, Sheffield S10 2TN, UK*

The unit of comparative plant ecology: an historical perspective

J. P. Grime, J. M. L. Mackey, S. H. Hillier & D. J. Read*

*Unit of Comparative Plant Ecology (NERC), Department of Botany,
The University, Sheffield and *Department of Botany, The University,
Sheffield S10 2TN, UK*

The role of soil heterogeneity, grazing and vesicular–arbuscular mycorrhizas in species diversity: a 12-month experiment using turf microcosms

P. L. Gupta & I. H. Rorison

*Unit of Comparative Plant Ecology (NERC), Department of Botany,
The University, Sheffield S10 2TN, UK*

Interspecific response by grasses to P in infertile and uncultivated soils

G. A. F. Hendry & K. J. Brocklebank

*Unit of Comparative Plant Ecology (NERC), Department of Botany,
The University, Sheffield S10 2TN, UK*

Manipulating oxygen radical metabolism in plants

G. A. F. HENDRY, K. J. BROCKLEBANK & J. G. HODGSON

Unit of Comparative Plant Ecology (NERC), Department of Botany, The University, Sheffield S10 2TN, UK

Herbivore nutrition and survival strategies

J. G. HODGSON & J. M. L. MACKEY

Unit of Comparative Plant Ecology (NERC), Department of Botany, The University, Sheffield S10 2TN, UK

Family specialization: its causes and ecological significance

R. HUNT

Unit of Comparative Plant Ecology (NERC), Department of Botany, The University, Sheffield S10 2TN, UK

Variation in relative growth rate

L. KAUTSKY

Department of Botany, University of Stockholm, Sweden

Aquatic macrophyte strategies: similarity and differences to terrestrial phanerograms

C. KÖRNER & U. RENHARDT

Institut für Botanik, Universität Innsbruck, Sternwartestrasse 15, A-6020 Innsbruck, Austria

Dry matter and nitrogen partitioning in low- and high-altitude plants

T. P. McGONIGLE

Department of Biology, University of York, York YO1 5DD, UK

Vesicular–arbuscular mycorrhizal function under field conditions

H. POORTER & S. POT

Department of Plant Ecology, University of Utrecht, Lange Nieuwstraat 106, 3512 PN Utrecht, The Netherlands

The influence of an increased pCO_2 on growth and photosynthesis of plants with different growth rates

M. J. POWELL, M. S. DAVIES & D. FRANCIS

Department of Plant Science, University College Cardiff, Cardiff CF1 1XL, UK

Cellular effects of Zn in a Zn-tolerant and a non-tolerant cultivar

A. H. PRICE & G. A. F. HENDRY

Unit of Comparative Plant Ecology (NERC), Department of Botany, The University, Sheffield S10 2TN, UK

The significance of vitamin E in stress survival in higher plants

D. L. RAYNAL*, J. P. GRIME & R. BOOT

**State University of New York, College of Environmental Science and Forestry, Syracuse, New York 13210, USA and Unit of Comparative Plant Ecology (NERC), Department of Botany, The University, Sheffield S10 2TN, UK*

A new method for the experimental droughting of plants

J. C. ROSE & I. H. RORISON

Unit of Comparative Plant Ecology (NERC), Department of Botany,
The University, Sheffield, S10 2TN, UK

Growth and nutrient uptake in relation to root temperature gradients

R. M. SIBLY* & J. P. GRIME

*Department of Pure and Applied Zoology, University of Reading,
Reading, RG6 2AJ and Unit of Comparative Plant Ecology (NERC),
Department of Botany, The University, Sheffield S10 2TN, UK

Strategies of resource capture by plants: evidence for
adversity selection

M. T. O. SILVA

Instituto de Botanica, Universide do Porto, Portugal

Food selection by terrestrial molluscs and its ecological consequences
for plant communities

N. SMIRNOFF, N. SHAD & G. R. STEWART*

Department of Biological Sciences, University of Exeter, Exeter and
*Department of Botany and Microbiology, University College, London

Photosynthesis and stomatal behaviour of the parasite,
Striga hermonthica

G. R. STEWART

Department of Botany and Microbiology, University College London,
Gower Street, London WC1E 6BT, UK

Nitrate acquisition and assimilation

S. WALDREN

School of Botany, Trinity College, Dublin 2, Republic of Ireland

Hybridization and waterlogging tolerance of *Geum rivale*
and *G. urbanum*

I. C. WISHEN & P. A. KEDDY

Department of Biology, University of Ottawa, Canada

Changes in species richness along standing crop gradients

J. E. YOUNG & P. S. JONES

Department of Plant Science, University College, PO Box 78,
Cardiff CF1 1XL, UK

Fern gametophyte development in relation to substrate

Index of Generic, Binomial and Vernacular Names*

* Prepared by Ms Judith Fletcher (UCPE).

* *Molinea* in text is incorrect.

Subject Index*

* Prepared by Ms Judith Fletcher (UCPE).